T0192181

Lecture Notes in Computer Science 14151

The series Lecture Notes in Computer Science (LNCS), including its subseries Lecture Notes in Artificial Intelligence (LNAI) and Lecture Notes in Bioinformatics (LNBI), has established itself as a medium for the publication of new developments in computer science and information technology research, teaching, and education.

LNCS enjoys close cooperation with the computer science R & D community, the series counts many renowned academics among its volume editors and paper authors, and collaborates with prestigious societies. Its mission is to serve this international community by providing an invaluable service, mainly focused on the publication of conference and workshop proceedings and postproceedings. LNCS commenced publication in 1973.

Benedek Nagy
Editor

Implementation and Application of Automata

27th International Conference, CIAA 2023
Famagusta, North Cyprus, September 19–22, 2023
Proceedings

 Springer

Editor
Benedek Nagy
Eastern Mediterranean University
Famagusta, North Cyprus

Eszterházy Károly Catholic University
Eger, Hungary

ISSN 0302-9743   ISSN 1611-3349 (electronic)
Lecture Notes in Computer Science
ISBN 978-3-031-40246-3   ISBN 978-3-031-40247-0 (eBook)
https://doi.org/10.1007/978-3-031-40247-0

This Springer imprint is published by the registered company Springer Nature Switzerland AG
The registered company address is: Gewerbestrasse 11, 6330 Cham, Switzerland

Preface

The 27th International Conference on Implementation and Application of Automata (CIAA 2023) was organized by the Department of Mathematics of the Eastern Mediterranean University (EMU). The conference took place during September 19–22, 2023, in Salamis Bay Conti Resort Hotel, Famagusta, North Cyprus, co-located with the 13th International Workshop on Non-Classical Models of Automata and Applications (NCMA 2023, September 18–19). This event forms part of the CIAA conference series, a major international venue for the exchange of new ideas in the field of automata theory. The previous 26 conferences were held in various locations all around the globe: Rouen (2022), Bremen (2021), Košice (2019), Charlottetown (2018), Marne-la-Vallée (2017), Seoul (2016), Umeå (2015), Giessen (2014), Halifax (2013), Porto (2012), Blois (2011), Winnipeg (2010), Sydney (2009), San Francisco (2008), Prague (2007), Taipei (2006), Nice (2005), Kingston (2004), Santa Barbara (2003), Tours (2002), Pretoria (2001), London, Ontario (2000), Potsdam (WIA 1999), Rouen (WIA 1998), and London, Ontario (WIA 1997 and WIA 1996). Due to the COVID-pandemic, CIAA 2020, planned to be held in Loughborough, was canceled. The CIAA conference series brings together researchers and allows the dissemination of results in the implementation, application, and theory of automata.

Last year the international community, not only the scientific one, was shocked by the Russian invasion of Ukraine, which is in violation of international laws and a crime against the Ukrainian people and is strongly condemned by the Steering Committee (SC) of CIAA. One year later, the war is still going on and becomes more brutal every day leading to unthinkable human suffering. We as scientists are used to collaborating across different beliefs, political systems, and cultural backgrounds. Scientific research is inherently an international endeavor and is based on the principle of a free exchange of ideas among the international scientific community. The editorial decision whether a paper is accepted or rejected should not be affected by the origins of a manuscript, including the nationality, ethnicity, political beliefs, race or religion of a paper's authors. This is in accordance with the guidelines of the Committee on Publication Ethics (COPE; https://publicationethics.org). According to John Steinbeck "All war is a symptom of man's failure as a thinking animal." The SC hopes that the suffering in Ukraine ends soon and peace will be restored. Peace is one of the UN's Sustainable Development Goals and a prerequisite for human progress.

This volume of Lecture Notes in Computer Science contains the scientific papers presented at CIAA 2023. The volume also includes two papers from the invited talks presented by Viliam Geffert (Binary Coded Unary Regular Languages) and Friedrich Otto (A Survey on Automata with Translucent Letters), and the abstract of the invited talk presented by Cem Say (Finite Automata as Verifiers), we wish to warmly thank all of them. The 20 regular papers were selected from 30 submissions covering various fields in the application, implementation, and theory of automata and related structures. Each paper was reviewed by three Program Committee members with the assistance of external referees and thoroughly discussed by the Program Committee (PC). Papers were submitted by authors from various countries: Belgium, Canada, Czech Republic, Denmark, Estonia, France, Germany, Israel, Italy, Japan, Latvia, Poland, Portugal, Russia, Slovakia, South Korea, Sweden, Turkey, and United Kingdom. We wish to thank everybody who contributed to the success of this conference: the authors for submitting their carefully prepared manuscripts, the PC members and external referees for their valuable evaluation of the submitted manuscripts, the invited speakers for their excellent presentations of topics related to the theme of the conference, the session chairs, the presenters, and the participants who made CIAA 2023 possible. We also thank the editorial staff at Springer, in particular Anna Kramer and Ronan Nugent, for their guidance and help during the publication process of this volume.

Last but not least, we would like to express our sincere thanks to the local organizers. We all are looking forward to the next CIAA in Japan.

June 2023

Markus Holzer
SC-Chair

Benedek Nagy
PC-Chair

Organization

Steering Committee

Markus Holzer (Chair)	Justus Liebig University, Germany
Oscar Ibarra	University of California, Santa Barbara, USA
Sylvain Lombardy	Université de Bordeaux, France
Nelma Moreira	University of Porto, Portugal
Kai T. Salomaa (Co-chair)	Queen's University, Canada
Hsu-Chun Yen	National Taiwan University, Taiwan

Program Committee

Marie-Pierre Beal	Université Paris-Est Marne-la-Vallée, France
Johanna Björklund	Umeå University, Sweden
Francine Blanchet-Sadri	University of North Carolina at Chapel Hill, USA
Cezar Câmpeanu	University of Prince Edward Island, Canada
Pascal Caron	Université de Rouen, France
Erzsébet Csuhaj-Varjú	Eötvös Loránd University, Hungary
Frank Drewes	Umeå University, Sweden
Ömer Egecioglu	University of California, Santa Barbara, USA
Attila Egri-Nagy	Akita International University, Japan
Szilárd Zsolt Fazekas	Akita University, Japan
Markus Holzer	Justus Liebig University, Germany
Szabolcs Iván	University of Szeged, Hungary
Galina Jirásková	Slovak Academy of Sciences, Košice, Slovakia
Jarkko Kari	University of Turku, Finland
Martin Kutrib	Justus Liebig University, Germany
Markus Lohrey	University of Siegen, Germany
Sylvain Lombardy	Université de Bordeaux, France
Andreas Malcher	Justus Liebig University, Germany
Andreas Maletti	Universität Leipzig, Germany
Florin Manea	Georg August University, Germany
Sebastian Maneth	University of Bremen, Germany
Ian McQuillan	University of Saskatchewan, Canada
Brink van der Merwe	Stellenbosch University, South Africa
Roland Meyer	TU Braunschweig, Germany
Nelma Moreira	University of Porto, Portugal

František Mráz	Charles University, Czech Republic
Benedek Nagy (Chair)	Eastern Mediterranean University, Famagusta, North Cyprus and Eszterházy Károly Catholic University, Eger, Hungary
Alexander Okhotin	St. Petersburg State University, Russia
Giovanni Pighizzini	University of Milan, Italy
Igor Potapov	University of Liverpool, UK
Daniel Reidenbach	Loughborough University, UK
Rogério Reis	University of Porto, Portugal
Michel Rigo	University of Liège, Belgium
Kai T. Salomaa	Queen's University, Canada
Hiroyuki Seki	Nagoya University, Japan
Shinnosuke Seki	University of Electro-Communications, Tokyo, Japan
György Vaszil	University of Debrecen, Hungary
Hsu-Chun Yen	National Taiwan University, Taiwan

Invited Speakers

Viliam Geffert	Pavol Jozef Šafárik University, Košice, Slovakia
Friedrich Otto	University of Kassel, Germany
Cem Say	Boğaziçi University, Istanbul, Turkey

Additional Reviewers

Duncan Adamson
Martin Berglund
Sabine Broda
Arnaud Carayol
Dimitry Chistikov
Zoltán Fülöp
Hermann Gruber
Christoph Haase

Carolin Hannusch
Eren Keskin
Matthew Kozma
Julien Leroy
Luca Prigioniero
Marek Szykuła
Martin Vu
Sören van der Wall

Finite Automata as Verifiers (Invited Talk)

A. C. Cem Say

Department of Computer Engineering,
Boğaziçi University, İstanbul, Turkey

Abstract. An interactive proof system is characterized by the computational abilities of its two components (called the "prover" and the "verifier") and the nature of the interaction between them. The verifier is the weaker party, who is supposed to be able to check the claims of the more powerful prover. We focus on the weakest possible kind of computational model, namely, the finite automaton, in the role of the verifier. We will provide an overview of the literature and talk about recent work involving new concepts, like constant-randomness machines and thermodynamic complexity considerations.

Contents

Invited Talks

Critical Tales

Binary Coded Unary Regular Languages

Viliam Geffert[✉]

Department of Computer Science, P. J. Šafárik University,
Jesenná 5, 04154 Košice, Slovakia
viliam.geffert@upjs.sk

Abstract. $\mathcal{L} \subseteq \{0,1\}^*$ is a *binary coded unary regular language*, if there exists a unary regular language $\mathcal{L}' \subseteq \{a\}^*$ such that a^x is in $\mathcal{L}'$ if and only if the binary representation of x is in $\mathcal{L}$. If a unary language $\mathcal{L}'$ is accepted by an optimal deterministic finite automaton (DFA) $\mathcal{A}'$ with n states, then its binary coded version $\mathcal{L}$ is regular and can be accepted by a DFA $\mathcal{A}$ using at most n states, but at least $1+\lceil \log n \rceil$ states. There are witness languages matching these upper and lower bounds exactly, for each n.

   More precisely, if $\mathcal{A}'$ uses $\sigma \geq 0$ states in the initial segment and $\mu \cdot 2^\ell$ states in the loop, where μ is odd and $\ell \geq 0$, then the optimal $\mathcal{A}$ for $\mathcal{L}$ consists of a preamble with at most σ but at least $\max\{1, 1 + \lceil \log \sigma \rceil - \ell\}$ states, except for $\sigma = 0$ with no preamble, and a kernel with at most $\mu \cdot 2^\ell$ but at least $\mu + \ell$ states. Also these lower bounds are matched exactly by witness languages, for each σ, μ, ℓ.

   The conversion in the opposite way is not always granted: there are binary regular languages the unary versions of which are not even context free.

   The corresponding conversion of a unary nondeterministic finite automaton (NFA) to a binary NFA uses $O(n^2)$ states and introduces a binary version of Chrobak normal form.

Keywords: finite automata · unary regular languages · state complexity

1 Introduction

One of the simplest language classes ever studied in theoretical computer science is the class of unary regular languages, corresponding to deterministic finite automata (DFAs) with a single-letter input alphabet. The structure of such automaton is very simple: it consists of an *initial segment* and a *loop*.

   Unary (tally) languages play an important role as languages with a very low information content, and many of their properties are quite different from the general or binary case. One of the earliest observations is the fact that every unary context-free language is also regular [11]. For other results on unary languages in general, the reader is referred, among others, to [1,6,8,12].

   In the case of regular languages, there are also substantial differences in state complexity. For example, removing nondeterminism in a nondeterministic

Supported by the Slovak grant contract VEGA 1/0177/21.

finite automaton (NFA) with n states may increase the number of states up to 2^n [16,18–20], while the corresponding cost for automata with unary input alphabet is only $e^{(1+o(1))\cdot\sqrt{n\cdot\ln n}}$ [2,7,17].

In this paper, we focus our attention on state complexity of binary coded unary regular languages. $\mathcal{L} \subseteq \{0,1\}^*$ is a *binary coded unary regular language*, if there exists a unary regular language $\mathcal{L}' \subseteq \{a\}^*$ such that a^x is in $\mathcal{L}'$ if and only if the binary representation of x is in $\mathcal{L}$. Our interest is motivated by the fact that most present-day computers store data in a binary coded form and that, quite recently [9], we have obtained some results on state complexity of binary coded non-unary regular languages.

We shall show that if a unary language $\mathcal{L}'$ is regular and can be accepted by a DFA $\mathcal{A}'$ with n states, then its binary coded counterpart $\mathcal{L} = \mathrm{bin}\,\mathcal{L}'$ is regular as well and can be accepted by a DFA $\mathcal{A}$ with at most n states. Moreover, if $\mathcal{A}'$ is optimal, then $\mathcal{A}$ must use at least $1+\lceil\log n\rceil$ states. We shall also provide witness languages matching these upper and lower bounds exactly, for each $n \geq 1$.

The converse does not hold. There exist binary regular languages the unary versions of which are not regular: consider the unary language

$$\mathcal{L}' = \{a^{2^k} : k \geq 0\}\,.$$

This language is not even context-free, which can be proved by the use of the Pumping Lemma [13, Sect. 7.2]. Nevertheless, its binary coded version $\mathcal{L} = \mathrm{bin}\,\mathcal{L}'$ consists of all binary strings that contain exactly one symbol "1", which can be tested by a DFA with 3 states or be given by the regular expression 0^*10^* (allowing leading zeros).

The conversion of a unary $\mathcal{A}'$ to its binary counterpart $\mathcal{A}$ depends on the partial factorization of λ—the length of the loop in $\mathcal{A}'$—into $\lambda = \mu\cdot 2^\ell$, where μ is odd. Namely, if $\mathcal{A}'$ uses σ states in the initial segment and $\mu\cdot 2^\ell$ states in the loop, where μ is odd, then the optimal $\mathcal{A}$ for the binary coded $\mathcal{L}(\mathcal{A}')$ consists of a preamble with at most σ but at least $\max\{1, 1 + \lceil\log\sigma\rceil\} - \ell$ states (except for $\sigma = 0$, with no preamble) and a kernel with at most $\mu\cdot 2^\ell$ but at least $\mu + \ell$ states. These lower bounds are matched by witness languages, for each σ, μ, ℓ.

After basic definitions in Sect. 2, the upper bounds are established in Sect. 3. Next, Sect. 4 gives a detailed analysis of a kernel that simulates the loop, ending by a lower bound on the size of this kernel. It turns out that, for loops of even length λ, the important role play pairs of *twin states*, located at opposite positions across the loop, i.e., at positions i_1 and $i_2 = (i_1 + \lambda/2) \bmod \lambda$. For example, if λ is the optimal length of a loop in a unary $\mathcal{A}'$ and this length is even, *there must exist* a pair of twin states such that one of them is accepting while the other one is rejecting. However, the corresponding kernel simulating this loop in the binary $\mathcal{A}$ is of the same size λ if and only if *all pairs* of twins in $\mathcal{A}'$ differ in acceptance. Next, Sect. 5 gives an analysis of a preamble simulating the initial segment, ending by a lower bound for this preamble. The lower and upper bounds on the total number of states are concentrated in Sect. 6. Finally, Sect. 7 presents related results for NFAs, among others, a binary variant of *Chrobak normal form* [2,5,7].

2 Preliminaries

Here we briefly fix some basic definitions and notation. For more details, we refer the reader to [13,14], or any other standard textbook.

A *deterministic finite state automaton* (DFA, for short) is a quintuple $\mathcal{A} = (Q, \Sigma, \delta, q_1, F)$, where Q denotes a finite set of states, Σ a finite set of input symbols, $\delta : Q \times \Sigma \to Q$ a transition function, $q_1 \in Q$ an initial state, and $F \subseteq Q$ a set of final (accepting) states. The states in the set $Q \backslash F$ will be called rejecting.

The transition function δ can be extended to $\delta^* : Q \times \Sigma^* \to Q$ in a natural way: $\delta^*(q, \varepsilon) = q$ and $\delta^*(q, \alpha a) = \delta(\delta^*(q, \alpha), a)$, for each $q \in Q$, $\alpha \in \Sigma^*$, and $a \in \Sigma$. The language accepted by $\mathcal{A}$ is $\mathcal{L}(\mathcal{A}) = \{w \in \Sigma^* : \delta^*(q_1, w) \in F\}$.

A *nondeterministic finite state automaton* (NFA, for short) uses the same components as DFA above, except for $\delta : Q \times \Sigma \to 2^Q$, allowing a set of several different transitions for the given $q \in Q$ and $a \in \Sigma$.

Two automata are *equivalent*, if they accept the same language. A DFA (NFA) $\mathcal{A}$ is *optimal*, if no DFA (NFA, respectively) with fewer states is equivalent to $\mathcal{A}$.

To simplify notation, a transition $\delta(q, a) = q'$ will sometimes be presented in a more compact form $q \xrightarrow{a} q'$. Similarly, $q \xrightarrow{a_1 \cdots a_n} q'$ displays a path consisting of n transitions, beginning in q, ending in q', and reading $a_1 \cdots a_n \in \Sigma^*$.

We fix the *unary input alphabet* to $\Sigma' = \{a\}$ and the *binary input alphabet* to $\Sigma = \{0, 1\}$. The correspondence between binary strings and nonnegative integers can be established by a function $\mathrm{num} : \{0, 1\}^* \to \mathbb{N}$, defined as follows:[1]

$$\mathrm{num}\,\text{``}\varepsilon\text{''} = 0\,,$$
$$\mathrm{num}\,\text{``}wb\text{''} = (\mathrm{num}\,\text{``}w\text{''}) \cdot 2 + b\,, \quad \text{for each } w \in \{0,1\}^* \text{ and } b \in \{0,1\}.$$

It is easy to show, by induction on $|w|$, the length of w, that $\mathrm{num}\,\text{``}vw\text{''} = (\mathrm{num}\,\text{``}v\text{''}) \cdot 2^{|w|} + \mathrm{num}\,\text{``}w\text{''}$, for each $v, w \in \{0,1\}^*$.

A mapping in the opposite direction is not so unambiguous, because of leading zeros. Moreover, we shall occasionally need to use a "proper" number of leading zeros. This leads to the following definition: for each x and r in $\mathbb{N}$, let

$\mathrm{bin}\,x = $ the shortest $w \in \{0,1\}^*$ such that $\mathrm{num}\,\text{``}w\text{''} = x$,
$\mathrm{bin}_r\,x = $ the shortest $w \in \{0,1\}^*$ *of length at least* r such that $\mathrm{num}\,\text{``}w\text{''} = x$.

As an example, $\mathrm{num}\,\text{``}000101\text{''} = \mathrm{num}\,\text{``}101\text{''} = 5$, while $\mathrm{bin}\,5 = \text{``}101\text{''}$. On the other hand, we have $\mathrm{bin}_2\,5 = \text{``}101\text{''}$ and $\mathrm{bin}_6\,5 = \text{``}000101\text{''}$.

The correspondence between unary languages and their binary counterparts is quite straightforward: let $\mathcal{L}' \subseteq \{a\}^*$ and $\mathcal{L} \subseteq \{0, 1\}^*$. Then

$\mathrm{bin}\,\mathcal{L}' = \{w \in \{0,1\}^* : a^{\mathrm{num}\,w} \in \mathcal{L}'\}$ and $\mathrm{num}\,\mathcal{L} = \{a^x \in \{a\}^* : \mathrm{bin}\,x \in \mathcal{L}\}$.

$\mathcal{L} \subseteq \{0,1\}^*$ is a *binary coded unary regular language*, if there exists a unary regular language $\mathcal{L}'$ such that $\mathcal{L} = \mathrm{bin}\,\mathcal{L}'$.

---

[1] To distinguish between a multiplication of integers and a concatenation of strings in formulas with mixed contents, strings are sometimes enclosed in quotation marks.

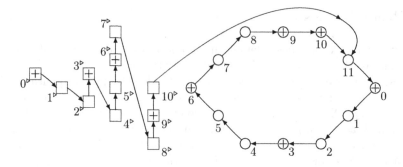

Fig. 1. The optimal one-way deterministic automaton $\mathcal{A}'$ accepting the unary language $\mathcal{L}' = \{a^k : k \bmod 12 \in \{0, 3, 6, 9, 10\}\} \setminus \{a^{10}\}$. The states in the initial segment are displayed as squares, those in the loop as circles. Accepting states are labeled by "+".

The structure of a *unary* DFA is very simple: such automaton $\mathcal{A}' = (Q', \Sigma', \delta', q_{\mathrm{I}}', F')$, with $\Sigma' = \{a\}$, consists of an *initial segment* $\mathbb{Z}_\sigma^\triangleright$ of length σ and a *loop* $\mathbb{Z}_\lambda$ of length λ, i.e., $Q' = \mathbb{Z}_\sigma^\triangleright \cup \mathbb{Z}_\lambda$. (See example in Fig. 1.) Formally,[2]

$$\mathbb{Z}_\sigma^\triangleright = \{0^\triangleright, 1^\triangleright, \ldots, (\sigma-1)^\triangleright\} \text{ and } \mathbb{Z}_\lambda = \{0, 1, \ldots, \lambda-1\}.$$

In the initial segment, $\mathcal{A}'$ counts the length of the input up to $\sigma - 1$ after which, in the loop, it counts modulo λ. Transitions are obvious:

$$\begin{aligned}
\delta'(j^\triangleright, a) &= (j+1)^\triangleright, && \text{for } j \in \{0, \ldots, \sigma-2\}, \\
\delta'((\sigma-1)^\triangleright, a) &= \sigma \bmod \lambda, && \\
\delta'(i, a) &= (i+1) \bmod \lambda, && \text{for } i \in \{0, \ldots, \lambda-1\}.
\end{aligned} \tag{1}$$

If $\sigma > 0$, the initial state is $q_{\mathrm{I}}' = 0^\triangleright$. However, if $\sigma = 0$, there is no initial segment, $\mathbb{Z}_\sigma^\triangleright = \varnothing$, and $q_{\mathrm{I}}' = 0$. Using (1), it is easy to see that, for each $x \geq 0$,

$$\delta'^*(q_{\mathrm{I}}', a^x) = \begin{cases} x^\triangleright, & \text{if } x < \sigma, \\ x \bmod \lambda, & \text{if } x \geq \sigma. \end{cases} \tag{2}$$

3   General Properties

Let us we begin with a "basic" construction:

Theorem 1. *If a unary language $\mathcal{L}'$ is accepted by a DFA $\mathcal{A}' = (Q', \{a\}, \delta', q_{\mathrm{I}}', F')$ using an initial segment $\mathbb{Z}_\sigma^\triangleright$ with σ states and a loop $\mathbb{Z}_\lambda$ with λ states, then its binary coded counterpart $\mathcal{L} = \mathrm{bin}\,\mathcal{L}'$ is regular and can be accepted by a DFA $\mathcal{A}^\diamond$ consisting of a preamble with σ states and a kernel with λ states.*

---

[2] Both in $\mathbb{Z}_\sigma^\triangleright$ and in $\mathbb{Z}_\lambda$, the states are associated with integers. To distinguish states in $\mathbb{Z}_\sigma^\triangleright$ from those in $\mathbb{Z}_\lambda$, the former are labeled by a triangle while the latter are not.

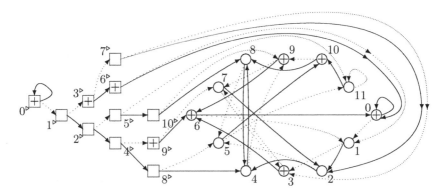

Fig. 2. The binary counterpart $\mathcal{A}^\diamond$ of the DFA $\mathcal{A}'$ presented in Fig. 1, constructed by the use of Theorem 1. Transitions reading the symbol "0" are displayed as solid arrows, those reading the symbol "1" as dotted arrows.

Proof (sketched). The new automaton $\mathcal{A}^\diamond = (Q', \{0,1\}, \delta^\diamond, q'_1, F')$ works with the same set of states $Q' = \mathbb{Z}_\sigma^\triangleright \cup \mathbb{Z}_\lambda$ as does $\mathcal{A}'$—see also Fig. 1—as well as with the same initial and accepting states, but with different transitions:

$$\delta^\diamond(j^\triangleright, b) = \begin{cases} (j{\cdot}2 + b)^\triangleright, & \text{if } j{\cdot}2 + b < \sigma, \\ (j{\cdot}2 + b) \bmod \lambda, & \text{if } j{\cdot}2 + b \geq \sigma, \end{cases} \tag{3}$$
$$\delta^\diamond(i, b) = (i{\cdot}2 + b) \bmod \lambda.$$

This holds for each $j \in \{0, \ldots, \sigma - 1\}$, each $i \in \{0, \ldots, \lambda - 1\}$, and each $b \in \{0,1\}$. The conversion is illustrated by an example shown in Fig. 2. Thus, instead of counting the length of a unary input, but using the same set of states $\mathbb{Z}_\sigma^\triangleright$, $\mathcal{A}^\diamond$ computes the numerical value num "w" for the prefix w which, so far, has been read from the given binary input. At the moment when this value exceeds $\sigma - 1$, $\mathcal{A}^\diamond$ starts to compute this value modulo λ, using the states in $\mathbb{Z}_\lambda$.

To prove that $\mathcal{L}(\mathcal{A}^\diamond) = \text{bin}\,\mathcal{L}'$, we need to show that, for each $w \in \{0,1\}^*$,

$$\delta^{\diamond*}(q'_1, w) = \begin{cases} (\text{num}\,w)^\triangleright, & \text{if num}\,w < \sigma, \\ (\text{num}\,w) \bmod \lambda, & \text{if num}\,w \geq \sigma. \end{cases} \tag{4}$$

This can be done by an induction on the length of w. Then, by comparing (4) with (2), for $x = \text{num}\,w$, it is easy to see that

$$\delta^{\diamond*}(q'_1, w) = \delta'^*(q'_1, a^{\text{num}\,w}), \quad \text{for each } w \in \{0,1\}^*. \tag{5}$$

Since $\mathcal{A}^\diamond$ agrees with $\mathcal{A}'$ in the set of accepting states, this means that $w \in \{0,1\}^*$ is accepted by $\mathcal{A}^\diamond$ if and only if $a^{\text{num}\,w} \in \{a\}^*$ is accepted by $\mathcal{A}'$. □

Before passing further, let us extend (3), presented in the construction in Theorem 1, from bits to arbitrary binary strings:

$$\delta^{\diamond*}(j^\triangleright, w) = \begin{cases} (j{\cdot}2^{|w|} + \text{num}\,w)^\triangleright, & \text{if } j{\cdot}2^{|w|} + \text{num}\,w < \sigma, \\ (j{\cdot}2^{|w|} + \text{num}\,w) \bmod \lambda, & \text{if } j{\cdot}2^{|w|} + \text{num}\,w \geq \sigma, \end{cases} \tag{6}$$
$$\delta^{\diamond*}(i, w) = (i{\cdot}2^{|w|} + \text{num}\,w) \bmod \lambda.$$

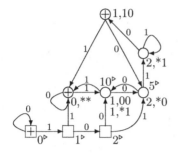

$t(q)$	$q \in \mathbb{Z}_\lambda \cup \mathbb{Z}_\sigma^\triangleright$		
0,**	0, 3, 6, 9,	$3^\triangleright, 6^\triangleright, 9^\triangleright$	
1,00 1,*1	1, 4, 7,	$4^\triangleright, 7^\triangleright,$	$10^\triangleright$
1,10	10		
2,*0	2, 8	$8^\triangleright,$	$5^\triangleright$
2,*1	5, 11		
$2^\triangleright$			$2^\triangleright$
$1^\triangleright$			$1^\triangleright$
$0^\triangleright$			$0^\triangleright$

Fig. 3. The optimal DFA $\mathcal{A}$ for $\mathcal{A}^\diamond$ presented in Fig. 2 and a table presenting the function t that maps the states of the original unary $\mathcal{A}'$ to the states in $\mathcal{A}$. The states are labeled as follows: "1,10" corresponds to the value i such that $i \bmod 3 = 1$ and the binary representation of i ends by "10" (i.e., $i \bmod 4 = 2$); similarly, "2,*1" stands for $i \bmod 3 = 2$ with the binary representation of i ending by "00" or "01", and "0,**" represents $i \bmod 3 = 0$ with all four possible bit combinations. The remaining labels should be obvious; the only exception is the path $0^\triangleright \xrightarrow{1} 1^\triangleright \xrightarrow{0} 2^\triangleright \xrightarrow{1} 5^\triangleright \xrightarrow{0} 10^\triangleright$.

This holds for each $j \in \{0, \ldots, \sigma - 1\}$, each $i \in \{0, \ldots, \lambda - 1\}$, and each $w \in \{0,1\}^*$. Also the argument for (6) uses induction on the length of w.

The binary DFA $\mathcal{A}^\diamond$ constructed in Theorem 1 is usually far from being optimal, even if the original unary DFA $\mathcal{A}'$ is. As an example, compare the binary DFA $\mathcal{A}^\diamond$ from Fig. 2 with its minimized version $\mathcal{A}$, presented by Fig. 3.

Theorem 2. *Let* $\mathcal{A}' = (Q', \{a\}, \delta', q'_\mathrm{I}, F')$ *be an optimal DFA accepting a unary language* $\mathcal{L}'$ *and let* $\mathcal{A} = (Q, \{0,1\}, \delta, q_\mathrm{I}, F)$ *be the optimal DFA accepting* $\mathcal{L} = \mathrm{bin}\, \mathcal{L}'$. *Then there exists a function* $t: Q' \to Q$ *preserving the machine's behavior, i.e., it preserves the initial states, up to binary coding the machine's transitions, and also acceptance/rejection:*

(I) $q_\mathrm{I} = t(q'_\mathrm{I})$,
(II) $\delta^*(q_\mathrm{I}, w) = t(\delta'^*(q'_\mathrm{I}, a^{\mathrm{num}\, w}))$, *for each* $w \in \{0,1\}^*$,
(III) $t(q) \in F$ *if and only if* $q \in F'$, *for each* $q \in Q'$.

Proof (sketched). For the given $\mathcal{A}'$, we first construct $\mathcal{A}^\diamond = (Q', \{0,1\}, \delta^\diamond, q'_\mathrm{I}, F')$ by the use of Theorem 1. This machine is not necessarily optimal, but it accepts $\mathcal{L} = \mathrm{bin}\, \mathcal{L}'$, using the same set of states, the same initial state, and the same accepting states. By (5), $\delta^{\diamond*}(q'_\mathrm{I}, w) = \delta'^*(q'_\mathrm{I}, a^{\mathrm{num}\, w})$, for each $w \in \{0,1\}^*$.

Now, since the optimal DFA is up to isomorphism always unique for each regular language, we can reconstruct all components in the optimal $\mathcal{A} = (Q, \{0,1\}, \delta, q_\mathrm{I}, F)$ by applying the standard procedure of minimization on $\mathcal{A}^\diamond$. (See, for example, [13, Sect. 4.4].) This procedure proceeds as follows.

First, we should eliminate states that are not be reachable from the initial state. However, using (2), we see that each $q \in Q'$ is reached in $\mathcal{A}'$ by reading a^x, for some $x \in \{0, \ldots, \sigma + \lambda - 1\}$. But then q is also reached in $\mathcal{A}^\diamond$, by reading $\mathrm{bin}\, x$, since $\delta^{\diamond*}(q'_\mathrm{I}, \mathrm{bin}\, x) = \delta'^*(q'_\mathrm{I}, a^{\mathrm{num\, bin}\, x}) = \delta'^*(q'_\mathrm{I}, a^x) = q$, by (5).

Second, partition Q' into blocks of equivalent states, i.e., $Q' = X_1 \cup \ldots \cup X_h$, so that these blocks are pairwise disjoint and any pair of states in Q' belongs to the same block if and only if these two states are equivalent in $\mathcal{A}^\diamond$. This establishes $Q = \{X_1, \ldots, X_h\}$, the state set for $\mathcal{A}$.

Third, we define the remaining components for $\mathcal{A}$. If $q_1' \in X$, for some $X \in Q$, then $q_1 \stackrel{\text{df.}}{=} X$. Transitions in $\mathcal{A}$ are established as follows. For each $X \in Q$ and $b \in \{0, 1\}$, let us take any $q \in X$. If $\delta^\diamond(q, b) \in X'$, for some $X' \in Q$, then $\delta(X, b) \stackrel{\text{df.}}{=} X'$. Finally, $X \in Q$ is included in F if and only if there is at least one state $q \in X$ such that $q \in F'$.

We are now ready to introduce $t : Q' \to Q$: if $q \in X$, then $t(q) \stackrel{\text{df.}}{=} X$. The fundamental property of this function is that

$$\delta^*(t(q), w) = t(\delta^{\diamond *}(q, w)), \quad \text{for each } q \in Q' \text{ and } w \in \{0, 1\}^*. \tag{7}$$

If $w = \varepsilon$, the above equation is trivial. For longer strings, the argument proceeds by induction on the length of w, beginning with $|w| = 1$.

Claim: For each $q \in Q'$, the state $t(q)$ in $\mathcal{A}$ is equivalent with q in $\mathcal{A}^\diamond$. More precisely, for each $u \in \{0, 1\}^*$, $X \in Q$, and $q \in Q'$ such that $q \in X$ (i.e., for each q satisfying $t(q) = X$), the state $\delta^*(X, u)$ is accepting in $\mathcal{A}$ if and only if $\delta^{\diamond *}(q, u)$ is accepting in $\mathcal{A}^\diamond$. This can be shown by an induction on the length of u.

It only remains to show that t satisfies (II) in the statement of the theorem; (I) and (III) follow directly from the definition of the initial and final states in $\mathcal{A}$.

First, for each $w, u \in \{0, 1\}^*$, the state $\delta^*(q_1, w \cdot u) = \delta^*(\delta^*(q_1, w), u)$ is accepting in $\mathcal{A}$ if and only if $\delta^{\diamond *}(q_1', w \cdot u) = \delta^{\diamond *}(\delta^{\diamond *}(q_1', w), u)$ is accepting in $\mathcal{A}^\diamond$, since $\mathcal{A}$ is equivalent to $\mathcal{A}^\diamond$. This gives that $\delta^*(q_1, w)$ is equivalent to $\delta^{\diamond *}(q_1', w)$, for each $w \in \{0, 1\}^*$.

However, using $q = \delta^{\diamond *}(q_1', w)$ in Claim above, we see that $\delta^{\diamond *}(q_1', w)$ is equivalent to $t(\delta^{\diamond *}(q_1', w))$. Thus, $\delta^*(q_1, w)$ is equivalent to $t(\delta^{\diamond *}(q_1', w))$. But $\mathcal{A}$ is optimal and hence $\delta^*(q_1, w) = t(\delta^{\diamond *}(q_1', w))$. Using (5), we then get $\delta^*(q_1, w) = t(\delta'^*(q_1', a^{\text{num } w}))$. □

The next theorem shows closure of DFAs accepting binary coded regular languages under the following nonstandard language operation: by modifying the set of accepting states (but keeping the structure of transitions), we obtain a DFA accepting a binary coded unary regular language again. This is later used to obtain some lower bounds.

Theorem 3. *Let $\mathcal{A} = (Q, \{0, 1\}, \delta, q_1, F)$ be an optimal DFA accepting a binary coded unary regular language and let $\mathcal{B}$ be a binary DFA that agrees with $\mathcal{A}$ in all components except for the set of accepting states. Then $\mathcal{B}$ accepts a binary coded unary regular language as well—but $\mathcal{B}$ is not necessarily optimal and $\mathcal{L}(\mathcal{B})$ may differ from $\mathcal{L}(\mathcal{A})$.*

More precisely, if $\mathcal{A}' = (Q', \{a\}, \delta', q_1', F')$ is the optimal unary DFA accepting $\text{num } \mathcal{L}(\mathcal{A})$ and t is the function preserving the machine's behavior, satisfying (I)–(III) in Theorem 2, then there exists a unary DFA $\mathcal{B}'$ accepting $\text{num } \mathcal{L}(\mathcal{B})$ that *agrees with $\mathcal{A}'$ in all components except for accepting states such that, for each $q \in Q'$, q is accepting in $\mathcal{B}'$ if and only if $t(q)$ is accepting in $\mathcal{B}$. Neither $\mathcal{B}'$ is necessarily optimal.*

Proof (sketched). Since $\mathcal{A}$ accepts a binary coded unary regular language, there must exist an optimal DFA $\mathcal{A}'$ for this unary language, with bin $\mathcal{L}(\mathcal{A}') = \mathcal{L}(\mathcal{A})$. But then, by Theorem 2, there must exist a function $t : Q' \to Q$ satisfying (I)–(III) in the statement of this theorem,[3] namely,

(I) $q_I = t(q'_I)$,
(II) $\delta^*(q_I, w) = t(\delta'^*(q'_I, a^{\mathrm{num}\, w}))$, for each $w \in \{0, 1\}^*$,
(III) $t(q) \in F$ if and only if $q \in F'$, for each $q \in Q'$.

Now, let $\mathcal{B} = (R, \{0, 1\}, \vartheta, r_I, G)$ be any DFA that agrees with $\mathcal{A}$ in all components except for accepting states, i.e., $R = Q$, $\vartheta = \delta$, and $r_I = q_I$, with G possibly different from F. This gives the following relation between $\mathcal{A}'$ and $\mathcal{B}$:

- $r_I = t(q'_I)$,
- $\vartheta^*(r_I, w) = t(\delta'^*(q'_I, a^{\mathrm{num}\, w}))$, for each $w \in \{0, 1\}^*$,

since (I) and (II) do not depend on accepting states.

Next, let $\mathcal{B}' = (R', \{a\}, \vartheta', r'_I, G')$ be a unary DFA that agrees with $\mathcal{A}'$ in all components except for accepting states, i.e., $R' = Q'$, $\vartheta' = \delta'$, and $r'_I = q'_I$; G' to be specified later. This $\mathcal{B}'$ is related to $\mathcal{B}$ as follows:

(I') $r_I = t(r'_I)$,
(II') $\vartheta^*(r_I, w) = t(\vartheta'^*(r'_I, a^{\mathrm{num}\, w}))$, for each $w \in \{0, 1\}^*$.

By fixing accepting states in $\mathcal{B}'$ properly, we can satisfy (III) in Theorem 2 as well: by definition, $q \in R'$ is included in G' if and only if $t(q) \in G$. Then

(III') $t(q) \in G$ if and only if $q \in G'$, for each $q \in R'$.

This gives that $\vartheta'^*(r'_I, a^{\mathrm{num}\, w}) \in G'$ if and only if $t(\vartheta'^*(r'_I, a^{\mathrm{num}\, w})) \in G$, that is, if and only if $\vartheta^*(r_I, w) \in G$. Thus, $a^{\mathrm{num}\, w}$ is accepted by $\mathcal{B}'$ if and only if w is accepted by $\mathcal{B}$, and hence $\mathcal{L}(\mathcal{B}) = \text{bin}\, \mathcal{L}(\mathcal{B}')$, with $\mathcal{L}(\mathcal{B}') \subseteq \{a\}^*$.   □

4   A Loop

Recall that a unary DFA $\mathcal{A}'$ uses an initial segment $\mathbb{Z}_\sigma^\triangleright$ and a loop $\mathbb{Z}_\lambda$, and hence, by Theorem 2, the state set of the optimal binary $\mathcal{A}$ accepting bin $\mathcal{L}(\mathcal{A}')$ is $t(\mathbb{Z}_\sigma^\triangleright \cup \mathbb{Z}_\lambda)$, where t is a function satisfying (I)–(III) in the statement of this theorem. That is, the state set of $\mathcal{A}$ consists of a *preamble* $t(\mathbb{Z}_\sigma^\triangleright) = \{t(q) : q \in \mathbb{Z}_\sigma^\triangleright\}$ and a *kernel* $t(\mathbb{Z}_\lambda) = \{t(q) : q \in \mathbb{Z}_\lambda\}$; these two sets are *not* necessarily disjoint.

---

[3] Such mapping t does exist even if the optimal DFA accepting bin $\mathcal{L}(\mathcal{A}')$ has not been obtained by the use of Theorem 2, since the optimal DFA $\mathcal{A}$ is always unique, up to isomorphism (see, e.g., [13]).

Let us express the length of the loop in $\mathcal{A}'$ in the form

$$\lambda = \mu \cdot 2^\ell, \text{ where } \mu \text{ is odd and } \ell \geq 0.$$

By the Chinese Remainder Theorem, each $i \in \mathbb{Z}_\lambda = \{0, \ldots, \lambda - 1\}$ is unambiguously determined by its *Chinese Residual Representation*, that is, it can be expressed in the form $i = \langle m, d \rangle$, where $m = i \bmod \mu$ and $d = i \bmod 2^\ell$. (See, e.g., [3,4,10].) In addition, the value d is fully determined by the last ℓ bits in the binary representation of i. Namely, if $i = $ num "$b_{k-1} \cdots b_{\ell-1} \cdots b_0$", then $d = (\sum_{j=0}^{k-1} b_j \cdot 2^j) \bmod 2^\ell = (\sum_{j=0}^{\ell-1} b_j \cdot 2^j) \bmod 2^\ell = (\text{num } "b_{\ell-1} \ldots b_0") \bmod 2^\ell = $ num "$b_{\ell-1} \ldots b_0$", since $2^j \bmod 2^\ell = 0$ for each $j \geq \ell$ and num "$b_{\ell-1} \ldots b_0$" $< 2^\ell$.

For this reason, $i \in \mathbb{Z}_\lambda$ can also be identified with $\langle m, \text{num } "b_{\ell-1} \ldots b_0" \rangle$. Thus, without actually changing the structure of transitions, we can replace computing modulo λ, as presented in (3), by computing just modulo μ but, in addition, we have to keep track about the latest ℓ bits of the prefix which, so far, has been read from the binary input. (See also Fig. 3.)

As an example, with $\lambda = 12 = 3 \cdot 2^2$, our DFA $\mathcal{A}^\diamond$ in Fig. 2 gets from $i_1 = 2$ to $i_2 = 5$ by the sequence of transitions $2 \xrightarrow{0} 4 \xrightarrow{0} 8 \xrightarrow{1} 17 \bmod 12 = 5$. Now, computing modulo $\mu = 3$ and keeping the latest $\ell = 2$ bits, this path can also be displayed in the form $\langle 2, \text{num } "10" \rangle \xrightarrow{0} \langle 1, \text{num } "00" \rangle \xrightarrow{0} \langle 2, \text{num } "00" \rangle \xrightarrow{1} \langle 2, \text{num } "01" \rangle$. Finally, by (7), the optimal $\mathcal{A}$ in Fig. 3 must follow the path $t(\langle 2, \text{num } "10" \rangle) \xrightarrow{0} t(\langle 1, \text{num } "00" \rangle) \xrightarrow{0} t(\langle 2, \text{num } "00" \rangle) \xrightarrow{1} t(\langle 2, \text{num } "01" \rangle)$, not excluding repetitions, i.e., $t(j') = t(j'')$ for some $j' \neq j''$.

Because of such "long jumps", it is not obvious at the first glance that the states in the kernel $t(\mathbb{Z}_\lambda)$ must always form a single strongly connected component. This is proved by the following lemma:

Lemma 1. *Let $\mathcal{A}'$ be an optimal unary DFA using $\lambda = \mu \cdot 2^\ell$ states in the loop, with μ odd, and let $\mathcal{A}$ be the optimal DFA accepting $\text{bin } \mathcal{L}(\mathcal{A}')$. Then the states in the kernel of $\mathcal{A}$ are strongly connected, that is, for each $i_1, i_2 \in \mathbb{Z}_\lambda$, there exists $w_{i_1,i_2} \in \{0,1\}^*$ such that $\mathcal{A}$ gets from $t(i_1)$ to $t(i_2)$ by reading w_{i_1,i_2}.*

Proof (sketched). Let $i_1, i_2 \in \mathbb{Z}_\lambda$ be given by their Chinese Residual Representation, as $i_1 = \langle m_1, d_1 \rangle$ and $i_2 = \langle m_2, d_2 \rangle$. Now, define:

$$i_\mu = \text{the smallest nonnegative integer satisfying } (2^{i_\mu} \cdot 2) \bmod \mu = 1,$$
$$r = 1 + \lfloor \log \mu \rfloor + \ell,$$
$$u_{i_1,i_2} = (d_2 + (m_2 - d_2) \cdot 2^{i_\mu \cdot \ell} \cdot 2^\ell - m_1 \cdot 2^r) \bmod (\mu \cdot 2^\ell),$$
$$w_{i_1,i_2} = \text{bin}_r \, u_{i_1,i_2}.$$

Since the value u_{i_1,i_2} is computed modulo $\mu \cdot 2^\ell$, it is smaller than $\mu \cdot 2^\ell$, and hence it can be written down by the use of at most $r = 1 + \lfloor \log \mu \rfloor + \ell$ bits. We can write it down by the use of exactly r bits, with leading zeros if necessary. This gives the binary string $w_{i_1,i_2} = \text{bin}_r \, u_{i_1,i_2}$ of length $|w_{i_1,i_2}| = r$.

Now, let $\mathcal{A}^\diamond$ be the binary counterpart of $\mathcal{A}'$ (not necessarily optimal) from Theorem 1. Using (6), we see that $\mathcal{A}^\diamond$, by reading the string w_{i_1,i_2} from the input, gets from $i_1 = \langle m_1, d_1 \rangle$ to the state $i' = (i_1 \cdot 2^{|w_{i_1,i_2}|} + \text{num } w_{i_1,i_2}) \bmod \lambda = $

$(i_1 \cdot 2^r + u_{i_1,i_2})$ mod λ. We are now going to show that $i' = i_2$ by showing that these two values agree in the Chinese Residual Representation.

To compute i' mod μ, we shall need: i_1 mod $\mu = m_1$, $(x$ mod $(\mu \cdot 2^\ell))$ mod $\mu = x$ mod μ for each $x \geq 0$, $(2^{i_\mu} \cdot 2)$ mod $\mu = 1$, and m_2 mod $\mu = m_2$:

$$\begin{aligned}
i' \bmod \mu &= ((i_1 \cdot 2^r + u_{i_1,i_2}) \bmod (\mu \cdot 2^\ell)) \bmod \mu = (m_1 \cdot 2^r + u_{i_1,i_2}) \bmod \mu \\
&= (m_1 \cdot 2^r + d_2 + (m_2 - d_2) \cdot 2^{i_\mu \cdot \ell} \cdot 2^\ell - m_1 \cdot 2^r) \bmod \mu \\
&= (d_2 + (m_2 - d_2) \cdot (2^{i_\mu} \cdot 2)^\ell) \bmod \mu \\
&= (d_2 + (m_2 - d_2) \cdot 1) \bmod \mu = m_2 \bmod \mu = m_2 \,.
\end{aligned}$$

Similarly, using $(x$ mod $(\mu \cdot 2^\ell))$ mod $2^\ell = x$ mod 2^ℓ for each $x \geq 0$, 2^r mod $2^\ell = 0$ (this follows from $r \geq \ell$), and d_2 mod $2^\ell = d_2$, we get

$$\begin{aligned}
i' \bmod 2^\ell &= ((i_1 \cdot 2^r + u_{i_1,i_2}) \bmod (\mu \cdot 2^\ell)) \bmod 2^\ell = u_{i_1,i_2} \bmod 2^\ell \\
&= (d_2 + (m_2 - d_2) \cdot 2^{i_\mu \cdot \ell} \cdot 2^\ell - m_1 \cdot 2^r) \bmod 2^\ell = d_2 \bmod 2^\ell = d_2 \,.
\end{aligned}$$

Summing up, by reading the string w_{i_1,i_2}, the DFA $\mathcal{A}^\diamond$ gets from $i_1 = \langle m_1, d_1 \rangle$ to the state $i' = \langle m_2, d_2 \rangle = i_2$, and hence, by (7), the optimal DFA $\mathcal{A}$ has the corresponding computation path $t(i_1) \xrightarrow{w_{i_1,i_2}} t(i_2)$. □

Lemma 2. *Let $\mathcal{A}'$ be an optimal unary DFA using $\lambda = \mu \cdot 2^\ell$ states in the loop, with μ odd, and let $\mathcal{A}$ be the optimal DFA accepting* bin $\mathcal{L}(\mathcal{A}')$. *Then $t(j_1) \neq t(j_2)$, for each $j_1, j_2 \in \mathbb{Z}_\lambda$ such that j_1 mod $\mu \neq j_2$ mod μ. In addition, $\delta(t(j_1), b) \neq \delta(t(j_2), b)$, for some $b \in \{0, 1\}$.*

The argument uses similar techniques as in Lemma 1; from the given pair of states that differ in the related residues modulo μ, we can reach a pair of states that differ in acceptance/rejection, by a carefully constructed string. The situation is different for pairs that agree in the related residues modulo μ:

Lemma 3. *Let $\mathcal{A}'$ be an optimal unary DFA using $\lambda = \mu \cdot 2^\ell$ states in the loop, with μ odd, and let $\mathcal{A}^\diamond, \mathcal{A}$ be the DFAs accepting* bin $\mathcal{L}(\mathcal{A}')$, *constructed in Theorems 1 and 2. Then, for each $i_1, i_2 \in \mathbb{Z}_\lambda$ such that i_1 mod $\mu = i_2$ mod μ and each $w \in \{0, 1\}^*$ of length at least ℓ, $\delta^{\diamond*}(i_1, w) = \delta^{\diamond*}(i_2, w)$ and $\delta^*(t(i_1), w) = \delta^*(t(i_2), w)$.*

Proof (sketched). Let $i_1, i_2 \in \mathbb{Z}_\lambda$ be given by their Chinese Residual Representation, as $i_1 = \langle m, d_1 \rangle$ and $i_2 = \langle m, d_2 \rangle$, and let w be an arbitrary binary string of length at least ℓ. Therefore, $2^{|w|}$ mod $2^\ell = 0$. But then

$$\begin{aligned}
(i_1 \cdot 2^{|w|} + \text{num } w) \bmod \mu &= (m \cdot 2^{|w|} + \text{num } w) \bmod \mu = i_2 \cdot 2^{|w|} + \text{num } w) \bmod \mu \,, \\
(i_1 \cdot 2^{|w|} + \text{num } w) \bmod 2^\ell &= (\text{num } w) \bmod 2^\ell = (i_2 \cdot 2^{|w|} + \text{num } w) \bmod 2^\ell.
\end{aligned}$$

Thus, $\delta^{\diamond*}(i_1, w), \delta^{\diamond*}(i_2, w)$ agree in residues modulo μ and modulo 2^ℓ, and hence $\delta^{\diamond*}(i_1, w) = \delta^{\diamond*}(i_2, w)$, which gives $\delta^*(t(i_1), w) = \delta^*(t(i_2), w)$, by (7). □

Let us now concentrate on a unary DFA $\mathcal{A}'$ with a loop of even length, i.e., with $\lambda = \mu \cdot 2^\ell$, where μ is odd and $\ell \geq 1$. It turns out that here the important role is played by pairs of states that are located at the opposite positions across the loop, that is, $i_1, i_2 \in \mathbb{Z}_\lambda$ such that $i_2 = (i_1 + \lambda/2)$ mod $\lambda = (i_1 + \mu \cdot 2^{\ell-1})$ mod λ. We shall call such pairs *twin states*, or *twins*, for short.

Lemma 4. *Let $\lambda = \mu\cdot 2^\ell$, where μ is odd and $\ell \geq 1$. Then $i_1, i_2 \in \mathbb{Z}_\lambda$ are twins if and only if they agree in their residues modulo μ and the binary representations of their residues modulo 2^ℓ differ only in the ℓ-th bit from the right. That is, given by Chinese Residual Representation, $i_1 = \langle m, \text{num} \text{``} b_1\beta''\rangle$ and $i_2 = \langle m, \text{num} \text{``} b_2\beta''\rangle$, for some m and some bits b_1, b_2 satisfying $b_1 \neq b_2$, with $|\beta| = \ell - 1$.*

Consider now behavior of $t(i_1), t(i_2)$ in $\mathcal{A}$, the optimal DFA for $\text{bin}\,\mathcal{L}(\mathcal{A}')$, if i_1, i_2 are twins. First, by (3), we easily obtain that $\delta^\diamond(\langle m, \text{num} \text{``} b_1\beta''\rangle, b) = \langle m\cdot 2 + b, \text{num} \text{``} \beta b''\rangle = \delta^\diamond(\langle m, \text{num} \text{``} b_2\beta''\rangle, b)$. That is, after reading any symbol $b \in \{0, 1\}$ from the input, the distinguishing bit disappears: $\delta^\diamond(i_1, b) = \delta^\diamond(i_2, b)$. But then, using (7), we get that $\delta(t(i_1), b) = t(\delta^\diamond(i_1, b)) = t(\delta^\diamond(i_2, b)) = \delta(t(i_2), b)$. This gives:

$$\delta^*(t(i_1), w) = \delta^*(t(i_2), w)\,, \quad \text{for each twin pair } i_1, i_2 \text{ and each } w \neq \varepsilon. \qquad (8)$$

Consequently, $t(i_1), t(i_2)$ are not equivalent (and hence not equal) if and only if one of them is accepting while the other one is rejecting.

Theorem 4. *Let $\mathcal{A}'$ be an optimal unary DFA using $\lambda = \mu\cdot 2^\ell$ states in the loop, with μ odd and $\ell \geq 1$. Then the optimal DFA $\mathcal{A}$ accepting $\text{bin}\,\mathcal{L}(\mathcal{A}')$ uses exactly λ states in its kernel if and only if, for each pair of twin states $i_1, i_2 \in \mathbb{Z}_\lambda$, one of i_1, i_2 is accepting while the other one is rejecting.*

Proof (sketched). Assume first that, for each pair of twin states $i_1, i_2 \in \mathbb{Z}_\lambda$, one of i_1, i_2 is accepting while the other one is rejecting in $\mathcal{A}'$. Suppose also, for contradiction, that the loop $\mathbb{Z}_\lambda$ contains some different states j_1, j_2 such that $t(j_1) = t(j_2)$. There are now two subcases:

First, if $j_1 \bmod \mu \neq j_2 \bmod \mu$, then, by Lemma 2, the states $t(j_1), t(j_2)$ in the kernel of $\mathcal{A}$ are different, a contradiction.

Second, if $j_1 \bmod \mu = j_2 \bmod \mu$, these values must differ in residues modulo 2^ℓ, and hence the binary representations of these residues must differ in at least one bit. Thus, given by their Chinese Residual Representation, $j_1 = \langle m, \text{num} \text{``} \beta_1 b_1\beta''\rangle$ and $j_2 = \langle m, \text{num} \text{``} \beta_2 b_2\beta''\rangle$, for some $m \in \{0, \ldots, \mu-1\}$, some $\beta_1, \beta_2, \beta \in \{0, 1\}^*$, and some bits $b_1, b_2 \in \{0, 1\}$, satisfying $|\beta_1 b_1\beta| = |\beta_2 b_2\beta| = \ell$ and $b_1 \neq b_2$. Here b_1, b_2 represent the first pair of different bits from the right.

But then, using (6) and (7), we see that, by reading the string $0^{\ell-|\beta|-1}$ from the input, $\mathcal{A}$ gets from $t(j_1)$ to $t(i_1) = t(\langle m\cdot 2^{\ell-|\beta|-1}, \text{num} \text{``} b_1 30^{\ell-|\beta|-1}''\rangle)$ and from $t(j_2)$ to the state $t(i_2) = t(\langle m\cdot 2^{\ell-|\beta|-1}, \text{num} \text{``} b_2 30^{\ell-|\beta|-1}''\rangle)$. Clearly, if $t(j_1) = t(j_2)$, then $t(i_1) = t(i_2)$ as well.

However, i_1, i_2 agree in residues modulo μ and their residues modulo 2^ℓ differ only in the ℓ-th bit from the right, $b_1 \neq b_2$. Thus, by Lemma 4, i_1, i_2 are twins and hence, by assumption, one of i_1, i_2 is accepting while the other one is rejecting in $\mathcal{A}'$. But then, by (III) in Theorem 2, one of the states $t(i_1), t(i_2)$ in $\mathcal{A}$ is accepting while the other one is rejecting, which contradicts $t(i_1) = t(i_2)$.

Thus, if each pair of twin states $i_1, i_2 \in \mathbb{Z}_\lambda$ is such that one of them is accepting while the other one is rejecting, there are no different states $j_1, j_2 \in \mathbb{Z}_\lambda$ such that $t(j_1) = t(j_2)$ in $\mathcal{A}$, and hence $t(\mathbb{Z}_\lambda)$ contains λ states, all different.

Assume now that there exists a pair of twin states $i_1, i_2 \in \mathbb{Z}_\lambda$ such that either both of them are accepting or both of them are rejecting in $\mathcal{A}'$.

But, by (III) in Theorem 2, either both $t(i_1)$ and $t(i_2)$ are accepting or both of them are rejecting in $\mathcal{A}$. Taking into account that $\delta^*(t(i_1), w) = \delta^*(t(i_2), w)$ for each $w \neq \varepsilon$, by (8), we see that $t(i_1), t(i_2)$ are equivalent, that is, $t(i_1) = t(i_2)$. Therefore, the kernel $t(\mathbb{Z}_\lambda)$ contains at most $\lambda - 1$ different states. $\square$

The corresponding condition for unary DFAs is simpler: if each pair of twin states i_1, i_2 in a unary DFA $\mathcal{A}'$ is such that either both of them are accepting or both of them are rejecting, then $\mathcal{A}'$ can be replaced by an equivalent DFA with $\lambda/2$ states in its loop, and hence it is not optimal.

We are now ready to present the lower bound on the size of the kernel:

Theorem 5. *Let $\mathcal{A}'$ be an optimal unary DFA using $\lambda = \mu \cdot 2^\ell$ states in the loop, with μ odd, and let $\mathcal{A}$ be the optimal DFA accepting* $\mathrm{bin}\, \mathcal{L}(\mathcal{A}')$. *Then the kernel of $\mathcal{A}$ must contain at least $\mu + \ell$ states.*

Proof (sketched). Let us begin with the special case of $\ell = 0$. In this case $\lambda = \mu$, with μ odd. By Lemma 2, for each $j_1, j_2 \in \mathbb{Z}_\lambda$ such that they do not agree in their residues modulo μ, the states $t(j_1), t(j_2)$ in the kernel of $\mathcal{A}$ are different. This implies that $t(0), \ldots, t(\mu - 1)$ is a sequence of $\mu = \mu + \ell$ different states.

Consider a loop of even length $\lambda = \mu \cdot 2^\ell$. Since the unary $\mathcal{A}'$ is optimal, there exists a pair of twins $i_1, i_2 \in \mathbb{Z}_\lambda$ such that one of them is accepting while the other one is rejecting, or else $\mathcal{A}'$ could be replaced by an equivalent DFA with only $\lambda/2$ states in its loop. But then, by (III) in Theorem 2, one of $t(i_1), t(i_2)$ is accepting while the other one is rejecting in $\mathcal{A}$, and hence $t(i_1) \neq t(i_2)$.

Now we proceed by induction on ℓ. First, let $\ell = 1$. As shown above, the sequence $t(0), \ldots, t(\mu-1)$ consists of μ different states. In addition, by Lemma 4, the twins i_1, i_2 agree in residues modulo μ, that is, $i_1 \bmod \mu = i_2 \bmod \mu = j$, for some $j \in \{0, \ldots, \mu-1\}$. But then $t(0), \ldots, t(j-1), t(i_1), t(i_2), t(j+1), \ldots, t(\mu-1)$ is a sequence consisting of $\mu + 1 = \mu + \ell$ different states.

Next, let $\ell \geq 2$ and let the statement of the theorem hold for $\ell' = \ell - 1$ by induction. For contradiction, assume that the kernel of $\mathcal{A}$ contains less than $\mu + \ell$ states. To complete the proof, it is enough to show that this leads to existence of an optimal unary DFA $\mathcal{B}'''$ with $\mu \cdot 2^{\ell-1}$ states in the loop, such that kernel of the optimal binary counterpart of $\mathcal{B}'''$ contains less than $\mu + \ell - 1$ states. Note that $\mathcal{L}(\mathcal{B}''')$ may differ from $\mathcal{L}(\mathcal{A}')$.

Fixing Special States. Using Lemma 4 and $\ell \geq 2$, the Chinese Residual Representation of the twins i_1, i_2 can be expressed as $i_1 = \langle m, \mathrm{num}\,``b_1 \beta b''\rangle$ and $i_2 = \langle m, \mathrm{num}\,``b_2 \beta b''\rangle$, for some m, β, and some bits b_1, b_2, b, satisfying $b_1 \neq b_2$.

Now, there exists $m' \in \{0, \ldots, \mu - 1\}$ such that $(m' \cdot 2 + b) \bmod \mu = m$. Using (3) and (7), we see that $t(i_1), t(i_2)$ can be reached by the following transitions:

$$\left.\begin{array}{l} t(j_{0,1}) \overset{\text{df.}}{=} t(\langle m', \mathrm{num}\,``0b_1\beta''\rangle) \\ t(j_{1,1}) \overset{\text{df.}}{=} t(\langle m', \mathrm{num}\,``1b_1\beta''\rangle) \end{array}\right\} \xrightarrow{\ b\ } t(\langle m, \mathrm{num}\,``b_1\beta b''\rangle) = t(i_1),$$

$$\left.\begin{array}{l} t(j_{0,2}) \overset{\text{df.}}{=} t(\langle m', \mathrm{num}\,``0b_2\beta''\rangle) \\ t(j_{1,2}) \overset{\text{df.}}{=} t(\langle m', \mathrm{num}\,``1b_2\beta''\rangle) \end{array}\right\} \xrightarrow{\ b\ } t(\langle m, \mathrm{num}\,``b_2\beta b''\rangle) = t(i_2).$$

By a rather long but not too difficult argument, one can show that

(I) either $t(j_{e_1,1}) \neq_1 t(i_1)$ and $t(j_{e_1,1}) \neq_1 t(i_2)$, for each $e_1 \in \{0,1\}$,
(II) or $t(j_{e_2,2}) \neq_1 t(i_1)$ and $t(j_{e_2,2}) \neq_1 t(i_2)$, for each $e_2 \in \{0,1\}$, or both.

Here "$q =_1 r$" means that $\delta(q, b') = \delta(r, b')$ for each $b' \in \{0,1\}$. This relation is reflexive, symmetric, and transitive, but $q =_1 r$ does not imply that $q = r$. Thus, "$q \neq_1 r$" means that $\delta(q, b') \neq \delta(r, b')$, for some bit $b' \in \{0,1\}$.
From this point forward, our reasoning proceeds on the assumption that (I) holds true; the case of (II) is symmetrical and hence omitted.

Fixing a Pair of Smaller Machines. By changing accepting states in the binary $\mathcal{A} = (Q, \{0,1\}, \delta, q_1, F)$, let us construct a new binary DFA $\mathcal{B}$ that agrees with $\mathcal{A}$ in all components, but the new set of accepting states is

$$G = \{t(k) \in t(\mathbb{Z}_\lambda) : t(k) =_1 t(j_{0,1})\}. \tag{9}$$

By (I), we have $t(i_1) \neq_1 t(j_{0,1})$ and $t(i_2) \neq_1 t(j_{0,1})$, and hence, in the new DFA $\mathcal{B}$, both $t(i_1)$ and $t(i_2)$ are rejecting. Recall that $t(i_1) \neq t(i_2)$ and i_1, i_2 are twins. But then $\delta^*(t(i_1), w) = \delta^*(t(i_2), w)$ for each $w \neq \varepsilon$, by (8). Thus, in $\mathcal{B}$, the states $t(i_1), t(i_2)$ are different but equivalent. Consequently, the new DFA $\mathcal{B}$ is not optimal: by minimization, we can save at least one state and hence *the kernel of the optimal DFA accepting $\mathcal{L}(\mathcal{B})$ contains less than $\mu + \ell - 1$ states.*

Next, by Theorem 3, $\mathcal{B}$ accepts a binary coded unary regular language and there exists a unary DFA $\mathcal{B}'$ that agrees with $\mathcal{A}'$ in all components (thus, with the same loop $\mathbb{Z}_\lambda$), except for accepting states, with $\text{bin}\,\mathcal{L}(\mathcal{B}') = \mathcal{L}(\mathcal{B})$. Moreover, for each $k \in \mathbb{Z}_\lambda$, k is accepting in $\mathcal{B}'$ if and only if $t(k)$ is accepting in $\mathcal{B}$.

It can be shown that each pair of twin states $k_1', k_2' = k_1' + \mu \cdot 2^{\ell-1}$ is equivalent in $\mathcal{B}$: by (8), we have $\delta^*(t(k_1'), w) = \delta^*(t(k_2'), w)$ for each $w \neq \varepsilon$, and hence also $t(k_1') =_1 t(k_2')$. But then, by (9), either both of them are accepting in $\mathcal{B}$, or both of them are rejecting in $\mathcal{B}$.

This implies that $\mathcal{B}'$ can be replaced by an equivalent $\mathcal{B}''$ with the same initial segment, but using the loop $\mathbb{Z}_{\lambda/2} = \{0, \ldots, \lambda/2 - 1\}$ instead of $\mathbb{Z}_\lambda$, counting modulo $\lambda/2 = \mu \cdot 2^{\ell-1}$. The new set of accepting states is $G \cap \mathbb{Z}_{\lambda/2}$. By a rather lengthy argument, one can show that *the size of the loop* in $\mathcal{B}''$ is optimal, i.e., it cannot be replaced by a shorter loop. But then *the optimal DFA accepting $\mathcal{L}(\mathcal{B}'')$ uses exactly $\mu \cdot 2^{\ell-1}$ states in its loop.* This optimal automaton, denoted here by $\mathcal{B}'''$, may differ from $\mathcal{B}''$ in a smaller initial segment.

Conclusion. If there exists an optimal unary DFA $\mathcal{A}'$ with $\lambda = \mu \cdot 2^\ell$ states in its loop such that the optimal DFA accepting $\text{bin}\,\mathcal{L}(\mathcal{A}')$ uses less than $\mu + \ell$ states in its kernel, then there must exist an optimal unary DFA $\mathcal{B}'''$ with $\mu \cdot 2^{\ell-1}$ states in its loop such that the optimal DFA accepting $\text{bin}\,\mathcal{L}(\mathcal{B}''') = \text{bin}\,\mathcal{L}(\mathcal{B}'') = \text{bin}\,\mathcal{L}(\mathcal{B}') = \mathcal{L}(\mathcal{B})$ uses less than $\mu + \ell - 1$ states in its kernel. But this contradicts the induction hypothesis, and hence such DFA $\mathcal{A}'$ does not exist. $\qquad\square$

5   An Initial Segment

Now we shall concentrate on $t(\mathbb{Z}_\sigma^\triangleright) \setminus t(\mathbb{Z}_\lambda)$, that is, on states in the preamble but outside the kernel. As can be seen from the example in Fig. 3, the sets $t(\mathbb{Z}_\sigma^\triangleright), t(\mathbb{Z}_\lambda)$ are not necessarily disjoint. In general, each state in $t(\mathbb{Z}_\sigma^\triangleright) \cap t(\mathbb{Z}_\lambda)$ can be expressed as $q = t(j^\triangleright) = t(i)$, for some $j^\triangleright \in t(\mathbb{Z}_\sigma^\triangleright)$ and $i \in t(\mathbb{Z}_\lambda)$. In this case, the value j must respect the corresponding residue modulo μ:

Lemma 5. *Let $\mathcal{A}'$ be an optimal unary DFA using σ states in the initial segment and $\lambda = \mu \cdot 2^\ell$ states in the loop, with μ odd, and let $\mathcal{A}$ be the optimal DFA accepting* bin $\mathcal{L}(\mathcal{A}')$. *Then $j \bmod \mu = i \bmod \mu$, for each $j \in \{0, \ldots, (\sigma-1)\}$ and each $i \in \{0, \ldots, (\lambda-1)\}$ such that $t(j^\triangleright) = t(i)$.*

Using the above lemma and the fact that the binary coding allow leading zeros—i.e., $w \in \mathcal{L}(\mathcal{A})$ if and only if $0^\ell w \in \mathcal{L}(\mathcal{A})$—one can show:

Lemma 6. *Let $\mathcal{A}'$ be an optimal unary DFA using σ states in the initial segment and let $\mathcal{A}$ be the optimal DFA accepting* bin $\mathcal{L}(\mathcal{A}')$. *Then the initial state of $\mathcal{A}$ is not in the kernel, if $\sigma > 0$.*

We are now ready to present the lower bound on the size of the preamble:

Theorem 6. *Let $\mathcal{A}'$ be an optimal unary DFA using σ states in the initial segment and $\lambda = \mu \cdot 2^\ell$ states in the loop, with μ odd, and let $\mathcal{A}$ be the optimal DFA accepting* bin $\mathcal{L}(\mathcal{A}')$. *Then $\mathcal{A}$ must use at least $\max\{1, 1 + \lceil \log \sigma \rceil - \ell\}$ states in the preamble but outside the kernel, except for $\sigma = 0$, with no preamble at all.*

Proof (sketched). Assume first that $\sigma \geq 2$. By (4) and (7), the binary DFA $\mathcal{A}$ gets from $q_\mathrm{I} = t(0^\triangleright)$ to $t((\sigma-1)^\triangleright)$ by reading $\mathrm{bin}(\sigma-1)$, with $1 + \lceil \log \sigma \rceil$ states along this path, including $t(0^\triangleright)$ and $t((\sigma-1)^\triangleright)$. There can be at most ℓ states in the kernel $t(\mathbb{Z}_\lambda)$ along this path:

Supposing for contradiction that at least $\ell + 1$ states are in $t(\mathbb{Z}_\lambda)$, this path can be split into two segments, namely, $t(0^\triangleright) \xrightarrow{\beta'} t(j^\triangleright)$ and $t(j^\triangleright) \xrightarrow{\beta} t((\sigma-1)^\triangleright)$, for some $t(j^\triangleright) \in t(\mathbb{Z}_\lambda)$, with $\beta'\beta = \mathrm{bin}(\sigma-1)$ and $|\beta| = \ell$. But then $t(j^\triangleright) = t(i')$, for some $i' \in \mathbb{Z}_\lambda$, and hence the computation path $t(j^\triangleright) \xrightarrow{\beta} t((\sigma-1)^\triangleright)$ can also be expressed as $t(i') \xrightarrow{\beta} t(i)$, for some $t(i) \in t(\mathbb{Z}_\lambda)$, with $t(i) = t((\sigma-1)^\triangleright)$. Moreover, $i = (\sigma-1) \bmod \lambda$: these values agree in residues modulo μ, by Lemma 5, and they agree in residues modulo 2^ℓ as well, because the same string β represents the last ℓ bits in the binary representations of both $i \bmod \lambda$ and $\sigma - 1$. Now, using $i = (\sigma-1) \bmod \lambda$, $t(i) = t((\sigma-1)^\triangleright)$, and (III) in Theorem 2, we get that $(\sigma-1) \bmod \lambda$ is accepting in $\mathcal{A}'$ if and only if $(\sigma-1)^\triangleright$ is accepting in $\mathcal{A}'$. But then the states $(\sigma-1) \bmod \lambda$, $(\sigma-1)^\triangleright$ are equivalent in $\mathcal{A}'$, which contradicts our assumption that $\mathcal{A}'$ is optimal.

Summing up, the path from $q_\mathrm{I} = t(0^\triangleright)$ to $t((\sigma-1)^\triangleright)$ consists of $1 + \lceil \log \sigma \rceil$ states, at most ℓ of them are in $t(\mathbb{Z}_\lambda)$, and all states outside $t(\mathbb{Z}_\lambda)$ are different, since the preamble $t(\mathbb{Z}_\sigma^\triangleright)$ is loop-free, except for the trivial loop reading "0" in q_I. Therefore, the preamble in $\mathcal{A}$ must contain at least $1 + \lceil \log \sigma \rceil - \ell$ states outside the kernel. Moreover, by Lemma 6, $q_\mathrm{I} = t(0^\triangleright)$ is never in the kernel, if $\sigma > 0$, which gives the lower bound $\max\{1, 1 + \lceil \log \sigma \rceil - \ell\}$. This covers also the special case of $\sigma = 1$, when $\max\{1, 1 + \lceil \log \sigma \rceil - \ell\} = \max\{1, 1 - \ell\} = 1$. □

6   Total Number of States

Here we begin by showing that the bounds presented in Theorems 5 and 6 are best possible, by presenting witness languages matching these bounds.

Theorem 7. *For each $\sigma \geq 0$, each odd $\mu \geq 1$, and each $\ell \geq 0$, there exists a unary language $\mathcal{L}'_{\sigma,\mu,\ell}$ such that the optimal DFA accepting $\mathcal{L}'_{\sigma,\mu,\ell}$ uses exactly σ states in the initial segment and exactly $\lambda = \mu \cdot 2^\ell$ states in the loop, while be the optimal DFA accepting bin $\mathcal{L}'_{\sigma,\mu,\ell}$ uses exactly $\max\{1, 1 + \lceil \log \sigma \rceil - \ell\}$ states in the preamble but outside the kernel (except for $\sigma = 0$, with no preamble) and exactly $\mu + \ell$ states in the kernel.*

We do not present the lengthy argument about the state complexities of $\mathcal{L}'_{\sigma,\mu,\ell}$ and bin $\mathcal{L}'_{\sigma,\mu,\ell}$ here—this will be done in a journal version of the paper—but content ourselves with presenting the witness language. For the given $\sigma \geq 0$, odd $\mu \geq 1$, and $\ell \geq 0$, let $\lambda = \mu \cdot 2^\ell$, $s = \max\{1, \lfloor (\sigma - 1)/2^{\ell-1} \rfloor\}$, and let

$$\mathcal{L}'_{\sigma,\mu,\ell} = \{a^k : k \bmod \lambda = (\sigma - 1) \bmod \lambda\} \setminus \{a^k : k \leq \sigma - 1 \text{ and}$$
$$\text{one of the strings bin } s, \text{bin } k \text{ is a prefix of the other}\}.$$

Note also that for $\sigma = 0$ we obtain $\mathcal{L}'_{\sigma,\mu,\ell} = \{a^k : k \bmod \lambda = (-1) \bmod \lambda\} \setminus \emptyset$, since there is no $a^k \in \{a\}^*$ with $k \leq -1$.

Now we shall consider the total number of states for the conversion. Let us begin with the upper bound:

Theorem 8. (i) *If a unary language $\mathcal{L}'$ is accepted by an optimal DFA $\mathcal{A}'$ using n states, then the optimal DFA $\mathcal{A}$ accepting bin $\mathcal{L}'$ uses at most n states.* (ii) *For each $n \geq 1$, there exists a unary language $\mathcal{L}'_n$ matching this bound, i.e., both the optimal $\mathcal{A}'_n$ for $\mathcal{L}'_n$ and the optimal $\mathcal{A}_n$ for bin $\mathcal{L}'_n$ use exactly n states.*

Proof (sketched). The universal upper bound in (i) follows trivially from Theorem 1. To provide a witness language for (ii), define:

$$\mathcal{L}'_n = \{a^k : k \bmod n \geq n/2\}.$$

The unary $\mathcal{A}'_n$ for $\mathcal{L}'_n$ counts modulo n, without any initial segment. By definition, the state j is accepting if and only if $j \bmod n \geq n/2$. It is easy to see that this DFA has no pair of different but equivalent states.

To see that the optimal $\mathcal{A}_n$ for bin $\mathcal{L}'_n$—constructed in Theorem 2—cannot use less than n states, we consider the following cases:

If n is odd, the loop in $\mathcal{A}'_n$ is of length $n = \lambda = \mu \cdot 2^0 = \mu$, where μ is odd. Clearly, all values in the sequence $0, \ldots, \mu - 1$ differ in their residues modulo μ. Hence, by Lemma 2, $t(0), \ldots, t(\mu - 1)$ is a sequence of $\mu = n$ different states.

If n is even, then j and $j + n/2$ form a pair of twin states, for each $j \in \{0, \ldots, n/2 - 1\}$. Moreover, j is rejecting and $j + n/2$ is accepting. But then, by Theorem 4, $\mathcal{A}_n$ uses exactly $\lambda = n$ states. $\square$

Theorem 9. (i) *If a unary language $\mathcal{L}'$ is accepted by an optimal* DFA $\mathcal{A}'$ *using n states, then the optimal* DFA $\mathcal{A}$ *accepting* bin $\mathcal{L}'$ *uses at least $1 + \lceil \log n \rceil$ states.* (ii) *For each $n \geq 1$, there exists a unary language $\mathcal{L}''_n$ matching this bound, i.e., the optimal* DFA *for $\mathcal{L}''_n$ uses exactly n states but the optimal* DFA *for* bin $\mathcal{L}''_n$ *exactly $1 + \lceil \log n \rceil$ states.*

Proof (sketched). (i) If $\mathcal{A}'$ uses n states, then $n = \sigma + \mu \cdot 2^\ell$, where $\sigma \geq 0$ denotes the length of the initial segment and $\lambda = \mu \cdot 2^\ell$ the length of the loop, with odd $\mu \geq 1$ and $\ell \geq 0$. If $\mathcal{A}'$ is optimal, then, by Theorems 6 and 5, the binary $\mathcal{A}$ must use at least $N = \max\{1, 1 + \lceil \log \sigma \rceil - \ell\} + (\mu + \ell)$ states, except for the case of $\sigma = 0$, when there is no preamble and $N = \mu + \ell$. There are now three cases, depending on whether $\sigma > 2^\ell$, $0 < \sigma \leq 2^\ell$, or $\sigma = 0$. For all of them, a more detailed evaluation of the formula reveals that $N \geq 1 + \lceil \log n \rceil$.

(ii) To provide a witness language, let us take ℓ, the unique integer satisfying $2^\ell \leq n < 2^{\ell+1}$. Clearly, n can be expressed as $n = \sigma + \mu \cdot 2^\ell$, for some $\sigma < n/2$ and $\mu = 1$. Now, let $\mathcal{L}''_n = \mathcal{L}'_{\sigma,\mu,\ell}$, where $\mathcal{L}'_{\sigma,\mu,\ell}$ is the unary language presented in Theorem 7. Note that $\sigma < n/2 < 2^\ell$. Therefore, the optimal DFA for $\mathcal{L}''_n$ uses exactly $\sigma + \mu \cdot 2^\ell = n$ states, while the optimal DFA for bin $\mathcal{L}''_n$ exactly $N = 1 + (\mu + \ell) = 2 + \ell$ states, if $\sigma > 0$, but exactly $N = \mu + \ell = 1 + \ell$ states, if $\sigma = 0$.

If $\sigma > 0$, then $n \geq 2$. But then $2^\ell = n - \sigma \leq n - 1$, which gives $N = 2 + \ell \leq 2 + \log(n-1)$, and hence $N \leq 2 + \lfloor \log(n-1) \rfloor$, since N is an integer. Now, using the fact that $\lceil \log n \rceil = 1 + \lfloor \log(n-1) \rfloor$ for each $n \geq 2$, we get $N \leq 1 + \lceil \log n \rceil$.

If $\sigma = 0$, then $N = 1 + \ell = 1 + \log n \leq 1 + \lceil \log n \rceil$. □

7   Other Properties

Let us now return to Theorem 8. By using 2^n instead of n, it states that, for each $n \geq 0$, there exists a unary language $\mathcal{L}'_{2^n}$ such that both the optimal DFA for $\mathcal{L}'_{2^n}$ and the optimal DFA for bin $\mathcal{L}'_{2^n}$ use exactly 2^n states. The witness language for this was $\mathcal{L}'_{2^n} = \{a^k : k \bmod 2^n \geq 2^{n-1}\}$. However, since $k \bmod 2^n$ is determined by the last n bits in the binary representation of k, it is easy to see that the binary coded version of $\mathcal{L}'_{2^n}$ can also be established as

$$\text{bin } \mathcal{L}'_{2^n} = \{w \in \{0,1\}^* : \text{the } n\text{-th bit from the end of} w \text{ is "1"}\},$$

which is actually the most popular witness example of an exponential blowup for removing nondeterminism in NFAs (see, e.g., [13, Sect. 2.3.6]).

It is well known that there exists a simple *nondeterministic* automaton for this language: this NFA moves along the given binary input w, nondeterministically guesses the position of the n-th bit from the end, after which it verifies that w ends by "$1 \cdot v$", for some v of length $n - 1$. Clearly, $n + 1$ states are sufficient for this task. This many states are also necessary, since no state along the segment "10^{n-1}" at the end of "$0^n 10^{n-1}$" can be repeated. This gives:

Corollary 1. *For each $n \geq 1$, there exists a binary coded unary language $\mathcal{L}$ such that the optimal* NFA *for $\mathcal{L}$ uses exactly n states, but both the optimal* DFA *for $\mathcal{L}$ and the optimal* DFA *for the unary version of $\mathcal{L}$ use exactly 2^{n-1} states.*

The conversion of a *unary* NFA to a binary DFA is less expensive: each unary NFA with n states can be replaced by an equivalent DFA with $e^{(1+o(1))\cdot\sqrt{n\cdot\ln n}}$ states—which is the growth rate of Landau's function [2,15,17], and then converted to a DFA for the binary version of the original language. by Theorem 1. The conversion of a unary NFA to a binary NFA is quadratic:

Theorem 10 (Binary version of Chrobak normal form [2,5,7]). *If a unary language $\mathcal{L}'$ is accepted by an NFA with n states, then its binary coded counterpart $\mathcal{L} = \mathrm{bin}\,\mathcal{L}'$ can be accepted by an NFA $\mathcal{A}$ consisting of:*

- *A deterministic preamble $\mathbb{Z}_\sigma^\triangleright$ using $\sigma \leq n^2 - n$ states, without any loops, except for the trivial loop reading the symbol "0" in the initial state.*
- *Deterministic strongly connected components $\mathbb{Z}_{\lambda_1}, \ldots, \mathbb{Z}_{\lambda_m}$ containing, respectively, $\lambda_1, \ldots, \lambda_m$ states, with $\lambda_1 + \cdots + \lambda_m \leq n - 1$. These components are disjoint and there are no transitions from any component to another.*
- *This NFA makes at most one nondeterministic choice during the course of the entire computation, by a single-step transition from $\mathbb{Z}_\sigma^\triangleright$ to some $\mathbb{Z}_{\lambda_i}$.*

In the special case of empty preamble, there is only one component $\mathbb{Z}_{\lambda_1}$, of size $\lambda_1 \leq n$, and $\mathcal{A}$ is deterministic.

It is easy to see that $\mathrm{bin}\,\mathcal{L}'_1 \cup \mathrm{bin}\,\mathcal{L}'_2 = \mathrm{bin}(\mathcal{L}'_1 \cup \mathcal{L}'_2)$, and hence the class of binary coded unary regular languages is closed under union; the same holds for intersection and complement. To illustrate different closure properties from binary regular languages in general, we point out that this class is not closed under concatenation. To see this, consider two finite unary languages, namely, $\mathcal{L}'_{\{1\}} = \{a^1\} = \{a\}$ and $\mathcal{L}'_{\{0\}} = \{a^0\} = \{\varepsilon\}$. Their binary coded versions can be given by regular expressions 0*1 and 0*, respectively, which gives 0*10* for concatenation. But $(\mathrm{bin}\,\mathcal{L}'_{\{1\}})\cdot(\mathrm{bin}\,\mathcal{L}'_{\{0\}})$ is binary coded $\{a^{2^k} : k \geq 0\}$.

We do not know whether the gap 2^{n-1} in Corollary 1 cannot be increased; however, the cost of removing nondeterminism cannot exceed 2^n, given by the standard subset construction. In this context, a promising line of research is studying nondeterministic automata; we are convinced that the quadratic cost in Theorem 10 can be improved. The representation of binary coded unary languages by two-way automata is worth studying as well.

Sometimes, e.g., in [8], a one-to-one mapping between integers and binary strings is obtained by making leading zeros significant, stating by convention that the most significant bit is hidden. That is, the binary inputs $\varepsilon, 0, 1, 00, 01, \ldots$ are interpreted as $1, 2, 3, 4, 5, \ldots$. This ensures that the same integer cannot be represented by two different strings, but we exclude a binary representation for zero. Conversion from a binary DFA working with leading zeros to a DFA working with the hidden leftmost bit is simple: if q_I is the original initial state, use $\delta(q_I, 1)$ as a new initial state. This imitates reading the hidden most significant bit.

The special role of the number 2 in formulas for the number of states follows from the fact that integers are represented in *binary* here. We expect different number of states for representations based on values different from 2.

References

1. Bertoni, A., Mereghetti, C., Pighizzini, G.: An optimal lower bound for nonregular languages. Inform. Process. Lett. **50**, 289–292 (1994). (Corr. ibid., **52**, p. 339, 1994)
2. Chrobak, M.: Finite automata and unary languages. Theoret. Comput. Sci. **47**, 149–158 (1986). (Corr. ibid., **302**, 497–498, 2003)
3. Davida, G.I., Litow, B.: Fast parallel arithmetic via modular representation. SIAM J. Comput. **20**, 756–765 (1991)
4. Dietz, P.F., Macarie, I.I., Seiferas, J.I.: Bits and relative order from residues, space efficiently. Inform. Process. Lett. **50**, 123–127 (1994)
5. Gawrychowski, P.: Chrobak normal form revisited, with applications. In: Bouchou-Markhoff, B., Caron, P., Champarnaud, J.-M., Maurel, D. (eds.) CIAA 2011. LNCS, vol. 6807, pp. 142–153. Springer, Heidelberg (2011). https://doi.org/10.1007/978-3-642-22256-6_14
6. Geffert, V.: Space hierarchy theorem revised. Theoret. Comput. Sci. **295**, 171–187 (2003)
7. Geffert, V.: Magic numbers in the state hierarchy of finite automata. Inform. Comput. **205**, 1652–1670 (2007)
8. Geffert, V.: Unary coded PSPACE-complete languages in ASPACE(log log n). Theory Comput. Syst. **63**, 688–714 (2019)
9. Geffert, V., Pališínová, D., Szabari, A.: State complexity of binary coded regular languages. In: Han, Y.S., Vaszil, G. (eds.) DCFS 2022. LNCS, vol. 13439, pp. 72–84. Springer, Cham (2022). https://doi.org/10.1007/978-3-031-13257-5_6
10. Geffert, V., Pardubská, D.: Factoring and testing primes in small space. RAIRO Inform. Théor. Appl. **47**, 241–259 (2013)
11. Ginsburg, S., Rice, H.G.: Two families of languages related to ALGOL. J. Assoc. Comput. Mach. **9**, 350–371 (1962)
12. Hartmanis, J., Immerman, N., Sewelson, W.: Sparse sets in NP−P: EXPTIME versus NEXPTIME. Inform. Control **65**, 158–181 (1985)
13. Hopcroft, J., Motwani, R., Ullman, J.: Introduction to Automata Theory, Languages, and Computation. Addison-Wesley, Boston (2001)
14. Hopcroft, J. E., Ullman, J. D.: Introduction to Automata Theory, Languages, and Computation. Addison-Wesley, Boston (1979)
15. Ljubič, Ju. I.: Ocenki dlja optimal'noj determinizacii nedeterminirovannyh avtonomnyh avtomatov. Sibirsk. Mat. Zh. **V/2**, 337–355 (1964). (In Russian)
16. Lupanov, O. B.: Über den Vergleich zweier Typen endlicher Quellen. Probleme der Kybernetik **6**, 329–335 (1966). (Akademie-Verlag, Berlin, in German)
17. Mereghetti, C., Pighizzini, G.: Optimal simulations between unary automata. SIAM J. Comput. **30**, 1976–1992 (2001)
18. Moore, F.R.: On the bounds for state-set size in the proofs of equivalence between deterministic, nondeterministic, and two-way finite automata by deterministic automata. IEEE Trans. Comput. **C-20**, 1211–1214 (1971)
19. Rabin, M., Scott, D.: Finite automata and their decision problems. IBM J. Res. Develop. **3**, 114–125 (1959)
20. Salomaa, A., Wood, D., Yu, S.: On the state complexity of reversals of regular languages. Theoret. Comput. Sci. **320**, 315–329 (2004)

A Survey on Automata with Translucent Letters

Friedrich Otto[(✉)]

Fachbereich Elektrotechnik/Informatik, Universität Kassel, 34109 Kassel, Germany
f.otto@uni-kassel.de

Abstract. In this survey we present the various types of automata with translucent letters that have been studied in the literature. These include the finite automata and the pushdown automata with translucent letters, which are obtained as reinterpretations of certain cooperating distributed systems of a restricted type of restarting automaton, the linear automaton with translucent letters, and the visibly pushdown automaton with translucent letters. For each of these types of automata with translucent letters, it has been shown that they accept those trace languages which are obtained from the class of languages that is accepted by the corresponding type of automaton without translucent letters.

Keywords: finite automaton · pushdown automaton · translucent letter · trace language · language class · inclusion · incomparability

1 Introduction

While a 'classical' finite automaton or pushdown automaton reads its input strictly from left to right, letter by letter, many types of automata have been studied in the literature that process their inputs in a different, non-sequential way. Under this aspect, the *jumping finite automaton* of A. Meduna and P. Zemek [21] is the most 'extreme,' as it processes a given input in an arbitrary order. Other types of automata that do not process their inputs in a strictly sequential order include the right one-way jumping finite automaton [3,10], the restarting automaton [16,17], the right-revolving finite automaton [4], the non-deterministic linear automaton [18], and the input-reversal pushdown automaton [8]. Here, however, we concentrate on a different type of automaton that does not read its input sequentially from left to right: the automaton with *translucent letters*.

A finite automaton with translucent letters is obtained from a 'classical' finite automaton by adding an end-of-tape marker and a so-called *translucency mapping* τ. This mapping assigns, to each state q, a subset $\tau(q)$ of input letters that the automaton cannot see when it is in state q. Accordingly, in state q, the automaton reads (and deletes) the first letter from the left that is *not* translucent for the current state. If there is no such letter, then the automaton sees the end-of-tape marker and it halts, either accepting or rejecting. Although this

© The Author(s), under exclusive license to Springer Nature Switzerland AG 2023
B. Nagy (Ed.): CIAA 2023, LNCS 14151, pp. 21–50, 2023.
https://doi.org/10.1007/978-3-031-40247-0_2

extension of the model of finite automaton looks quite natural, it was only discovered through a reinterpretation of cooperating distributed systems of a certain type of restarting automaton. The restarting automaton was introduced in [16] to model the *analysis by reduction*, which is a technique from linguistics that can be applied to analyze sentences of natural languages with a free word-order. By placing restrictions on the operations that a restarting automaton may execute, characterizations have been obtained for various classes of formal languages (see, e.g., [42,43]).

The weakest type of restarting automaton is the stateless deterministic R-automaton with a window of size one, abbreviated as stl-det-R(1)-automaton, which only accepts a rather restricted class of regular languages. In [36], several stl-det-R(1)-automata are combined into a cooperating distributed system inspired by the cooperating distributed grammar systems as presented in [11]. As observed in [34], these systems can be interpreted as NFAs with translucent letters. Other types of cooperating distributed systems of stl-det-R(1)-automata give rise to the DFA with translucent letters and the (deterministic and non-deterministic) pushdown automaton with translucent letters. In addition, linear automata with translucent letters have been studied in [40]. Interestingly, it has turned out that each nondeterministic type of automaton with translucent letters accepts those trace languages that are obtained from the class of languages that is accepted by the corresponding type of automaton without translucent letters. Finally, in [23], an extension of the NFA and the DFA with translucent letters is considered that, after reading and deleting a letter, does not return its head to the left end of its tape but that, instead, continues from the position of the letter just deleted.

In the current survey, we present the above development in short. Accordingly, the paper is structured as follows. After establishing notation at the end of this section, we introduce several classes of trace languages in Sect. 2. In the following section, we present the restarting automaton, concentrating on the stl-det-R(1)-automaton, and in Sect. 4, we recall the various types of cooperating distributed systems of stl-det-R(1)-automata. In the next two sections, we present finite automata with translucent letters and pushdown automata with translucent letters. Finally, in Sect. 7, we consider the non-returning NFA and DFA with translucent letters, comparing them to various other types of automata that do not read their inputs sequentially from left to right. In the concluding section, we list a number of questions and open problems for future work.

Notation. We use $\mathbb{N}$ to denote the set of non-negative integers. For a finite alphabet Σ, Σ^* denotes the set of all words over Σ, including the empty word λ, and Σ^+ denotes the set of all non-empty words. For a word $w \in \Sigma^*$, $|w|$ denotes the length of w and, for $a \in \Sigma$, $|w|_a$ is the a-length of w, that is, the number of occurrences of the letter a in w. The concatenation (or product) of two words u and v is written as $u \cdot v$ or simply as uv. Moreover, for a set S, we use $\mathcal{P}(S)$ to denote the power set of S. Finally, we shall encounter the classes REG of regular languages, LIN of linear context-free languages, DCFL of deterministic context-free languages, VPL of visibly pushdown languages, CFL of context-free languages, GCSL of growing context-sensitive languages, and CSL of context-sensitive languages. In addition, NFA (DFA) is used to denote a nondeterministic (deterministic) finite automaton (see, e.g., [1,12,15]).

2   Trace Languages

As we concentrate on types of automata that do not process their inputs strictly from left to right, we need some language classes in addition to those of the classical Chomsky hierarchy. Specifically, we consider various classes of *trace languages*.

Starting with A. Mazurkiewicz's seminal paper [20], the theory of traces has become an important part of the theory of concurrent systems. A *trace* is an equivalence class of words over a given finite alphabet with respect to a partial commutativity relation [9]. Informally speaking, if the letters of an alphabet Σ are interpreted as *atomic actions*, then a word w over Σ stands for a finite sequence of such actions. If some of these atomic actions, say a and b, are *independent* of each other, then it does not matter in which order they are executed, that is, the sequences of actions ab and ba yield the same result. A reflexive and symmetric binary relation D on Σ is called a *dependency relation* on Σ, and $I_D = (\Sigma \times \Sigma) \smallsetminus D$ is the corresponding *independence relation*. Obviously, the relation I_D is also symmetric. The equivalence relation $\equiv_D$ that is induced by the pairs

$$\{ (xaby, xbay) \mid (a,b) \in I_D,\, x,y \in \Sigma^* \}$$

is the *partial commutativity relation* induced by D. By collecting all words (sequences) that are equivalent to a given word w into a class $[w]_D = \{ z \in \Sigma^* \mid z \equiv_D w \}$, one abstracts from the order between independent actions. These equivalence classes are called *traces*, and the set $M(D) = \{ [w]_D \mid w \in \Sigma^* \}$ of all traces is the *trace monoid* $M(D)$ that is presented by the pair (Σ, D) (see, e.g., [13]).

Now we restate the definitions of various classes of trace languages in short.

Definition 1. *Let Σ be a finite alphabet. A language $L \subseteq \Sigma^*$ is called*

(a) a rational trace language *if there exist a dependency relation D on Σ and a regular language $R \subseteq \Sigma^*$ such that $L = \bigcup_{w \in R}[w]_D$, that is, $L = \{w \in \Sigma^* \mid \exists z \in R : w \equiv_D z \}$;*

(b) a linear context-free trace language *if there exist a dependency relation D on Σ and a linear context-free language $R \subseteq \Sigma^*$ such that $L = \bigcup_{w \in R}[w]_D$;*

(c) a context-free trace language *if there exist a dependency relation D on Σ and a context-free language $R \subseteq \Sigma^*$ such that $L = \bigcup_{w \in R}[w]_D$;*

(d) a visibly pushdown trace language *if there exist a dependency relation D on Σ and a visibly pushdown language $R \subseteq \Sigma^*$ such that $L = \bigcup_{w \in R}[w]_D$.*

$\mathcal{LRAT}$ *($\mathcal{LLIN}$, $\mathcal{LCF}$, $\mathcal{LVP}$) is used to denote the class of all rational (linear context-free, context-free, visibly pushdown) trace languages.*

Context-free trace languages have been considered in [5,6], while visibly push-down trace languages have been studied in [7]. However, in that paper, it is assumed that the push and pop operations of any process are independent of all the push and pop operations of any other process.

The relation $D = \{\,(a,b) \mid a,b \in \Sigma\,\}$ is the trivial dependency relation on Σ, as $x \equiv_D y$ if and only if $x = y$. This shows that the free monoid Σ^* is a trace monoid, which in turn implies that REG $\subseteq \mathcal{LRAT}$, LIN $\subseteq \mathcal{LLIN}$, CFL $\subseteq \mathcal{LCF}$, and VPL $\subseteq \mathcal{LVP}$. In fact, all these inclusions are proper as is shown by the following simple example.

Example 2. Let $\Sigma = \{a,b,c\}$ and $D = \{(a,a),(b,b),(c,c)\}$. Then $\equiv_D$ is the *commutation relation* on Σ^*, that is, the trace monoid presented by (Σ, D) is the free abelian monoid generated by Σ. Now $R = \{\,(abc)^n \mid n \geq 1\,\}$ is a regular language and $L = \bigcup_{w \in R}[w]_D$ is a rational trace language. However, $L = \{\,w \in \Sigma^* \mid |w|_a = |w|_b = |w|_c \geq 1\,\}$, which is not even context-free. ∎

3   Restarting Automata

Motivated by the analysis by reduction, P. Jančar, F. Mráz, M. Plátek, and J. Vogel presented the first type of the *restarting automaton* at the FCT 1995 in Dresden [16]. As defined in that paper, a restarting automaton consists of a finite-state control, a finite tape that initially contains the input and which is bordered on the left by the sentinel $\triangleright$ and on the right by the sentinel $\triangleleft$, and a read/write window of a fixed finite size $k \geq 1$. This type of automaton differs from the Turing machine (or, rather, the *linear-bounded automaton*) in three important aspects:

- A restarting automaton works in *cycles*, where in each cycle, it scans the tape from left to right until it detects a position at which it can apply a delete/restart operation. A delete/restart operation removes one or more letters from the contents of the window and restarts the automaton, which means that the window is repositioned at the left end of the tape.
- The tape of a restarting automaton is *flexible*, that is, it adjusts to the length of the tape inscription. Thus, when one or more letters are removed during a delete/restart step, then the tape is shortened accordingly.
- By a delete/restart step, the finite-state control is reset to the initial state, that is, all the information that the restarting automaton may have collected in its finite-state control while scanning the tape from left to right is forgotten during a delete/restart step.

Despite these limitations, the deterministic variant of the (above type of) restarting automaton, called an R-*automaton*, accepts a proper superset of the deterministic context-free languages, while the nondeterministic variant is even more expressive. However, the latter is not sufficient to accept all context-free languages [17]. Therefore, in subsequent papers, P. Jančar, F. Mráz, M. Plátek, and J. Vogel extended the restarting automaton in several ways (see, e.g., the survey papers [42] and [43]).

Here we are interested in the weakest possible type of restarting automaton, the *stateless deterministic R-automaton of window size one*, as this automaton serves as the main building block for the various types of automata with translucent letters. This particular type of restarting automaton is defined as follows.

Definition 3 ([36]). *A stateless deterministic R-automaton of window size one, or a* stl-det-R(1)-*automaton for short, is defined through a four-tuple* $M = (\Sigma, \triangleright, \triangleleft, \delta)$, *where* Σ *is a finite input alphabet,* $\triangleright, \triangleleft \notin \Sigma$ *are special letters, called* sentinels, *that are used to mark the left and right border of the tape, and* $\delta : (\Sigma \cup \{\triangleright, \triangleleft\}) \to \{\lambda, \mathsf{MVR}, \mathsf{Accept}\}$ *is a partial transition function. Here it is required that* $\delta(\triangleright) \neq \lambda \neq \delta(\triangleleft)$.

For a word $w \in \Sigma^*$, the corresponding initial configuration of M is written as $\underline{\triangleright} \cdot w \cdot \triangleleft$, where the underlined letter indicates the position of the read/write window of size one. A general configuration of M has the form $\triangleright \cdot u\underline{a}v \cdot \triangleleft$ for any words $u, v \in \Sigma^*$ and a letter $a \in \Sigma$. The stl-det-R(1)-automaton M induces the following single-step computation relation $\vdash_M$ on its set of configurations $\triangleright \cdot \Sigma^* \cdot \triangleleft \cup \{\mathsf{Accept}, \mathsf{Reject}\}$, where $u, v \in \Sigma^*$ and $a, b \in \Sigma$:

$$\triangleright \cdot u\underline{a}bv \cdot \triangleleft \vdash_M \begin{cases} \triangleright \cdot ua\underline{b}v \cdot \triangleleft, & \text{if } \delta(a) = \mathsf{MVR}, \\ \underline{\triangleright} \cdot ubv \cdot \triangleleft, & \text{if } \delta(a) = \lambda, \\ \mathsf{Accept}, & \text{if } \delta(a) = \mathsf{Accept}, \\ \mathsf{Reject}, & \text{if } \delta(a) = \emptyset. \end{cases}$$

The language $L(M)$ accepted by M is defined as

$$L(M) = \{\, w \in \Sigma^* \mid \underline{\triangleright} \cdot w \cdot \triangleleft \vdash_M^* \mathsf{Accept} \,\},$$

where $\vdash_M^*$ denoted the reflexive transitive closure of the relation $\vdash_M$.

The following simple example illustrates the way in which a stl-det-R(1)-automaton works.

Example 4. Let $M = (\Sigma, \triangleright, \triangleleft, \delta)$ be the stl-det-R(1)-automaton that is specified through $\Sigma = \{a, b, c\}$ and $\delta(\triangleright) = \mathsf{MVR}, \delta(a) = \mathsf{MVR}, \delta(b) = \lambda, \delta(c) = \emptyset, \delta(\triangleleft) = \mathsf{Accept}$. Then

$$\underline{\triangleright} \cdot abc \cdot \triangleleft \vdash_M \triangleright \cdot \underline{a}bc \cdot \triangleleft \vdash_M \triangleright \cdot a\underline{b}c \cdot \triangleleft \vdash_M \underline{\triangleright} \cdot ac \cdot \triangleleft \vdash_M \triangleright \cdot \underline{a}c \cdot \triangleleft \vdash_M \triangleright \cdot a\underline{c} \cdot \triangleleft \vdash_M \mathsf{Reject},$$

as $\delta(c)$ is undefined, while

$$\underline{\triangleright} \cdot ba \cdot \triangleleft \vdash_M \triangleright \cdot \underline{b}a \cdot \triangleleft \vdash_M \underline{\triangleright} \cdot a \cdot \triangleleft \vdash_M \triangleright \cdot \underline{a} \cdot \triangleleft \vdash_M \triangleright \cdot a \cdot \underline{\triangleleft} \vdash_M \mathsf{Accept}.$$

Thus, $ba \in L(M)$, but $abc \notin L(M)$. ∎

As we see, a stl-det-R(1)-automaton works in cycles: starting at the left sentinel, it performs a number of move-right steps until it either accepts, rejects, or deletes a letter and restarts its computation on a shortened tape. Accordingly, each configuration in which the head of the automaton is on the left sentinel is called a *restarting configuration*. The execution of a complete cycle is expressed by the relation $\vdash^c_M$. For example, in Example 4, we have $\rhd \cdot abc \cdot \lhd \vdash^c_M \rhd \cdot ac \cdot \lhd$ and $\rhd \cdot ba \cdot \lhd \vdash^c_M \rhd \cdot a \cdot \lhd$.

In order to characterize the languages that are accepted by stl-det-R(1)-automata, we introduce the following notation.

Definition 5. *Let $M = (\Sigma, \rhd, \lhd, \delta)$ be a stateless deterministic R(1)-automaton. Then we can partition the alphabet Σ into four disjoint subalphabets:*

(1) $\Sigma_1 = \{\, a \in \Sigma \mid \delta(a) = \mathsf{MVR}\,\}$, (3) $\Sigma_3 = \{\, a \in \Sigma \mid \delta(a) = \mathsf{Accept}\,\}$,
(2) $\Sigma_2 = \{\, a \in \Sigma \mid \delta(a) = \lambda\,\}$, (4) $\Sigma_4 = \{\, a \in \Sigma \mid \delta(a) = \emptyset\,\}$.

Thus, Σ_1 is the set of letters that M just moves across, Σ_2 is the set of letters that M deletes, Σ_3 is the set of letters that cause M to accept, and Σ_4 is the set of letters on which M gets stuck, that is, which cause M to reject.

Then we obtain the following characterization.

Proposition 6 ([36]). *Let $M = (\Sigma, \rhd, \lhd, \delta)$ be a stateless deterministic R(1)-automaton, and assume that the subalphabets $\Sigma_1, \Sigma_2, \Sigma_3, \Sigma_4$ are defined as above. Then the language $L(M)$ can be characterized as follows:*

$$L(M) = \begin{cases} \emptyset, & \text{if } \delta(\rhd) = \emptyset, \\ \Sigma^*, & \text{if } \delta(\rhd) = \mathsf{Accept}, \\ (\Sigma_1 \cup \Sigma_2)^* \cdot \Sigma_3 \cdot \Sigma^*, & \text{if } \delta(\rhd) = \mathsf{MVR} \text{ and } \delta(\lhd) \neq \mathsf{Accept}, \\ (\Sigma_1 \cup \Sigma_2)^* \cdot ((\Sigma_3 \cdot \Sigma^*) \cup \{\lambda\}), & \text{if } \delta(\rhd) = \mathsf{MVR} \text{ and } \delta(\lhd) = \mathsf{Accept}. \end{cases}$$

Thus, we see that the class of languages $\mathcal{L}(\text{stl-det-R}(1))$ that are accepted by stl-det-R(1)-automata is just a very special proper subclass of the regular languages. It is easily seen that a stateless finite automaton with input alphabet Σ accepts a language of the form Σ_0^*, where Σ_0 is a subalphabet of Σ. This yields the following consequence.

Corollary 7. *A language L is accepted by a stateless deterministic R(1)-automaton that only accepts on reaching the right sentinel $\lhd$, if and only if L is accepted by a stateless finite automaton.*

By increasing the window size, we obtain the family of language classes $(\mathcal{L}(\text{stl-det-R}(k)))_{k \geq 1}$, which form a strictly ascending hierarchy that is incomparable to the regular languages with respect to inclusion [36]. On the other hand, by admitting states, we obtain the language class $\mathcal{L}(\text{det-R}(1))$, which coincides with the class of regular languages [22]. The diagram in Fig. 1 summarizes the corresponding inclusion relations, where each arrow denotes a proper inclusion. In addition, if there is no sequence of oriented edges between two classes, then these classes are incomparable with respect to inclusion.

For obtaining automata with translucent letters, the stateless deterministic R(1)-automaton is generalized in a different way by combining finitely many of these automata into a cooperating distributed system.

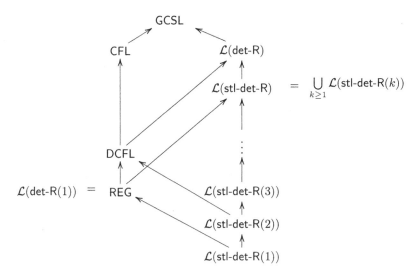

Fig. 1. Taxonomy of language classes accepted by stateless deterministic R-automata.

4   CD-Systems of Restarting Automata

Following [36] (see also [30]), we now combine several stateless deterministic R(1)-automata into a cooperating distributed system.

Definition 8 ([36]). *A cooperating distributed system of stl-det-R(1)-auto-mata, or a stl-det-local-CD-R(1)-system for short, is specified by a triple $\mathcal{M} = ((M_i)_{i \in I}, I_0, \delta)$, where I is a finite set of indices, $(M_i)_{i \in I}$ is a collection of stl-det-R(1)-automata $M_i = (\Sigma, \triangleright, \triangleleft, \delta_i)$ $(i \in I)$ that all have the same input alphabet Σ, $I_0 \subseteq I$ is the set of initial indices, and $\delta : \bigcup_{i \in I}(\{i\} \times \Sigma_2^{(i)}) \to \mathcal{P}(I)$ is a global successor function. This function assigns, to each pair $(i, a) \in \{i\} \times \Sigma_2^{(i)}$, a set of possible successor automata, where $\Sigma_2^{(i)}$ denotes the set of letters that the automaton M_i can delete. Here it is required that $I_0 \neq \emptyset$, that $\delta_i(\triangleright) = \mathsf{MVR}$ for all $i \in I$, and that $\delta(i, a) \neq \emptyset$ for all $i \in I$ and all $a \in \Sigma_2^{(i)}$.*

For a given input $w \in \Sigma^*$, the computation of $\mathcal{M}$ proceeds as follows. First, an index $i_0 \in I_0$ is chosen nondeterministically. Then the stl-det-R(1)-automaton M_{i_0} starts the computation with the initial configuration $\triangleright \cdot w \cdot \triangleleft$, executing one cycle. Thereafter, an index $i_1 \in \delta(i_0, a)$ is chosen nondeterministically, where a is the letter an occurrence of which has been deleted by M_{i_0} in the above cycle, and M_{i_1} continues the computation by executing one cycle. This continues until, for some $j \geq 0$, the automaton M_{i_j} accepts. Should at some stage the chosen automaton M_{i_j} be unable to execute a cycle, then the computation fails.

In order to describe computations of stl-det-local-CD-R(1)-systems in a more transparent way, we encode restarting configurations through pairs of the form $(i, \triangleright \cdot w \cdot \triangleleft)$, where i is the index of the currently active component automaton and $\triangleright \cdot w \cdot \triangleleft$ is the current tape contents. Then $\mathcal{L}(\text{stl-det-local-CD-R}(1))$ denotes the class of languages that are accepted by stl-det-local-CD-R(1)-systems. The next example shows that stl-det-local-CD-R(1)-systems accept some important non-regular languages.

Example 9. Let $\mathcal{M}_1 = ((M_i)_{i \in I}, I_0, \delta)$, where $I = \{a, b, +\}$, $I_0 = \{a, +\}$, M_a, M_b, and M_+ are the stateless deterministic R(1)-automata on $\Sigma = \{a, b\}$ that are given by the following transition functions:

$$M_a : (1)\ \delta_a(\triangleright) = \text{MVR},\ (2)\ \delta_a(a) = \lambda,$$
$$M_b : (3)\ \delta_b(\triangleright) = \text{MVR},\ (4)\ \delta_b(a) = \text{MVR},\quad (5)\ \delta_b(b) = \lambda,$$
$$M_+ : (6)\ \delta_+(\triangleright) = \text{MVR},\ (7)\ \delta_+(\triangleleft) = \text{Accept},$$

and the global successor function δ is defined through

$$\delta(a, a) = \{b\} \text{ and } \delta(b, b) = \{a, +\}.$$

Here we assume that these transition functions are undefined for all other values. Then $w \in L(\mathcal{M}_1)$ iff $|w|_a = |w|_b$ and $|u|_a \geq |u|_b$ for all prefixes u of w. Hence, $L(\mathcal{M}_1)$ is the semi-Dyck language D_1. ∎

A major feature of these stl-det-local-CD-R(1)-systems is the fact that, although all their component automata are deterministic, the computations of these CD-systems are not, as in each computation, the initial component automaton and the successor component automata are still chosen nondeterministically. In [38] (see also [35]), a completely deterministic variant of these CD-systems has been introduced.

Definition 10. *Let $((M_i)_{i \in I}, I_0, \delta)$ be a CD-system of stl-det-R(1)-automata over Σ. This system is called a stl-det-global-CD-R(1)-system, if $|I_0| = 1$ and if $|\delta(i, a)| = 1$ for all $i \in I$ and all $a \in \Sigma_2^{(i)}$.*

If $w = uav$ for some word $u \in \Sigma_1^{(i)*}$ and a letter $a \in \Sigma_2^{(i)}$, and if $\delta(i, a) = j$, then $(i, \triangleright \cdot uav \cdot \triangleleft) \vdash_{\mathcal{M}}^c (j, \triangleright \cdot uv \cdot \triangleleft)$. It follows that, for each input word $w \in \Sigma^*$, the system $\mathcal{M} = ((M_i)_{i \in I}, I_0, \delta)$ has a unique computation that starts from the initial configuration corresponding to input w, that is, $\mathcal{M}$ is completely deterministic. We use $\mathcal{L}(\text{stl-det-global-CD-R}(1))$ to denote the class of all languages that are accepted by stl-det-global-CD-R(1)-systems.

We illustrate this definition by presenting a stl-det-global-CD-R(1)-system for the language $L_{abc} = \{w \in \{a, b, c\}^* \mid |w|_a = |w|_b = |w|_c \geq 0\}$, which is not even context-free.

Example 11. Let $\mathcal{M}_2 = ((M_i)_{i \in I}, I_0, \delta)$, where $I = \{0, 1, 2, 3, 4, 5, 6\}$, $I_0 = \{0\}$, and $M_0, M_1, \ldots, M_6$ are the stateless deterministic R(1)-automata on $\Sigma =$

$\{a, b, c\}$ that are given by the following transition functions:

M_0 : (1) $\delta_0(\rhd)$ = MVR,
         (2) $\delta_0(a)$ = λ,
         (3) $\delta_0(b)$ = λ,
         (4) $\delta_0(c)$ = λ,
         (5) $\delta_0(\lhd)$ = Accept,

M_1 : (6) $\delta_1(\rhd)$ = MVR, M_2 : (10) $\delta_2(\rhd)$ = MVR,
         (7) $\delta_1(a)$ = MVR,   (11) $\delta_2(a)$ = MVR,
         (8) $\delta_1(b)$ = λ,   (12) $\delta_2(b)$ = MVR,
         (9) $\delta_1(c)$ = MVR,   (13) $\delta_2(c)$ = λ,

M_3 : (14) $\delta_3(\rhd)$ = MVR, M_4 : (18) $\delta_4(\rhd)$ = MVR,
        (15) $\delta_3(a)$ = λ,   (19) $\delta_4(a)$ = MVR,
        (16) $\delta_3(b)$ = MVR,   (20) $\delta_4(b)$ = MVR,
        (17) $\delta_3(c)$ = MVR,   (21) $\delta_4(c)$ = λ,

M_5 : (22) $\delta_5(\rhd)$ = MVR, M_6 : (26) $\delta_6(\rhd)$ = MVR,
        (23) $\delta_5(a)$ = λ,   (27) $\delta_6(a)$ = MVR,
        (24) $\delta_5(b)$ = MVR,   (28) $\delta_6(b)$ = λ,
        (25) $\delta_5(c)$ = MVR,   (29) $\delta_6(c)$ = MVR,

and $\delta_i(\lhd) = \emptyset$ for all $i = 1, 2, \ldots, 6$.

Finally, the global transition function δ is defined as follows:

$$\delta(0, a) = 1,\ \delta(1, b) = 2,\ \delta(2, c) = 0,$$
$$\delta(0, b) = 3,\ \delta(3, a) = 4,\ \delta(4, c) = 0,$$
$$\delta(0, c) = 5,\ \delta(5, a) = 6,\ \delta(6, b) = 0.$$

Then $w \in L(\mathcal{M}_2)$ iff $|w|_a = |w|_b = |w|_c$, that is, $L(\mathcal{M}_2) = L_{abc}$. ∎

In [26], also CD-systems of stateless deterministic R-automata of window size 2 are studied. These systems accept all linear context-free languages, and they even accept some languages that are not semi-linear.

During the computation of a stl-det-local-CD-R(1)-system, the successor of an active component automaton is chosen based on the index of that component automaton and the letter that has been deleted in the current cycle. In [32] (see also [31]), CD-systems of stl-det-R(1)-automata are introduced that use an additional external pushdown store to determine the successor of the currently active component automaton. These systems, called pushdown CD-systems of stateless deterministic R(1)-automata, are formally defined as follows.

Definition 12 ([32]). *A pushdown CD-system of stateless deterministic R(1)-automata, a PD-CD-R(1)-system for short, consists of a CD-system of stateless deterministic R(1)-automata and an external pushdown store. It is defined through a tuple*

$$\mathcal{M} = (I, \Sigma, (M_i)_{i \in I}, \Gamma, \bot, I_0, \delta),$$

where I is a finite set of indices, Σ is a finite input alphabet, for all $i \in I$, M_i is a stateless deterministic R(1)-automaton on Σ, Γ is a finite pushdown alphabet,

$\perp \notin \Gamma$ is the bottom marker of the pushdown store, $I_0 \subseteq I$ is the set of initial indices, and

$$\delta : (I \times \Sigma \times (\Gamma \cup \{\perp\})) \to \mathcal{P}(I \times (\Gamma \cup \{\perp\})^{\leq 3})$$

is the successor relation. For each $i \in I$, $a \in \Sigma$, and $A \in \Gamma$, $\delta(i, a, A)$ is a subset of $I \times \Gamma^{\leq 2}$, and $\delta(i, a, \perp)$ is a subset of $I \times (\perp \cdot \Gamma^{\leq 2})$, where $\Gamma^{\leq 2}$ denotes the set of all words over Γ of length at most 2.

A *configuration* of $\mathcal{M}$ is a triple of the form $(i, \omega, \perp\alpha)$, where $i \in I$, $\omega \in (\triangleright \cdot \Sigma^* \cdot \triangleleft) \cup \{\mathsf{Accept}\}$, and $\alpha \in \Gamma^*$. A configuration $(i, \triangleright \cdot w \cdot \triangleleft, \perp\alpha)$ describes the situation that the component automaton M_i has just been activated, $\underline{\triangleright} \cdot w \cdot \triangleleft$ is the corresponding restarting configuration of M_i, and the word $\perp\alpha$ is the current contents of the pushdown store with the last letter of α at the top. An *initial configuration* of $\mathcal{M}$ on input $w \in \Sigma^*$ has the form $(i_0, \triangleright \cdot w \cdot \triangleleft, \perp)$ for any $i_0 \in I_0$, and an *accepting configuration* has the form $(i, \mathsf{Accept}, \perp)$ for any $i \in I$.

To exclude some trivial cases, we assume that each automaton M_i ($i \in I$) performs a move-right step on the left sentinel $\triangleright$. Furthermore, for each $i \in I$, let $\Sigma_1^{(i)}$, $\Sigma_2^{(i)}$, and $\Sigma_3^{(i)}$ denote the subsets of Σ that correspond to the automaton M_i according to Definition 5. Then the *single-step computation relation* $\Rightarrow_{\mathcal{M}}$ that $\mathcal{M}$ induces on its set of configurations is defined by the following three rules, where $i \in I$, $w \in \Sigma^*$, $\alpha \in \Gamma^*$, and $A \in \Gamma$:

(1) $(i, \triangleright \cdot w \cdot \triangleleft, \perp\alpha A) \Rightarrow_{\mathcal{M}} (j, \triangleright \cdot w' \cdot \triangleleft, \perp\alpha\eta)$ if $\exists u \in \Sigma_1^{(i)*}, a \in \Sigma_2^{(i)}, v \in \Sigma^*$:
$$w = uav, w' = uv, \text{ and } (j, \eta) \in \delta(i, a, A);$$

(2) $(i, \triangleright \cdot w \cdot \triangleleft, \perp) \Rightarrow_{\mathcal{M}} (j, \triangleright \cdot w' \cdot \triangleleft, \perp\eta)$ if $\exists u \in \Sigma_1^{(i)*}, a \in \Sigma_2^{(i)}, v \in \Sigma^*$:
$$w = uav, w' = uv, \text{ and } (j, \perp\eta) \in \delta(i, a, \perp);$$

(3) $(i, \triangleright \cdot w \cdot \triangleleft, \perp) \Rightarrow_{\mathcal{M}} (i, \mathsf{Accept}, \perp)$ if $\exists u \in \Sigma_1^{(i)*}, a \in \Sigma_3^{(i)}, v \in \Sigma^*$:
$$w = uav, \text{ or } w \in \Sigma_1^{(i)*} \text{ and } \delta_i(\triangleleft) = \mathsf{Accept}.$$

Notice that the contents of the pushdown store is always a word of the form $\perp\alpha$ for some $\alpha \in \Gamma^*$, that is, the bottom marker $\perp$ cannot be removed from the pushdown store. Then $\Rightarrow_{\mathcal{M}}^*$ denotes the *computation relation* of $\mathcal{M}$, which is the reflexive and transitive closure of the relation $\Rightarrow_{\mathcal{M}}$. The language $L(\mathcal{M})$ accepted by $\mathcal{M}$ consists of all words for which $\mathcal{M}$ has an accepting computation, that is,

$$L(\mathcal{M}) = \{ w \in \Sigma^* \mid \exists i_0 \in I_0 \, \exists i \in I : (i_0, \triangleright \cdot w \cdot \triangleleft, \perp) \Rightarrow_{\mathcal{M}}^* (i, \mathsf{Accept}, \perp) \}.$$

Now $\mathcal{L}(\text{PD-CD-R}(1))$ denotes the class of languages that are accepted by PD-CD-R(1)-systems.

The PD-CD-R(1)-system $\mathcal{M}$ accepts if and when both of the following conditions are satisfied simultaneously: the currently active component automaton M_i executes an accepting computation starting from the current tape contents $\triangleright \cdot w \cdot \triangleleft$, and the pushdown store just contains the bottom marker $\perp$. Observe that the contents of the pushdown store of $\mathcal{M}$ is manipulated only in steps of the forms (1) and (2), and that during each step of either of these forms, a component automaton of $\mathcal{M}$ executes a cycle, that is, an input letter is being erased. Thus, there is no way that $\mathcal{M}$ can manipulate its pushdown store without reading (that is, deleting) input letters.

We illustrate the PD-CD-R(1)-systems through a simple example. Let

$$L = \{ a^n v \mid v \in \{b, c\}^*, |v|_b = |v|_c = n, n \geq 0 \}.$$

As $L \cap (\{a\}^* \cdot \{b\}^* \cdot \{c\}^*) = \{ a^n b^n c^n \mid n \geq 0 \}$ is not context-free, we see that L itself is not context-free.

Example 13. The language L is accepted by the PD-CD-R(1)-system

$$\mathcal{M} = (I, \Sigma, (M_i)_{i \in I}, \Gamma, \bot, I_0, \delta)$$

that is specified by $I = \{a, b, c, +\}$, $I_0 = \{a, +\}$, $\Sigma = \{a, b, c\}$, and $\Gamma = \{C\}$, by defining M_a, M_b, M_c, and M_+ by the following transition functions:

(1) $\delta_a(\triangleright) = $ MVR,   (5) $\delta_b(\triangleright) = $ MVR,   (8) $\delta_c(\triangleright) = $ MVR,
(2) $\delta_a(a) = \lambda$,   (6) $\delta_b(b) = \lambda$,   (9) $\delta_c(c) = \lambda$,
(3) $\delta_+(\triangleright) = $ MVR,   (7) $\delta_b(c) = $ MVR,   (10) $\delta_c(b) = $ MVR,
(4) $\delta_+(\triangleleft) = $ Accept,

where $\delta_x(y)$ $(x \in I, y \in \Sigma \cup \{\triangleleft\})$ is undefined for all other cases, and by defining the global transition function δ as follows:

(1) $\delta(a, a, \bot) = \{(a, \bot C), (b, \bot C)\}$,   (3) $\delta(b, b, C) = \{(c, C)\}$,
(2) $\delta(a, a, C) = \{(a, CC), (b, CC)\}$,   (4) $\delta(c, c, C) = \{(b, \lambda), (+, \lambda)\}$.

The component automaton M_+ just accepts the empty word, and it gets stuck on all other words. The component automaton M_a just deletes the first letter, if it is an a, otherwise, it gets stuck. The component automaton M_b reads across c's and deletes the first b it encounters, and analogously, the component automaton M_c reads across b's and deletes the first c it encounters. Thus, from the global transition function, we see that $\mathcal{M}$ only accepts certain words of the form $a^m v$ such that $v \in \{b, c\}^*$. However, when M_a deletes an a, then a letter C is pushed onto the pushdown store, and when M_c deletes a c, then a letter C is popped from the pushdown store. In fact, it can be checked that $L(\mathcal{M}) = L$. ∎

In [37] (see also [33]), a deterministic variant of the PD-CD-R(1)-system has been introduced.

Definition 14. *A PD-CD-R(1)-system $\mathcal{M} = (I, \Sigma, (M_i)_{i \in I}, \Gamma, \bot, I_0, \delta)$ is called (globally) deterministic, if $|I_0| = 1$ and if $|\delta(i, a, A)| \leq 1$ for all $i \in I$, $a \in \Sigma$, and $A \in \Gamma \cup \{\bot\}$. The class of all (globally) deterministic PD-CD-R(1)-systems is denoted by* det-PD-CD-R(1), *and* $\mathcal{L}$(det-PD-CD-R(1)) *denotes the class of languages that are accepted by det-PD-CD-R(1)-systems.*

A pushdown automaton A is called a *one-counter automaton* if its pushdown alphabet contains only one letter in addition to the bottom marker. By putting the corresponding restriction on PD-CD-R(1)-systems and their deterministic variants, the OC-CD-R(1)-systems and the det-OC-CD-R(1)-systems are obtained [32,37].

By combining the concept of a visibly pushdown automaton with the concept of trace languages, a variant of the *visibly pushdown trace languages* is used in [7] to analyze concurrent recursive programs. In that paper, the visibly pushdown trace languages are realized by so-called *concurrent visibly pushdown automata* that combine the concept of a visibly pushdown automaton with that of W. Zielonka's asynchronous automaton [46]. However, for the concurrent pushdown alphabets considered in [7], it is assumed that the push and pop operations of any process are independent of all the push and pop operations of any other process. In [44], visibly pushdown trace languages are studied that are obtained without enforcing any such restriction, that is, the dependency relation D on a given pushdown alphabet Σ is completely independent of the actual partitioning of Σ into call, return, and internal symbols. The resulting systems are called *visibly pushdown-CD-R(1)-systems* or VPD-CD-R(1)-systems for short, and $\mathcal{L}(\mathsf{VPD\text{-}CD\text{-}R}(1))$ is the class of all languages that are accepted by these systems.

5   Finite Automata with Translucent Letters

Let $\mathcal{M} = ((M_i)_{i \in I}, I_0, \delta)$ be a stl-det-local-CD-R(1)-system on an alphabet Σ, and for each $i \in I$, let $(\Sigma_1^{(i)}, \Sigma_2^{(i)}, \Sigma_3^{(i)}, \Sigma_4^{(i)})$ be the partitioning of Σ associated with M_i according to Definition 5. We can actually assume that $\Sigma_3^{(i)} = \emptyset$ for all $i \in I$ by introducing an additional component automaton. Now we can present the system $\mathcal{M}$ by a diagram that contains a vertex for each component automaton M_i and a special vertex 'Accept.' For each $i \in I$ and $a \in \Sigma_2^{(i)}$, M_i deletes the left-most occurrence of the letter a, provided it is preceded only by a word from $\Sigma_1^{(i)^*}$. Accordingly, the diagram contains an edge labelled $(\Sigma_1^{(i)^*}, a)$ from vertex i to vertex j for all $j \in \delta(i, a)$ (see Fig. 2). Furthermore, if $\delta_i(\lhd) = \mathsf{Accept}$, then M_i accepts all words from the set $\Sigma_1^{(i)^*}$, and accordingly, there is an edge labelled $\Sigma_1^{(i)^*}$ from vertex i to the vertex 'Accept' (see Fig. 3). In addition, vertex i is specifically marked for all initial indices $i \in I_0$. We illustrate this way of describing a stl-det-local-CD-R(1)-system by an example.

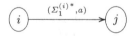

Fig. 2. A read/delete transition for $a \in \Sigma_2^{(i)}$ and $j \in \delta(i, a)$.

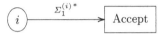

Fig. 3. An accept transition.

Example 15. Let $\mathcal{M} = ((M_i)_{i\in I}, I_0, \delta)$ be the following system, where $I = \{1, 2, 3, 4, +\}$, $I_0 = \{1\}$, the various R-automata are given by the following transition functions:

$$
\begin{aligned}
M_1 &: \delta_1(\triangleright) = \mathsf{MVR}, & M_+ &: \delta_+(\triangleright) = \mathsf{MVR}, \\
& \;\;\, \delta_1(a) = \lambda, & & \;\;\, \delta_+(\triangleleft) = \mathsf{Accept}, \\
M_2 &: \delta_2(\triangleright) = \mathsf{MVR}, & M_3 &: \delta_3(\triangleright) = \mathsf{MVR}, \\
& \;\;\, \delta_2(a) = \lambda, & & \;\;\, \delta_3(a) = \mathsf{MVR}, \\
& \;\;\, \delta_2(b) = \mathsf{MVR}, & & \;\;\, \delta_3(b) = \lambda, \\
& \;\;\, \delta_2(c) = \mathsf{MVR}, & & \;\;\, \delta_3(c) = \mathsf{MVR}, \\
M_4 &: \delta_4(\triangleright) = \mathsf{MVR}, & & \;\;\, \delta_4(b) = \mathsf{MVR}, \\
& \;\;\, \delta_4(a) = \mathsf{MVR}, & & \;\;\, \delta_4(c) = \lambda,
\end{aligned}
$$

where these functions are undefined for all other cases, and the global successor function δ is defined though

$$\delta(1, a) = \{1, 2\}, \delta(2, a) = \{3\}, \delta(3, b) = \{4\}, \delta(4, c) = \{2, +\}.$$

Using the component automaton M_1, $\mathcal{M}$ deletes a positive number of a's, and then using component automata M_2, M_3, and M_4, it deletes an equal number of a's, b's, and c's, before it accepts the empty word by component automaton M_+. Thus,

$$L_{=1}(\mathcal{M}) = \{\, a^n w \mid n \geq 1, \; w \in \{a, b, c\}^+ \text{ satisfying } |w|_a = |w|_b = |w|_c \,\}.$$

Now this CD-system of stateless R-automata of window size 1 can be described more compactly by the diagram given in Fig. 4. ∎

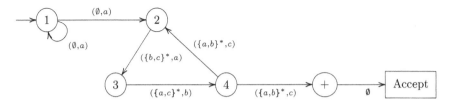

Fig. 4. The stl-det-local-CD-R(1)-system $\mathcal{M}$ from Example 15.

Motivated by this graphical representation, a stl-det-local-CD-R(1)-system can be interpreted as a nondeterministic finite automaton with translucent letters.

Definition 16 ([34]). *A* finite automaton with translucent letters, *or an* NFAwtl *for short, is defined as a* 7-*tuple* $A = (Q, \Sigma, \vartriangleleft, \tau, I, F, \delta)$, *where* Q *is a finite set of internal states,* Σ *is a finite alphabet of input letters,* $\vartriangleleft \notin \Sigma$ *is a special letter that is used as an* end-of-tape marker, $\tau : Q \to \mathcal{P}(\Sigma)$ *is a translucency mapping,* $I \subseteq Q$ *is a set of initial states,* $F \subseteq Q$ *is a set of final states, and* $\delta : Q \times \Sigma \to \mathcal{P}(Q)$ *is a transition relation. Here we require that, for each state* $q \in Q$ *and each letter* $a \in \Sigma$, *if* $a \in \tau(q)$, *then* $\delta(q, a) = \emptyset$. *For each state* $q \in Q$, *the letters from the set* $\tau(q)$ *are* translucent *for* q, *that is, in state* q, *the automaton* A *does not see these letters.*

Example 15 *(cont.)* The NFAwtl $A = (Q, \Sigma, \vartriangleleft, \tau, I, F, \delta)$ that corresponds to the stl-det-local-CD-R(1)-system $\mathcal{M} = ((M_i)_{i \in I}, I_0, \delta)$ is specified through $Q = \{q_1, q_2, q_3, q_4, q_+\}$, $\Sigma = \{a, b, c\}$, $I = \{q_1\}$, $F = \{q_+\}$, and the relations

$$\tau(q_1) = \emptyset, \tau(q_2) = \{b, c\}, \tau(q_3) = \{a, c\}, \tau(q_4) = \{a, b\}, \tau(q_+) = \emptyset,$$

and

$$\delta(q_1, a) = \{q_1, q_2\}, \delta(q_2, a) = \{q_3\}, \delta(q_3, b) = \{q_4\}, \delta(q_4, c) = \{q_2, q_+\}. \quad \blacksquare$$

An NFAwtl $A = (Q, \Sigma, \vartriangleleft, \tau, I, F, \delta)$ works as follows. For an input word $w \in \Sigma^*$, it starts in a nondeterministically chosen initial state $q_0 \in I$ with the word $w \cdot \vartriangleleft$ on its tape. This configuration is denoted by $q_0 w \cdot \vartriangleleft$. Assume that A is in a configuration $qw \cdot \vartriangleleft$, where $q \in Q$ and $w = a_1 a_2 \cdots a_n$ for some $n \geq 1$ and $a_1, a_2, \ldots, a_n \in \Sigma$. Then A looks for the first occurrence from the left of a letter that is not translucent for state q, that is, if $w = uav$ such that $u \in (\tau(q))^*$, $a \in (\Sigma \smallsetminus \tau(q))$, and $v \in \Sigma^*$, then A nondeterministically chooses a state $q_1 \in \delta(q, a)$, erases the letter a from the tape, thus producing the tape contents $uv \cdot \vartriangleleft$, and sets its internal state to q_1. In case $\delta(q, a) = \emptyset$, A halts without accepting. Finally, if $w \in (\tau(q))^*$, then A sees the end-of-tape marker $\vartriangleleft$ and the computation halts. In this case, A accepts if q is a final state; otherwise, it does not accept. Thus, A executes the following computation relation on its set $Q \cdot \Sigma^* \cdot \vartriangleleft \cup \{\mathsf{Accept}, \mathsf{Reject}\}$ of configurations:

$$qw \cdot \vartriangleleft \vdash_A \begin{cases} q'uv \cdot \vartriangleleft, & \text{if } w = uav, \ u \in (\tau(q))^*, \ a \notin \tau(q), \text{ and } q' \in \delta(q, a), \\ \mathsf{Reject}, & \text{if } w = uav, \ u \in (\tau(q))^*, \ a \notin \tau(q), \text{ and } \delta(q, a) = \emptyset, \\ \mathsf{Accept}, & \text{if } w \in (\tau(q))^* \text{ and } q \in F, \\ \mathsf{Reject}, & \text{if } w \in (\tau(q))^* \text{ and } q \notin F. \end{cases}$$

A word $w \in \Sigma^*$ is *accepted by* A if there exists an initial state $q_0 \in I$ and a computation $q_0 w \cdot \vartriangleleft \vdash_A^* \mathsf{Accept}$, where $\vdash_A^*$ denotes the reflexive transitive closure of the single-step computation relation $\vdash_A$. Now $L(A) = \{w \in \Sigma^* \mid w \text{ is accepted by } A\}$ is the *language accepted by* A and $\mathcal{L}(\mathsf{NFAwtl})$ denotes the class of all languages that are accepted by NFAwtls.

Definition 17 ([34]). *An NFAwtl* $A = (Q, \Sigma, \vartriangleleft, \tau, I, F, \delta)$ *is a* deterministic finite automaton with translucent letters, *abbreviated as* DFAwtl, *if* $|I| = 1$ *and if* $|\delta(q, a)| \leq 1$ *for all* $q \in Q$ *and all* $a \in \Sigma$. *Then* $\mathcal{L}(\mathsf{DFAwtl})$ *denotes the class of all languages that are accepted by DFAwtls.*

It is easily seen that the DFAwtl corresponds to the stl-det-global-CD-R(1)-system. Concerning the language classes $\mathcal{L}(\mathsf{NFAwtl})$ and $\mathcal{L}(\mathsf{DFAwtl})$, the following results have been obtained.

Theorem 18 ([36]). *Each language* $L \in \mathcal{L}(\mathsf{NFAwtl})$ *contains a regular sublanguage* E *such that the Parikh images* $\pi(E)$ *and* $\pi(L)$ *coincide.*

Essentially, an NFA for E is obtained from an NFAwtl for L by removing the translucency mapping. As a consequence, we see that $\mathcal{L}(\mathsf{NFAwtl})$ only contains semi-linear languages. Moreover, it follows that the language $\{\, a^n b^n \mid n \geq 0 \,\}$ is not accepted by any NFAwtl, that is, we have the following incomparability result.

Corollary 19 ([36]). *The language class* $\mathcal{L}(\mathsf{NFAwtl})$ *is incomparable to the classes* DLIN, LIN, DCFL, *and* CFL *with respect to inclusion.*

Let Σ be a finite alphabet. A language $L \subseteq \Sigma^*$ is a *rational trace language* if there exist a dependency relation D on Σ and a regular language $R \subseteq \Sigma^*$ such that $L = \bigcup_{w \in R} [w]_D$ (see Definition 1). Now let A be an NFA for R. We can easily turn A into an equivalent NFA $B = (Q, \Sigma, I, F, \delta_B)$ such that, for each state $q \in Q$, there is only a single letter $a(q) \in \Sigma$ such that $\delta(q, a(q)) \neq \emptyset$. Essentially, when executing a transition, B must already guess the next letter that it will have to process. From B, we obtain an NFAwtl C by adding an end-of-tape marker $\lhd$ and by defining a translucency mapping τ by taking, for each state q, $\tau(q) = \{\, b \in \Sigma \mid (a(q), b) \notin D \,\}$. Thus, in state q, C cannot see all those letters that are independent of the letter $a(q)$. Then it can be shown that C accepts the rational trace language L. Hence, we have the following result, as by Example 9, $D_1 \in \mathcal{L}(\mathsf{NFAwtl})$, but D_1 is not a rational trace language.

Theorem 20 ([36]). $\mathcal{LRAT} \subsetneqq \mathcal{L}(\mathsf{NFAwtl})$.

In fact, in [36], a subclass of NFAwtls is presented that characterizes the class of rational trace languages. However, the DFAwtl is not sufficiently expressive to accept all rational trace languages.

Proposition 21 ([38]). *The rational trace language*

$$L_\vee = \{\, w \in \{a, b\}^* \mid \exists n \geq 0 : |w|_a = n \text{ and } |w|_b \in \{n, 2n\} \,\}$$

is not accepted by any DFAwtl.

This yields the following result.

Corollary 22 ([38]).

(a) REG $\subsetneqq \mathcal{L}(\mathsf{DFAwtl}) \subsetneqq \mathcal{L}(\mathsf{NFAwtl})$.
(b) *The language class* $\mathcal{L}(\mathsf{DFAwtl})$ *is incomparable to the classes* $\mathcal{LRAT}$ *and* CFL *with respect to inclusion.*

Concerning closure and non-closure properties the following results have been obtained.

Theorem 23 ([36,38]).

(a) *The language class $\mathcal{L}$(NFAwtl) is closed under union, product, Kleene star, inverse projections, disjoint shuffle, and the operation of taking the commutative closure, but it is neither closed under intersection (with regular sets), nor under complementation, nor under non-erasing morphisms.*

(b) *The language class $\mathcal{L}$(DFAwtl) is closed under complementation, but it is not closed under any of the following operations: union, intersection (with regular sets), product, Kleene star, reversal, alphabetic morphisms, and commutative closure.*

However, it is still open whether $\mathcal{L}$(NFAwtl) or $\mathcal{L}$(DFAwtl) is closed under inverse morphisms.

Finally, we turn to decision problems for NFAwtls and DFAwtls. Each step of an NFAwtl can be simulated in linear time by a Turing machine. Hence, it follows that the membership problem for an NFAwtl can be solved nondeterministically in quadratic time using linear space. Analogously, the membership problem for a DFAwtl can be solved deterministically in quadratic time using linear space. However, in [29], B. Nagy and L. Kovács derived an interesting improvement on the complexity of the membership problem for DFAwtls.

Let $A = (Q, \Sigma, \lhd, \tau, q_0, F, \delta)$ be a DFAwtl, where $\Sigma = \{a_1, a_2, \ldots, a_m\}$, and let $w = w_1 w_2 \cdots w_n$ be an input word, where $n \geq 1$ and $w_1, w_2, \ldots, w_n \in \Sigma$. For each state $q \in Q$, let $\psi(q) = \Sigma \smallsetminus \tau(q)$ be the set of letters that are non-translucent for this state. Assume that A is in state q. Then in order to execute the next step of the computation of A on the tape contents $w \cdot \lhd$, the first occurrence w_j of a letter from the set $\psi(q)$ must be found in w. Instead of writing the input $w = w_1 w_2 \cdots w_n$ as a continuous sequence of letters, B. Nagy and L. Kovács suggest a preprocessing stage that allows more efficiency. For each letter $a_i \in \Sigma$, $1 \leq i \leq m$, a linked list L_i is built that contains those indices $j \in \{1, 2, \ldots, n\}$ in increasing order for which $w_j = a_i$ holds. These lists can be constructed in time $\mathcal{O}(n \cdot \log n)$. Then these lists are used to simulate the computation of A on input w, which takes another $\mathcal{O}(n \cdot \log n)$ steps. This yields the following result.

Theorem 24 ([29]). *The membership problem for a DFAwtl is decidable in time $\mathcal{O}(n \cdot \log n)$.*

For an NFAwtl, the above considerations give a nondeterministic time bound of $\mathcal{O}(n \cdot \log n)$. The proof of Theorem 18 yields an effective construction of an NFA B from an NFAwtl A such that the language $E = L(B)$ is a subset of the language $L = L(A)$ that is letter-equivalent to L. Hence, E is non-empty if and only if L is non-empty, and E is infinite if and only if L is infinite. As the emptiness problem and the finiteness problem are decidable for finite automata, this immediately yields the following decidability results.

Proposition 25 ([36]). *The emptiness problem and the finiteness problem are effectively decidable for NFAwtls.*

As by Theorem 23, the class $\mathcal{L}(\mathsf{DFAwtl})$ is (effectively) closed under complementation, the decidability of the emptiness problem yields the following result.

Corollary 26 ([38]). *The universality problem is effectively decidable for DFAwtls, that is, it is decidable whether $L(A) = \Sigma^*$ for a given DFAwtl A on Σ.*

On the other hand, it is proved in [36] that regularity, inclusion, and equivalence are undecidable for NFAwtls. For DFAwtls, the inclusion problem is undecidable [38], but it is still open whether regularity and equivalence are undecidable for DFAwtls, too.

While the above types of automata with translucent letters are obtained as reinterpretations of certain types of CD-systems of stl-det-R(1)-automata, another type of automaton with translucent letters has been introduced and studied in [40] (see also [39]): the *nondeterministic linear automaton with translucent letters* (NLAwtl). It is obtained by extending the nondeterministic linear automaton (NLA) of [18] with translucent letters. A nondeterministic linear automaton has two heads that start from the two ends of a given input word, one reading the word from left to right and the other reading it from right to left, halting when the two heads meet. This automaton characterizes the class of linear context-free languages. It corresponds to the $5' \to 3'$ *sensing Watson-Crick automaton* studied in [25,28] and to a class of 2-head finite automata for linear context-free languages considered in [27].

By removing the translucency mapping from an NLAwtl A, one obtains an NLA A' such that $L(A')$ is a sublanguage of $L(A)$ that is letter-equivalent to $L(A)$. This implies immediately that emptiness and finiteness are decidable for NLAwtls. Furthermore, each linear context-free trace language is accepted by an NLAwtl. In fact, a subclass of NLAwtls can be specified that characterizes the class $\mathcal{LCLN}$ of linear context-free trace languages. Concerning closure and non-closure properties, it is known that the language class $\mathcal{L}(\mathsf{NLAwtl})$ is closed under union, but it is neither closed under intersection with regular sets nor under complementation.

On the other hand, the corresponding deterministic class $\mathcal{L}(\mathsf{DLAwtl})$ is closed under complementation, but not under intersection with regular sets, and hence, it is neither closed under intersection nor under union. In addition, this class is not closed under alphabetic morphisms. These closure and non-closure properties show that $\mathcal{L}(\mathsf{DLAwtl})$ is a proper subclass of $\mathcal{L}(\mathsf{NLAwtl})$. Moreover, both these classes are not closed under product, but the class $\mathcal{L}(\mathsf{NLAwtl})$ is closed under product with regular sets. However, it still remains open whether any of the classes $\mathcal{L}(\mathsf{DLAwtl})$ and $\mathcal{L}(\mathsf{NLAwtl})$ is closed under Kleene star or inverse morphisms, and whether the class $\mathcal{L}(\mathsf{NLAwtl})$ is closed under alphabetic morphisms. Finally, for NLAwtls, regularity, linear context-freeness, inclusion, and equivalence are undecidable, but it is still open whether any of these problems is decidable for DLAwtls.

In [19], a Mealey automaton M is combined with a DFAwtl A in such a way that a given input word w is first transformed by M into a word $M(w)$, and this word is then processed by A. The language accepted by the pair (M, A) is $L(M, A) = \{ w \in \Sigma^* \mid M(w) \in L(A) \}$. It is shown in [19] that the non-context-free languages $\{ a^n b^n c^n \mid n \geq 0 \}$, $\{ a^m b^n c^m d^n \mid m, n \geq 1 \}$, and $\{ wcw \mid w \in \{a, b\}^* \}$ are accepted by such pairs.

6   Pushdown Automata with Translucent Letters

In the same way as stl-det-local-CD-R(1)-systems give rise to NFAwtls, PD-CD-R(1)-systems lead to pushdown automata with translucent letters.

Definition 27. *A* pushdown automaton with translucent letters, *or a PDAwtl for short, is defined as a 9-tuple* $P = (Q, \Sigma, \triangleleft, \tau, \Gamma, \bot, Q_{\mathrm{in}}, Q_{\mathrm{fi}}, \delta)$, *where Q is a finite set of states, Σ is a finite input alphabet, $\triangleleft \notin \Sigma$ is a special letter that is used as an* end-of-tape marker, $\Gamma \cup \{\bot\}$ *is the pushdown alphabet, where $\bot$ is a special letter that is used as a bottom marker, $Q_{\mathrm{in}} \subseteq Q$ is the set of initial states, $Q_{\mathrm{fi}} \subseteq Q$ is the set of final states, $\tau : Q \to \mathcal{P}(\Sigma)$ is a* translucency mapping, *and*

$$\delta : (Q \times \Sigma \times (\Gamma \cup \{\bot\})) \to \mathcal{P}(Q \times (\Gamma \cup \{\bot\})^{\leq 3})$$

is the transition relation. For each $q \in Q$, $a \in \Sigma$, and $A \in \Gamma$, $\delta(q, a, A)$ is a subset of $Q \times \Gamma^{\leq 2}$, and $\delta(q, a, \bot)$ is a subset of $Q \times (\bot \cdot \Gamma^{\leq 2})$. In addition, we may require that, for each $q \in Q$, $a \in \Sigma$, and $A \in \Gamma \cup \{\bot\}$, if $a \in \tau(q)$, then $\delta(q, a, A) = \emptyset$.

A configuration of P has the form $(q, w \cdot \triangleleft, \bot\alpha)$. It describes the situation that the PDAwtl P is in state q, its input tape contains the word $w \cdot \triangleleft$, and the word $\bot\alpha$ is the current contents of the pushdown store with the last letter of α at the top. An *initial configuration* of P on input $w \in \Sigma^*$ has the form $(q, w \cdot \triangleleft, \bot)$ for any $q \in Q_{\mathrm{in}}$, and an *accepting configuration* has the form $(q, u \cdot \triangleleft, \bot)$ for any $q \in Q_{\mathrm{fi}}$ and $u \in (\tau(q))^*$.

The *single-step computation relation* $\vdash_P$ that P induces on its set of configurations is defined by the following two rules, where $q \in Q$, $w \in \Sigma^*$, $\alpha \in \Gamma^*$, and $A \in \Gamma$:

(1) $(q, w \cdot \triangleleft, \bot\alpha A) \vdash_P (q', w' \cdot \triangleleft, \bot\alpha\eta)$ if $\exists u \in (\tau(q))^*, a \in \Sigma \smallsetminus \tau(q), v \in \Sigma^*$:
$$w = uav, w' = uv, \text{ and } (q', \eta) \in \delta(q, a, A);$$
(2) $(q, w \cdot \triangleleft, \bot) \vdash_P (q', w' \cdot \triangleleft, \bot\eta)$ if $\exists u \in (\tau(q))^*, a \in \Sigma \smallsetminus \tau(q), v \in \Sigma^*$:
$$w = uav, w' = uv, \text{ and } (q', \bot\eta) \in \delta(q, a, \bot).$$

The language $L(P)$ accepted by P consists of all words for which P has an accepting computation, that is,

$$L(P) = \{ w \in \Sigma^* \mid \exists q \in Q_{\mathrm{in}}, q' \in Q_{\mathrm{fi}}, u \in (\tau(q'))^* : (q, w \cdot \triangleleft, \bot) \vdash_P^* (q', u \cdot \triangleleft, \bot) \}.$$

In fact, the PDAwtl corresponds to the PD-CD-R(1)-system. Let $\mathcal{M} = (I, \Sigma, (M_i)_{i \in I}, \Gamma, \bot, I_0, \delta)$ be a PD-CD-R(1)-system. From $\mathcal{M}$, we obtain a PDAwtl $P = (Q, \Sigma, \triangleleft, \tau, \Gamma, \bot, Q_{\mathrm{in}}, Q_{\mathrm{fi}}, \delta_P)$ as follows:

- $Q = \{q_i \mid i \in I\}$, $Q_{in} = \{q_i \mid i \in I_0\}$, and $Q_{fi} = \{q_i \mid \delta_i(\vartriangleleft) = \mathsf{Accept}\}$,
- the translucency mapping $\tau : Q \to \mathcal{P}(\Sigma)$ is defined through

$$\tau(q_i) = \{a \in \Sigma \mid \delta_i(a) = \mathsf{MVR}\}$$

for all $i \in I$,
- and $\delta_P = \delta$ is the transition relation.

It is easily seen that the PDAwtl P simulates the computations of $\mathcal{M}$. Conversely, from a PDAwtl, one can easily derive a PD-CD-R(1)-system that simulates the computations of the PDAwtl. We illustrate the above definition through a simple example.

Example 28. From the PD-CD-R(1)-system $\mathcal{M}$ from Example 13, we obtain the PDAwtl $P = (Q, \Sigma, \vartriangleleft, \tau, \Gamma, \bot, Q_{in}, Q_{fi}, \delta)$ that looks as follows:

- $Q = \{q_a, q_b, q_c, q_+\}$, $Q_{in} = \{q_a, q_+\}$, and $Q_{fi} = \{q_+\}$,
- $\Sigma = \{a, b, c\}$ and $\Gamma = \{C\}$,
- the translucency mapping τ is specified through

$$\tau(q_a) = \emptyset, \tau(q_b) = \{c\}, \tau(q_c) = \{b\}, \tau(q_+) = \emptyset,$$

- and the transition function δ is defined as follows:

(1) $\delta(q_a, a, \bot) = \{(q_a, \bot C), (q_b, \bot C)\}$, (3) $\delta(q_b, b, C) = \{(q_c, C)\}$,
(2) $\delta(q_a, a, C) = \{(q_a, CC), (q_b, CC)\}$, (4) $\delta(q_c, c, C) = \{(q_b, \lambda), (q_+, \lambda)\}$,

where this function is undefined for all other cases.

On input $w = aaccbb$, P can execute the following accepting computation:

$$(q_a, aaccbb \cdot \vartriangleleft, \bot) \vdash_P (q_a, accbb \cdot \vartriangleleft, \bot C) \vdash_P (q_b, ccbb \cdot \vartriangleleft, \bot CC)$$
$$\vdash_P (q_c, ccb \cdot \vartriangleleft, \bot CC) \vdash_P (q_b, cb \cdot \vartriangleleft, \bot C)$$
$$\vdash_P (q_c, c \cdot \vartriangleleft, \bot C) \qquad \vdash_P (q_+, \vartriangleleft, \bot).$$

In fact, $L(P) = \{a^n v \mid v \in \{b, c\}^*, |v|_b = |v|_c = n, n \geq 0\} = L(\mathcal{M})$. ∎

Observe that the bottom marker $\bot$ cannot be removed from the pushdown of a PDAwtl and that this marker can only occur as the bottom-most letter in the pushdown. Moreover, the push operations of a PDAwtl are restricted, and a PDAwtl accepts simultaneously by final state and by (almost) empty pushdown. In addition, a PDAwtl does not admit any λ-transitions. Despite all these restrictions, the PDAwtls accept all context-free languages, as each context-free language is generated by a context-free grammar in quadratic Greibach normal form [14,45], and as the left-most derivations of such a grammar can be simulated by a (stateless) pushdown automaton that accepts with empty pushdown. Accordingly, we have the following proper inclusion.

Theorem 29 ([32]). $\mathcal{L}(\mathsf{NFAwtl}) \cup \mathsf{CFL} \subsetneq \mathcal{L}(\mathsf{PDAwtl})$.

The results that have been obtained for PDAwtls correspond essentially to those for NFAwtls.

Theorem 30 ([32]). *Each language $L \in \mathcal{L}(\mathsf{PDAwtl})$ contains a context-free sublanguage E that is letter-equivalent to L. In fact, a pushdown automaton for E can be constructed effectively from a PDAwtl for L.*

It follows that $\mathcal{L}(\mathsf{PDAwtl})$ only contains semi-linear languages. In addition, we see that the semi-linear language $\{\, a^n b^n c^n \mid n \geq 0 \,\}$ is not accepted by any PDAwtl.

If L is a context-free trace language on Σ, then there exist a dependency relation D on Σ and a context-free language $R \subseteq \Sigma^*$ such that $L = \bigcup_{w \in R} [w]_D$ (see Definition 1). From a PDA for R, one can construct a PDAwtl B for the language L by introducing a corresponding translucency mapping. Moreover, let $\Sigma = \{a, b, c\}$ and let $L' = \{\, wa^m \mid |w|_a = |w|_b = |w|_c \geq 1, m \geq 1 \,\}$. It can be shown that this language is accepted by an NFAwtl, but as proved in [32], this language is not a context-free trace language. Thus, we get the following consequences.

Theorem 31 ([32]).

(a) $\mathcal{LCF} \subsetneqq \mathcal{L}(\mathsf{PDAwtl})$.

(b) $\mathcal{LCF}$ *is incomparable under inclusion to* $\mathcal{L}(\mathsf{NFAwtl})$.

In analogy to the situation for rational trace languages, a subclass of PDAwtls can be determined that characterizes the context-free trace languages. In addition, it is shown in [32] that the language class $\mathcal{L}(\mathsf{PDAwtl})$ is closed under union and commutative closure, but it is not closed under intersection or complementation. However, it is still open whether this class is closed under product, Kleene-star, reversal, λ-free morphisms, or inverse morphisms.

A deterministic variant of the PDAwtl is obtained from the (globally) deterministic PD-CD-R(1)-systems considered in [37].

Definition 32. *A PDAwtl $P = (Q, \Sigma, \lhd, \tau, \Gamma, \bot, Q_{\mathrm{in}}, Q_{\mathrm{fi}}, \delta)$ is deterministic or a DPDAwtl for short, if $|Q_{\mathrm{in}}| = 1$ and if $|\delta(q, a, A)| \leq 1$ for all $q \in Q$, $a \in \Sigma$, and $A \in \Gamma \cup \{\bot\}$.*

As seen easily, $\mathcal{L}(\mathsf{DFAwtl}) \subsetneqq \mathcal{L}(\mathsf{DPDAwtl})$, but due to the absence of λ-steps and the restrictive acceptance conditions, $\mathsf{DCFL} \not\subseteq \mathcal{L}(\mathsf{DPDAwtl})$. In fact, it is shown in [37] that the deterministic context-free language $\{\, a^n c a^n, a^n bc \mid n \geq 1 \,\}$ is not accepted by any DPDAwtl. In addition, the rational trace language $L_\vee$ (see Proposition 21) is not accepted by any DPDAwtl, either. On the other hand, the language

$$L_p = \{\, w \in \Sigma^* \mid |w|_a = |w|_b = |w|_c \geq 0 \text{ and } \forall uv = w : |u|_a \geq |u|_b \geq |u|_c \,\}$$

on $\Sigma = \{a, b, c\}$ belongs to the difference set $\mathcal{L}(\mathsf{DFAwtl}) \setminus \mathcal{LCF}$ [37]. Hence, already the class $\mathcal{L}(\mathsf{DFAwtl})$ contains languages that are not context-free trace

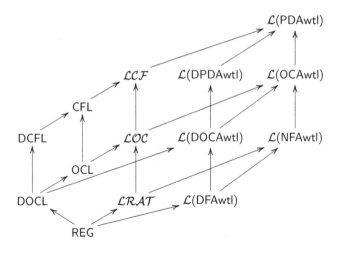

Fig. 5. Hierarchy of language classes accepted by various types of automata with translucent letters. Each arrow represents a proper inclusion, and classes that are not connected by a sequence of arrows are incomparable under inclusion.

languages, and hence, this class is incomparable under inclusion to the class of context-free trace languages.

A pushdown automaton A is called a *one-counter automaton* if its pushdown alphabet contains only one letter in addition to the bottom marker. Then OCL (DOCL) denotes the class of languages that are accepted by (deterministic) one-counter automata. It is well-known that

$$\mathsf{REG} \subsetneq \mathsf{DOCL} \subsetneq \mathsf{OCL} \subsetneq \mathsf{CFL}$$

holds (see, e.g., [5]). By putting the corresponding restriction on PDAwtls and their deterministic variants, the classes of automata OCAwtl and the DOCAwtl are obtained in [32, 37]. The inclusion relations between the resulting language classes and those studied above are summarized in the diagram in Fig. 5, where $\mathcal{LOC}$ denotes the class of *one-counter trace languages*. Finally, it is proved in [37] that the language classes $\mathcal{L}(\mathsf{DOCAwtl})$ and $\mathcal{L}(\mathsf{DPDAwtl})$ are anti-AFLs, that is, they are not closed under any of the six AFL operations. In addition, these classes are not closed under complementation, reversal, and commutative closure, either. Concerning decision problems for PDAwtls, we see from Theorem 30 that emptiness and finiteness are decidable for PDAwtls. On the other hand, as inclusion is undecidable for DFAwtl, it follows immediately that inclusion is also undecidable for DOCAwtls and DPDAwtls. However, it remains open whether equivalence, universality, or regularity are decidable for these automata.

By reinterpreting the visibly pushdown-CD-R(1)-systems of [44], we obtain the *visibly pushdown automata with translucent letters*. The nondeterministic variant of these automata accept a proper superclass of the visibly pushdown trace languages $\mathcal{LVP}$. However, the deterministic visibly pushdown automata with translucent letters accept a class of languages that is incomparable to the class $\mathcal{LVP}$ with respect to inclusion.

To conclude this section, we remark that, in [41], pushdown automata with translucent pushdown alphabets are studied. Such a device is a pushdown automaton A that is equipped with a translucency mapping τ that, to each pair (q, a) consisting of an internal state q of A and an input letter a, assigns a subset of pushdown symbols that are translucent for state q and input letter a. Accordingly, in a corresponding transition step, A does not read (and delete) the topmost symbol on its pushdown store, but the topmost one that is only covered by translucent symbols, that is, by symbols from the set $\tau(q, a)$. As it turns out, the deterministic variant of this type of pushdown automaton already accepts all recursively enumerable languages.

7   Non-returning NFAs with Translucent Letters

In contrast to a DFA or an NFA, which read their input strictly from left to right, letter by letter, a DFAwtl or an NFAwtl just reads (and deletes) the first letter from the left which it can see, that is, the first letter which is not translucent for its current state. In [23], an extended variant of these types of automata is presented that, after reading and deleting a letter, does not return its head to the left end of its tape but, instead, continues from the position of the letter just deleted. When the end-of-tape marker is reached, this type of automaton either accepts, rejects, or continues with its computation, which means that it again reads the remaining tape contents from the beginning. This type of automaton, called a *non-returning finite automaton with translucent letters* or an *nrNFAwtl*, is strictly more expressive than the NFAwtl.

In the literature, many other types of automata have been studied that do not simply read their inputs letter by letter from left to right. Among them are the right one-way jumping finite automaton of [3,10] and the right-revolving finite automaton of [4]. In [24], these automata are compared to the nrNFAwtl and its deterministic variant, the nrDFAwtl. As it turns out, the nrDFAwtl can be interpreted as a right one-way jumping finite automaton that can detect the end of its input. Here we restate these results in short.

Definition 33 ([23]). *An nrNFAwtl A is specified through a 6-tuple $A = (Q, \Sigma, \lhd, \tau, I, \delta)$, where Q is a finite set of states, Σ is a finite alphabet, $\lhd \notin \Sigma$ is a special letter that is used as an end-of-tape marker, $\tau : Q \to \mathcal{P}(\Sigma)$ is a translucency mapping, $I \subseteq Q$ is a set of initial states, and*

$$\delta : Q \times (\Sigma \cup \{\lhd\}) \to (\mathcal{P}(Q) \cup \{\mathsf{Accept}\})$$

is a transition relation. Here it is required that, for each state $q \in Q$ and each letter $a \in \Sigma$, $\delta(q, a) \subseteq Q$, and if $a \in \tau(q)$, then $\delta(q, a) = \emptyset$. For each state $q \in Q$, the letters from the set $\tau(q)$ are translucent for q, that is, in state q, the automaton A does not see these letters.

An nrNFAwtl $A = (Q, \Sigma, \lhd, \tau, I, \delta)$ *is a* non-returning deterministic finite automaton with translucent letters, *abbreviated as* nrDFAwtl, *if* $|I| = 1$ *and if* $|\delta(q, a)| \leq 1$ *for all* $q \in Q$ *and all* $a \in \Sigma \cup \{\lhd\}$.

From the above definition, we see that $\delta(q, \lhd)$ is either a subset of Q or the operation Accept. Thus, on seeing the end-of-tape marker $\lhd$ in state q, the nrNFAwtl A has either the option to change its state or to accept. The nrNFAwtl $A = (Q, \Sigma, \lhd, \tau, I, \delta)$ works as follows. For an input word $w \in \Sigma^*$, A starts in a nondeterministically chosen initial state $q_0 \in I$ with the word $w \cdot \lhd$ on its tape. This situation is described by the configuration $q_0 w \cdot \lhd$. Now assume that A is in a configuration of the form $x q_1 w \cdot \lhd$, where $q_1 \in Q$ and $x, w \in \Sigma^*$, that is, A is in state q_1, the tape contains the word $xw \cdot \lhd$, and the head of A is on the first letter of the suffix $w \cdot \lhd$. Then A looks for the first occurrence from the left of a letter in w that is not translucent for state q_1, that is, if $w = uav$ such that $u \in (\tau(q_1))^*$, $a \in (\Sigma \smallsetminus \tau(q_1))$, and $v \in \Sigma^*$, then A nondeterministically chooses a state $q_2 \in \delta(q_1, a)$, erases the letter a from the tape, thus producing the tape contents $xuv \cdot \lhd$, sets its internal state to q_2, and continues the computation from the configuration $xu q_2 v \cdot \lhd$. In case $\delta(q_1, a) = \emptyset$, A halts without accepting. Finally, if $w \in (\tau(q_1))^*$, then A reaches the end-of-tape marker $\lhd$, and a transition from the set $\delta(q_1, \lhd)$ is applied. This transition is either an accept step or a state q_2 from Q. In the former case, A halts and accepts, while in the latter case, it continues the computation in state q_2 by reading its tape again from left to right, that is, from the configuration $q_2 xw \cdot \lhd$. Finally, if $\delta(q_1, \lhd)$ is undefined, A halts and rejects. Thus, the computation relation $\vdash_A$ that A induces on its set of configurations $\Sigma^* \cdot Q \cdot \Sigma^* \cdot \{\lhd\} \cup \{\text{Accept}, \text{Reject}\}$ is the reflexive and transitive closure $\vdash_A^*$ of the single-step computation relation $\vdash_A$ that is specified as follows, where x, u, v, w are words from Σ^* and a is a letter from Σ:

$$
xqw \cdot \lhd \vdash_A
\begin{cases}
xuq'v \cdot \lhd, & \text{if } w = uav,\ u \in (\tau(q))^*,\ a \notin \tau(q),\ \text{and } q' \in \delta(q, a), \\
\text{Reject}, & \text{if } w = uav,\ u \in (\tau(q))^*,\ a \notin \tau(q),\ \text{and } \delta(q, a) = \emptyset, \\
q'xw \cdot \lhd & \text{if } w \in (\tau(q))^* \text{ and } q' \in \delta(q, \lhd), \\
\text{Accept}, & \text{if } w \in (\tau(q))^* \text{ and } \delta(q, \lhd) = \text{Accept}, \\
\text{Reject}, & \text{if } w \in (\tau(q))^* \text{ and } \delta(q, \lhd) = \emptyset.
\end{cases}
$$

We illustrate this definition through a simple example.

Example 34. Let $A = (Q, \{a, b, c\}, \lhd, \tau, \{q_a\}, \delta)$ be the nrNFAwtl that is defined by taking $Q = \{q_a, q_b, q_c, q_r\}$,

$$
\tau(q_a) = \emptyset, \tau(q_b) = \{a\}, \tau(q_c) = \{b\}, \tau(q_r) = \{c\},
$$

and

$$
\delta(q_a, a) = \{q_b\}, \delta(q_b, b) = \{q_c\}, \delta(q_c, c) = \{q_r\}, \delta(q_r, \lhd) = \{q_a\}, \delta(q_a, \lhd) = \text{Accept},
$$

where this function is undefined for all other cases. Given the word $w = aabbcc$ as input, the automaton A proceeds as follows:

$$
q_a aabbcc \cdot \lhd \vdash_A q_b abbcc \cdot \lhd \vdash_A a q_c bcc \cdot \lhd \vdash_A ab q_r c \cdot \lhd
$$
$$
\vdash_A q_a abc \cdot \lhd \quad \vdash_A q_b bc \cdot \lhd \quad \vdash_A q_c c \cdot \lhd
$$
$$
\vdash_A q_r \cdot \lhd \quad\quad \vdash_A q_a \cdot \lhd \quad\quad \vdash_A \text{Accept},
$$

that is, A accepts on input $w = aabbcc$. Actually, it is easily seen that $L(A) = \{ a^n b^n c^n \mid n \geq 0 \}$. Furthermore, A is actually an nrDFAwtl. ∎

Recall from Theorem 30 that this language is not even accepted by any PDAwtl.

As defined above, an nrNFAwtl may run into an infinite computation. Moreover, it may accept without having read and deleted its input completely. However, the following result shows that these situations can be avoided.

Lemma 35 ([23]). *Each nrNFAwtl A can effectively be converted into an equivalent nrNFAwtl C that never gets into an infinite computation and accepts only after reading and deleting its input completely. In addition, if A is deterministic, then so is C.*

It is fairly easy to simulate an NFAwtl by an nrNFAwtl. Hence, the example above shows that the nrNFAwtl (nrDFAwtl) is strictly more expressive than the NFAwtl (DFAwtl).

Definition 36. [3,10] *A nondeterministic right one-way jumping finite automaton, or an NROWJFA, is given through a five-tuple $J = (Q, \Sigma, I, F, \delta)$, where Q is a finite set of states, Σ is a finite alphabet, $I \subseteq Q$ is the set of initial states, $F \subseteq Q$ is the set of final states, and $\delta : Q \times \Sigma \to \mathcal{P}(Q)$ is a transition relation. For each state $q \in Q$, $\Sigma_q = \{ a \in \Sigma \mid \delta(q, a) \neq \emptyset \}$ is the set of letters that J can read in state q.*

A configuration of the NROWJFA J is a word qw from the set $Q \cdot \Sigma^*$. The computation relation $\circlearrowright_J^*$ that J induces on its set of configurations is the reflexive and transitive closure of the right one-way jumping relation $\circlearrowright_J$ that is defined as follows, where $x, y \in \Sigma^*$, $a \in \Sigma$, and $q, q' \in Q$:

$$qxay \circlearrowright_J q'yx \text{ if } x \in (\Sigma \smallsetminus \Sigma_q)^* \text{ and } q' \in \delta(q, a).$$

Thus, being in state q, J reads and deletes the first letter to the right of the actual head position that it can actually read in that state, while the prefix that consists of letters for which J has no transitions in the current state is cyclically shifted to the end of the current tape contents. Then

$$L(J) = \{ w \in \Sigma^* \mid \exists q_0 \in I \, \exists q_f \in F : q_0 w \circlearrowright_J^* q_f \}$$

is the language accepted by the NROWJFA J.

The NROWJFA J is deterministic, *that is, a* right one-way jumping finite automaton *or a* ROWJFA, *if $|I| = 1$ and $|\delta(q, a)| \leq 1$ for all $q \in Q$ and $a \in \Sigma$.*

We illustrate the workings of an (N)ROWJFA with a simple example.

Example 37. Let $J = (\{p_0, q_0, q_1\}, \{a, b\}, \{p_0, q_0\}, \{p_0, q_0\}, \delta)$ be the NROWJFA that is specified by the following transition relation:

$$\delta(p_0, a) = \{p_0\}, \, \delta(q_0, a) = \{q_1\}, \, \delta(q_1, b) = \{q_0\},$$

where this function is undefined for all other cases. Starting from state p_0, J accepts on input $w \in \{a, b\}^*$ iff $|w|_b = 0$, while starting from state q_0, J accepts on input $w \in \{a, b\}^*$ iff $|w|_a = |w|_b$. Hence, $L(J) = \{a\}^* \cup \{ w \in \{a, b\}^* \mid |w|_a = |w|_b \} = L'_\vee$. ∎

It has been shown that the language $\{a^n b^n \mid n \geq 0\}$ is not accepted by any NROWJFA [2], while the permutation-closed language $L'_\vee$ above is not accepted by an ROWJFA [2]. Thus, the ROWJFA is strictly less expressive than the NROWJFA. Moreover, right one-way jumping finite automata can be simulated by non-returning finite automata with translucent letters.

Theorem 38 ([24]).

(a) $\mathcal{L}(\mathsf{ROWJFA}) \subsetneq \mathcal{L}(\mathsf{nrDFAwtl})$.

(b) $\mathcal{L}(\mathsf{NROWJFA}) \subsetneq \mathcal{L}(\mathsf{nrNFAwtl})$.

When looking at the definitions, we see that the ROWJFA differs from the nrDFAwtl in two aspects. The first of these is the partitioning of the alphabet. For each state q of an ROWJFA J, the alphabet of J is split into two disjoint subsets: the set Σ_q of letters that J can read in state q and the remaining letters. For a state p of an nrDFAwtl A, the alphabet of A is split into three disjoint subsets: the set of letters that A can read in state p, the set $\tau(p)$ of letters that are translucent for state p, and the set of letters that are neither translucent for state p nor can be read in state p. However, the third type of letters for state p can be avoided by defining, for each letter b of this type, $\delta_A(p, b) = \{q_{\mathrm{fail}}\}$, where q_{fail} is a new non-final state in which A cannot read any letter at all. The second aspect is the fact that an nrDFAwtl has an end-of-tape marker and that, whenever its head reaches that marker, A can execute an additional change of state. It is actually this feature that allows the nrDFAwtl to accept the language $\{a^n b^n \mid n \geq 0\}$. Thus, the increase in expressive capacity from the ROWJFA to the nrDFAwtl is due to this second feature.

Definition 39. [4] *A* right-revolving NFA, *or an* rr-NFA, *is given through a six-tuple* $A = (Q, \Sigma, q_0, F, \Delta, \delta)$, *where* Q *is a finite set of states,* Σ *is a finite alphabet,* $q_0 \in Q$ *is an initial state,* $F \subseteq Q$ *is the set of final states, and*

$$\Delta : Q \times \Sigma \to \mathcal{P}(Q) \ \text{and} \ \delta : Q \times \Sigma \to \mathcal{P}(Q)$$

are two transition relations.

A configuration *of the rr-NFA* A *is a word* qw, *where* $q \in Q$ *and* $w \in \Sigma^*$. *The* computation relation $\vdash_A$ *that* A *induces on its set of configurations is the reflexive and transitive closure* $\vdash_A^*$ *of the* single-step computation relation $\vdash_A$ *that is defined as follows, where* $a \in \Sigma$ *and* $q, q' \in Q$:

(1) *If* $q' \in \delta(q, a)$, *then* $qaw \vdash_A q'w$. *The transitions of this form are called* ordinary transitions.
(2) *If* $q' \in \Delta(q, a)$, *then* $qaw \vdash_A q'wa$. *The transitions of this form are called* right-revolving transitions.

Thus, an ordinary transition consumes the first letter of the current tape contents, while a right-revolving transition just shifts the first letter to the end of the tape. Observe that it is possible that $\delta(q, a) \neq \emptyset \neq \Delta(q, a)$. *In this case,* A *nondeterministically chooses whether to execute an ordinary or a right-revolving transition.* Then $L(A) = \{ w \in \Sigma^* \mid \exists q_f \in F : q_0w \vdash_A^* q_f \}$ *is the language accepted by the rr-NFA A.*

The rr-NFA A *is* deterministic, *or an* rr-DFA, *if, for all* $q \in Q$ *and all* $a \in \Sigma$, $|\Delta(q,a)| + |\delta(q,a)| \leq 1$.

Concerning the expressive capacity of right-revolving finite automata, it is shown in [4,10] that $\mathcal{L}(\text{ROWJFA}) \subsetneq \mathcal{L}(\text{rr-DFA})$, that the non-context-free language

$$L_{eq3} = \{ w \in \{a,b,c\}^* \mid |w|_a = |w|_b = |w|_c \}$$

is accepted by an rr-DFA, and that $L'_\vee \in \mathcal{L}(\text{rr-NFA}) \smallsetminus \mathcal{L}(\text{rr-DFA})$. Concerning the relationship between the language classes defined by the various types of automata considered here, we have the following result.

Theorem 40 ([24]). *All arrows in the diagram in Fig. 6 depict proper inclusions, and classes that are not connected by a sequence of directed arrows are incomparable under inclusion.*

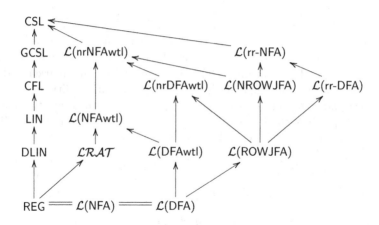

Fig. 6. The inclusion relations between the various types of jumping automata, right-revolving automata, and automata with translucent letters

For separating the various types of automata with translucent letters from the right-revolving automata, the following example languages are presented in [24]:

The language $L_c = \{ wc \mid w \in \{a,b\}^*, |w|_a = |w|_b \}$ is accepted by a DFAwtl but not by any rr-NFA, while the language $L_{cc} = \{ a^m ca^m c \mid m \geq 0 \}$ has the following properties:

– There exists a rr-DFA A such that $L(A) \cap (\{a\}^* \cdot \{c\} \cdot \{a\}^* \cdot \{c\}) = L_{cc}$,
– but there is no nrNFAwtl C such that $L(C) \cap (\{a\}^* \cdot \{c\} \cdot \{a\}^* \cdot \{c\}) = L_{cc}$.

The latter shows, in addition, that the deterministic linear language L_{cc} is not accepted by any nrNFAwtl.

8   Conclusion

Although finite automata with translucent letters have originally been obtained through a reinterpretation of certain cooperating distributed systems of restarting automata, the concept itself is quite straightforward. By adjoining a translucency mapping to a type of automaton, a corresponding automaton with translucent letters is obtained. We have seen this not only for finite automata, but also for linear automata, one-counter automata, and pushdown automata. From the results obtained, it should be clear that automata with translucent letters are an interesting variety of automata that do not read their input strictly from left to right. They are of particular interest because of their ability to accept (and even characterize) classes of trace languages.

Despite the fact that much work has already been done on automata with translucent letters, many questions are still unanswered, and many problems are still open. Here we list a few of them:

- Find further closure and non-closure properties for the various classes of languages accepted by automata with translucent letters!
- Determine the degree of complexity for those algorithmic problems that are decidable for the various types of automata with translucent letters!
- Which of the decision problems that are decidable for a certain type of automata remain decidable for the corresponding type of automata with translucent letters?
- What effect would the addition of λ-steps have on the expressive capacity of the various types of automata with translucent letters? For example, in [3], S. Beier and M. Holzer consider nondeterministic right one-way jumping finite automata with λ-transitions, that is, these automata can perform a change of state without reading an input letter. Actually, they study three different types of these automata based on the conditions that enable the execution of λ-transitions. On the other hand, in [4], λ-transitions are defined for rr-NFAs and rr-DFAs, but it is proved in that paper that λ-transitions do not increase the expressive capacity of right-revolving finite automata.
- How does the expressive capacity of a PDAwtl change if we abandon the requirement that the pushdown must only contain the bottom marker when the automaton accepts? For example, this change has already been adapted in the definition of the visibly pushdown-CD-R(1)-systems in [44].

Finally, the concept of non-returning finite automata with translucent letters can easily be extended to pushdown automata with translucent letters. The resulting non-returning PDAwtl will have a larger expressive capacity than the PDAwtl, but how do these automata compare to other types of pushdown automata that do not simply process their inputs from left to right like, e.g., the input-reversal pushdown automaton of [8].

References

1. Alur, R., Madhusudan, P.: Visibly pushdown languages. In: Babai, L. (ed.) STOC 2004, Proceedings, pp. 202–211. ACM Press, New York (2004)

2. Beier, S., Holzer, M.: Properties of right one-way jumping finite automata. Theoret. Comput. Sci. **798**, 78–94 (2019)
3. Beier, S., Holzer, M.: Nondeterministic right one-way jumping finite automata. Inf. Comput. **284**, 104687 (2022)
4. Bensch, S., Bordihn, H., Holzer, M., Kutrib, M.: On input-revolving deterministic and nondeterministic finite automata. Inf. Comput. **207**, 1140–1155 (2009)
5. Berstel, J.: Transductions and Context-Free Languages. Teubner, Stuttgart (1979)
6. Bertoni, A., Mauri, G., Sabadini, N.: Membership problems for regular and context-free trace languages. Inf. Comput. **82**, 135–150 (1989)
7. Bollig, B., Grindei, M.-L., Habermehl, P.: Realizability of concurrent recursive programs. In: de Alfaro, L. (ed.) FoSSaCS 2009. LNCS, vol. 5504, pp. 410–424. Springer, Heidelberg (2009). https://doi.org/10.1007/978-3-642-00596-1_29
8. Bordihn, H., Holzer, M., Kutrib, M.: Input reversals and iterated pushdown automata: a new characterization of Khabbaz geometric hierarchy of languages. In: Calude, C.S., Calude, E., Dinneen, M.J. (eds.) DLT 2004. LNCS, vol. 3340, pp. 102–113. Springer, Heidelberg (2004). https://doi.org/10.1007/978-3-540-30550-7_9
9. Cartier, P., Foata, D.: Problèmes Combinatoires de Commutation et Réarrangements. Lecture Notes in Mathematics, vol. 85. Springer, Heidelberg (1969)
10. Chigahara, H., Fazekas, S.Z., Yamamura, A.: One-way jumping finite automata. Int. J. Found. Comput. Sci. **27**, 391–405 (2016)
11. Csuhaj-Varjú, E., Dassow, J., Kelemen, J., Păun, G.: Grammar Systems - A Grammatical Approach to Distribution and Cooperation. Gordon and Breach, Newark (1994)
12. Dahlhaus, E., Warmuth, M.K.: Membership for growing context-sensitive grammars is polynomial. J. Comput. Syst. Sci. **33**, 456–472 (1986)
13. Diekert, V., Rozenberg, G. (eds.): The Book of Traces. World Scientific, Singapore (1995)
14. Greibach, S.A.: A new normal form theorem for context-free phrase structure grammars. J. Assoc. Comput. Mach. **12**, 42–52 (1965)
15. Hopcroft, J.E., Ullman, J.D.: Introduction to Automata Theory, Languages, and Computation. Addison-Wesley, Reading (1979)
16. Jančar, P., Mráz, F., Plátek, M., Vogel, J.: Restarting automata. In: Reichel, H. (ed.) FCT 1995. LNCS, vol. 965, pp. 283–292. Springer, Heidelberg (1995). https://doi.org/10.1007/3-540-60249-6_60
17. Jančar, P., Mráz, F., Plátek, M., Vogel, J.: On monotonic automata with a restart operation. J. Autom. Lang. Comb. **4**, 287–311 (1999)
18. Loukanova, R.: Linear context free languages. In: Jones, C.B., Liu, Z., Woodcock, J. (eds.) ICTAC 2007. LNCS, vol. 4711, pp. 351–365. Springer, Heidelberg (2007). https://doi.org/10.1007/978-3-540-75292-9_24
19. Madeeha, F., Nagy, B.: Transduced-input automata with translucent letters. C. R. Acad. Bulg. Sci. **73**, 33–39 (2020)
20. Mazurkiewicz, A.: Concurrent program schemes and their interpretations. DAIMI report PB 78, Aarhus University, Aarhus (1977)
21. Meduna, A., Zemek, P.: Jumping finite automata. Int. J. Found. Comput. Sci. **23**, 1555–1578 (2012)
22. Mráz, F.: Lookahead hierarchies of restarting automata. J. Autom. Lang. Comb. **6**, 493–506 (2001)

23. Mráz, F., Otto, F.: Non-returning finite automata with translucent letters. In: Bordihn, H., Horváth, G., Vaszil, G. (eds.) 12th International Workshop on Non-Classical Models of Automata and Applications (NCMA 2022). EPTCS, vol. 367, pp. 143–159 (2022)

24. Mráz, F., Otto, F.: Non-returning deterministic and nondeterministic finite automata with translucent letters (2023, submitted)

25. Nagy, B.: On $5' \to 3'$ sensing Watson-Crick finite automata. In: Garzon, M.H., Yan, H. (eds.) DNA 2007. LNCS, vol. 4848, pp. 256–262. Springer, Heidelberg (2008). https://doi.org/10.1007/978-3-540-77962-9_27

26. Nagy, B.: On CD-systems of stateless deterministic R(2)-automata. J. Autom. Lang. Comb. **16**, 195–213 (2011)

27. Nagy, B.: A class of 2-head finite automata for linear languages. Triangle **8** (Languages. Mathematical Approaches), 89–99 (2012)

28. Nagy, B.: On a hierarchy of $5' \to 3'$ sensing Watson-Crick finite automata languages. J. Log. Comput. **23**, 855–872 (2013)

29. Nagy, B., Kovács, L.: Finite automata with translucent letters applied in natural and formal language theory. In: Nguyen, N.T., Kowalczyk, R., Fred, A., Joaquim, F. (eds.) Transactions on Computational Collective Intelligence XVII. LNCS, vol. 8790, pp. 107–127. Springer, Heidelberg (2014). https://doi.org/10.1007/978-3-662-44994-3_6

30. Nagy, B., Otto, F.: CD-systems of stateless deterministic R(1)-automata accept all rational trace languages. In: Dediu, A.-H., Fernau, H., Martín-Vide, C. (eds.) LATA 2010. LNCS, vol. 6031, pp. 463–474. Springer, Heidelberg (2010). https://doi.org/10.1007/978-3-642-13089-2_39

31. Nagy, B., Otto, F.: An automata-theoretical characterization of context-free trace languages. In: Černá, I., et al. (eds.) SOFSEM 2011. LNCS, vol. 6543, pp. 406–417. Springer, Heidelberg (2011). https://doi.org/10.1007/978-3-642-18381-2_34

32. Nagy, B., Otto, F.: CD-systems of stateless deterministic R(1)-automata governed by an external pushdown store. RAIRO Theor. Inform. Appl. **45**, 413–448 (2011)

33. Nagy, B., Otto, F.: Deterministic pushdown-CD-systems of stateless deterministic R(1)-automata. In: Dömösi, P., Iván, S. (eds.) Automata and Formal Languages, AFL 2011, Proceedings, pp. 328–342. Institute of Mathematics and Informatics, College of Nyiregyháza (2011)

34. Nagy, B., Otto, F.: Finite-state acceptors with translucent letters. In: Bel-Enguix, G., Dahl, V., De La Puente, A.O. (eds.) BILC 2011: AI Methods for Interdisciplinary Research in Language and Biology, Proceedings, Portugal, pp. 3–13. SciTePress (2011)

35. Nagy, B., Otto, F.: Globally deterministic CD-systems of stateless R(1)-automata. In: Dediu, A.-H., Inenaga, S., Martín-Vide, C. (eds.) LATA 2011. LNCS, vol. 6638, pp. 390–401. Springer, Heidelberg (2011). https://doi.org/10.1007/978-3-642-21254-3_31

36. Nagy, B., Otto, F.: On CD-systems of stateless deterministic R-automata with window size one. J. Comput. Syst. Sci. **78**, 780–806 (2012)

37. Nagy, B., Otto, F.: Deterministic pushdown CD-systems of stateless deterministic R(1)-automata. Acta Informatica **50**, 229–255 (2013)

38. Nagy, B., Otto, F.: Globally deterministic CD-systems of stateless R-automata with window size 1. Int. J. Comput. Math. **90**(6), 1254–1277 (2013)

39. Nagy, B., Otto, F.: Two-head finite-state acceptors with translucent letters. In: Catania, B., Královič, R., Nawrocki, J., Pighizzini, G. (eds.) SOFSEM 2019. LNCS, vol. 11376, pp. 406–418. Springer, Cham (2019). https://doi.org/10.1007/978-3-030-10801-4_32

40. Nagy, B., Otto, F.: Linear automata with translucent letters and linear context-free trace languages. RAIRO Theor. Inform. Appl. **54**, 23 (2020). https://doi.org/10.1051/ita/2020002

41. Nagy, B., Otto, F., Vollweiler, M.: Pushdown automata with translucent pushdown symbols. In: Freund, R., Holzer, M., Mereghetti, C., Otto, F., Palano, B. (eds.) Third Workshop on Non-Classical Models of Automata and Applications (NCMA 2011), Proceedings, books@ocg.at, vol. 282, pp. 193–208. Österreichische Computer Gesellschaft, Wien (2011)

42. Otto, F.: Restarting automata and their relations to the Chomsky hierarchy. In: Ésik, Z., Fülöp, Z. (eds.) DLT 2003. LNCS, vol. 2710, pp. 55–74. Springer, Heidelberg (2003). https://doi.org/10.1007/3-540-45007-6_5

43. Otto, F.: Restarting automata. In: Ésik, Z., Martín-Vide, C., Mitrana, V. (eds.) Recent Advances in Formal Languages and Applications. Studies in Computational Intelligence, vol. 25, pp. 269–303. Springer, Heidelberg (2006). https://doi.org/10.1007/978-3-540-33461-3_11

44. Otto, F.: On visibly pushdown trace languages. In: Italiano, G.F., Margaria-Steffen, T., Pokorný, J., Quisquater, J.-J., Wattenhofer, R. (eds.) SOFSEM 2015. LNCS, vol. 8939, pp. 389–400. Springer, Heidelberg (2015). https://doi.org/10.1007/978-3-662-46078-8_32

45. Rosenkrantz, D.J.: Matrix equations and normal forms for context-free grammars. J. Assoc. Comput. Mach. **14**, 501–507 (1967)

46. Zielonka, W.: Note on finite asynchronous automata. RAIRO Theor. Inform. Appl. **21**, 99–135 (1987)

Contributed Papers

Earliest Query Answering for Deterministic Stepwise Hedge Automata

Antonio Al Serhali[(⊠)] and Joachim Niehren

Inria and University of Lille, Lille, France
`antonio.al-serhali@inria.fr`

Abstract. Earliest query answering (EQA) is the problem to enumerate certain query answers on streams at the earliest events. We consider *EQA* for regular monadic queries on hedges or nested words defined by deterministic stepwise hedge automata (dSHAs). We present an *EQA* algorithm for dSHAs that requires time $O(c\,m)$ per event, where m is the size of the automaton and c the concurrency of the query. We show that our EQA algorithm runs efficiently on regular XPath queries in practice.

1 Introduction

Streaming algorithms for hedges or nested words were widely studied for complex event processing [8,12] and for transforming *XML* documents in an online manner [5,17]. The open end of a stream can be instantiated continuously. Therefore, streams can be seen as incomplete databases, for which the notion of *certain query answers (CQAs)* was widely studied [10]. An element is a certain query answer if it is selected by all completions of the incomplete database. For instance, the XPATH query `following::a[following::b]` selects all *a*-elements of a nested word (modeling an *XML* document) that are followed eventually by some *b*-element. On the stream $aabaabaaaaa\ldots$ all *a*-positions before the last *b* are *CQAs* of this query, while those after are not.

Earliest query answering (*EQA*) is the problem of detecting *CQAs* on streams at the earliest event [7]. We study *EQA* for regular monadic queries which select nodes of trees or hedges. For this, we consider streams that elongate prefixes of nested words such as $aa\langle b\langle a\rangle\,\langle b\ldots$ to well-nested words. *EQA* requires to decide the existence of *CQAs* which is a computationally hard problem even for tiny fragments of regular XPATH queries [4], since CQA is a universality problem concerning all completions of the stream. Gauwin et al. [7] showed that *EQA* can be done in polynomial time for monadic queries defined by deterministic nested word automata (dNWAs) [3,16]. Their algorithm, however, requires time $O(c\,n^2)$ per event, where the concurrency c is the number of alive candidates of the query (not of the algorithm) and n is the number of the automaton states. In the worst case, c may be the length of the stream for monadic queries, so the overall complexity may be quadratic in the size of the stream too.

For complex event processing [8,12], *EQA* has often been avoided, by restricting the class of queries such that the certainty of an answer candidate depends

© The Author(s), under exclusive license to Springer Nature Switzerland AG 2023
B. Nagy (Ed.): CIAA 2023, LNCS 14151, pp. 53–65, 2023.
https://doi.org/10.1007/978-3-031-40247-0_3

only on the past of the stream and not on its future [8,12]. This rules out XPATH query with filters such as following::a[following::b] . Riveros et al. [13] proposed instead to enumerate the query answers late at the end of the stream, which however requires to buffer a large number of candidates. For XML streaming, EQA was often approximated [5,17] so that all $CQAs$ are eventually selected but not always earliest. Or else, as done by Saxon [9], only very restricted regular XPATH queries were permitted.

A major bottleneck for automata based EQA algorithms in practice [5] was the difficulty to compile regular path queries on nested words to reasonably small deterministic automata. This problem was solved recently [2] based on stepwise hedge automata (SHAS) [14] and schema-based determinization [15]. SHAS naturally generalize finite state automata for words and trees. They can recognize all regular languages of hedges equally to NWAS but without any explicit stack (such as tree automata). SHAS can always be determinized in a bottom-up and left-to-right manner. NWAS can also be determinized but differently, since their determinization has to deal with top-down processing via explicit stacks, often leading to a huge size increase.

The availability of dSHAS for regular path queries gave new hope for the feasibility of EQA in practice. For this, however, more efficient EQA algorithms are needed. In particular, the time per event should no more be quadratic in n and neither should the preprocessing time be cubic in n. Sakho [18] showed that EQA for boolean dSHA queries can be done in time $O(m)$ per event, where m is the overall size of the automaton. This improvement relies on the fact, that the set of accessible states of a dSHA can be computed in time $O(m)$, while for dNWAS it requires time $O(n^2)$ after $O(n^3)$ preprocessing, where n is the number of the states of the automaton.

In the present paper, we present a new EQA algorithm for monadic dSHA queries. Our approach is to adapt the general ideas of Gauwin from dNWAS to dSHAS. This yields an EQA algorithm in time $O(c\,m)$ per event where c is the concurrency of the query. Gauwin's quadratic factor n^2 is reduced to m while the cubic preprocessing in time $O(n^3)$ is removed. The algorithm obtained is more efficient than the best previous EQA algorithm, based on a reduction to Gauwin's EQA algorithm by compiling dSHAS to dNWAS in linear time.

We implemented our new EQA algorithm in the AStream tool and applied it to the regular XPath queries from the XPathMark collection [6] scaling to huge documents, and to the regular XPath queries extracted from practical XSLT programs by Lick and Schmitz [11] but on smaller documents. It turns out that AStream runs efficiently on huge XML documents (>100 GB) for all queries with low concurrency. Some queries can be answered in streaming mode where the best existing non earliest query answering algorithm failed to be earliest [5].

Outline. We start with preliminaries in Sect. 2, while 3 and 4 recall stepwise hedge automata and nested word automata, respectively. An earliest membership tester for dSHAS is presented in Sect. 5 and a late streaming algorithm for answering monadic queries in Sect. 6. We then present our new EQA algorithm in Sect. 7 and discuss experimental results with AStream in Sect. 8. Complete proofs and supplementary material are given in [1].

2   Preliminaries

Let A and B be sets. A partial function $f : A \hookrightarrow B$ is a relation $f \subseteq A \times B$ that is functional. The domain of a partial function is $dom(f) = \{a \in A \mid f(a) \in B\}$. A total function $f : A \to B$ is a partial function $f : A \hookrightarrow B$ with $dom(f) = A$. Let $\mathbb{N}$ be the set of natural numbers including 0.

Words. Let alphabet Ω be a set. The set of words over Ω is $\Omega^* = \cup_{n \in \mathbb{N}} \Omega^n$. A word $(a_1, \ldots, a_n) \in \Omega^n$ is written as $a_1 \ldots a_n$. We denote the empty word of length 0 by $\varepsilon \in \Omega^0$ and by $v_1 \cdot v_2 \in \Omega^*$ the concatenation of two words $v_1, v_2 \in \Omega^*$. For any word $v \in \Omega^*$ let $prefs(v) \subseteq \Omega^*$ be the set of its prefixes. For any $v \in \Omega^*$ and $a \in \Omega$ let $\#_a(v)$ be the number of occurrences of a in v.

Hedges. Hedges are sequences of letters and trees $\langle h \rangle$ with some hedge h. More formally, a hedge $h \in \mathcal{H}_\Omega$ has the following abstract syntax:

$$h, h' \in \mathcal{H}_\Omega ::= \varepsilon \quad | \quad a \quad | \quad \langle h \rangle \quad | \quad h \cdot h' \qquad \text{where } a \in \Omega$$

We assume $\varepsilon \cdot h = h \cdot \varepsilon = h$ and $(h \cdot h_1) \cdot h_2 = h \cdot (h_1 \cdot h_2)$. Therefore, we consider any word in Ω^* as a hedge in $\mathcal{H}_\Omega$, i.e., $\Omega^* \ni aab = a \cdot a \cdot b \in \mathcal{H}_\Omega$. For any $h \in \mathcal{H}_\Omega$ and $a \in \Omega$ let $\#_a(h)$ be the number of occurrences of a in h. The size $|h|$ is the number of letters and opening parenthesis of h. The nesting depth d of h is the maximal number of nested opening parenthesis of trees in h. The set of positions of a hedge $h \in \mathcal{H}_\Omega$ is $pos(h) = \{1, \ldots, |h|\}$.

Nested Words. Hedges can be identified with nested words, i.e., words over the alphabet $\hat{\Omega} = \Omega \cup \{\langle, \rangle\}$ in which all parentheses are well-nested. This is done by the function $nw(h) : \mathcal{H}_\Omega \to (\Omega \cup \{\langle, \rangle\})^*$ such that: $nw(\varepsilon) = \varepsilon$, $nw(\langle h \rangle) = \langle \cdot nw(h) \cdot \rangle$, $nw(a) = a$, and $nw(h \cdot h') = nw(h) \cdot nw(h')$. The set of nested word prefixes is $nwprefs_\Omega = prefs(nw(\mathcal{H}_\Omega)) \subseteq \hat{\Omega}^*$. Note that nested word prefixes may lack closing parenthesis, in which case they are not well-nested.

Monadic Queries. Let Σ be a set. A monadic query $\mathbf{Q}$ on hedges with signature Σ is a function mapping any hedge $h \in \mathcal{H}_\Sigma$ to a subset of its positions $\mathbf{Q}(h) \subseteq pos(h)$. We next relate monadic queries on hedges to hedge languages. For this, we fix a selection variable $x \notin \Sigma$ arbitrarily and consider hedge languages over signature $\Sigma^x = \Sigma \cup \{x\}$. For any $h \in \mathcal{H}_\Sigma$, let $\tilde{h} \in \mathcal{H}_{\Sigma \cup pos(h)}$ be its annotation with its positions. For instance $\widetilde{aa\langle\rangle a} = a1a2\langle 3 \rangle a4$. For any variable assignment $\alpha : \{x\} \hookrightarrow pos(h)$, we define the hedge $h * \alpha \in \mathcal{H}_{\Sigma^x}$ annotated with x by substituting in $\tilde{h}$ the position $\alpha(x)$ by x and removing all other positions. For instance, $aa\langle\rangle a * [x/2] = aax\langle\rangle a$. The monadic query on hedges with signature Σ defined by a hedge language $L \subseteq \mathcal{H}_{\Sigma^x}$ is $qry_L(h) = \{\alpha(x) \mid \alpha : \{x\} \to pos(h), \ h * \alpha \in L\}$.

3   Stepwise Hedge Automata

We define regular hedge languages by stepwise hedge automata (dSHAs).

Definition 1. *A dSHA is a tuple* $A = (\Omega, \mathcal{Q}, \delta, q_{init}, F)$ *where* Ω *and* $\mathcal{Q}$ *are finite sets,* $q_{init} \in \mathcal{Q}$, $F \subseteq \mathcal{Q}$, *and* $\delta = ((a^\delta)_{a \in \Omega}, \langle\rangle^\delta, @^\delta)$ *where:* $a^\delta : \mathcal{Q} \hookrightarrow \mathcal{Q}$, $\langle\rangle^\delta \in \mathcal{Q}$, *and* $@^\delta : \mathcal{Q} \times \mathcal{Q} \hookrightarrow \mathcal{Q}$.

There are states $q \in \mathcal{Q}$, the initial state is q_{init}, and final states in F. The transition rules in δ have three forms: If $a^\delta(q) = q'$ then we have an internal rule $q \xrightarrow{a} q'$, if $q@^\delta p = q'$ then an apply rule $q \xrightarrow{p} q'$, and if $q = \langle\rangle^\delta \in \mathcal{Q}$ then a tree initial rule $\xrightarrow{\langle\rangle} q$. We denote by $n = |\mathcal{Q}|$ the number of states of A, and by $m = n + |\Omega| + \sum_{a \in \Omega} |a^\delta| + |@^\delta| + |F| + 2$ its overall size. Note that $m \in O(n^2 + |\Omega| n)$ by determinism. For any hedge $h \in \mathcal{H}_\Sigma$ we define the transition $[\![h]\!]^\delta = [\![h]\!] : \mathcal{Q} \hookrightarrow \mathcal{Q}$ such that for all $q \in \mathcal{Q}$, $a \in \Omega$, and $h, h' \in \mathcal{H}_\Sigma$:

$$[\![\varepsilon]\!](q) = q \qquad\qquad [\![a]\!](q) = a^\delta(q)$$
$$[\![h \cdot h']\!](q) = [\![h']\!]([\![h]\!](q)) \qquad [\![\langle h \rangle]\!](q) = q@^\delta([\![h]\!](\langle\rangle^\delta))$$

A hedge is accepted if its transition from the initial state yields some final state. The language $\mathcal{L}(A)$ is the set of all accepted hedges: $\mathcal{L}(A) = \{h \in \mathcal{H}_\Omega \mid [\![h]\!](q_{init}) \in F\}$. We call a hedge language $L \subseteq \mathcal{H}_\Omega$ regular if it can be defined by some dSHA. A monadic query over hedges in $\mathcal{H}_\Sigma$ is called regular if it is equal to qry_L for some regular hedge language $L \subseteq \mathcal{H}_{\Sigma^x}$.

For example, let $\Omega = \Sigma^x$ where $\Sigma = \{a\}$. We draw in Fig. 1 the graph of a dSHA for the query on hedges in $\mathcal{H}_\Sigma$ that selects the positions $1, \ldots, n-1$ on hedges of the form $a^n \cdot \langle h \rangle \cdot h'$ if h does not start with letter "a" and position n otherwise. The drawing of dSHAs are similar to the usual finite state automata, except that now, edges may also be labeled by states and not only by letters.

A successful run of this automaton on the hedge $aaax\langle a \rangle a$ is given in Fig. 2. In state 5 the transition must suspend on the result of the evaluation of the subhedge, which is started by the tree initial rule $\xrightarrow{\langle\rangle} 1$. The two edges $5 \;\mathbin{\square\!\!\rightarrow}\; 9$ and $3 \;\mathbin{\text{-}\text{-}\square}\;$ are justified by the apply rule $5 \xrightarrow{3} 9$: the suspended computation in state 5 is resumed in state 9 when going up from the subtree in state 3.

The set of states that are accessible from a state $q \in \mathcal{Q}$ through some hedge is $acc^\delta(q) = \{q' \mid q' = [\![h]\!](q), h \in \mathcal{H}_\Omega\}$. For any $Q \subseteq \mathcal{Q}$, the set $acc^\delta(Q)$ can be computed in time $O(m)$ as well as $invacc^\delta(Q) = \{q' \mid q \in acc^\delta(q'), q \in Q\}$. A tree state is a state in $\mathcal{P} = acc^\delta(\langle\rangle^\delta)$. We call a set of transition rules δ complete if $@^\delta|_{\mathcal{Q} \times \mathcal{P}}$, as well as all a^δ with $a \in \Omega$, are total functions. For instance, the dSHA in Fig. 1 has the tree states $\mathcal{P} = \{1, 3, 7\}$. Note that δ is not complete since x^δ is not total. But its restriction to the letters in $\Sigma = \{a\}$ is complete due to the sink state 4.

4   Nested Word Automata

We define streaming algorithms for dSHAs by infinitary deterministic nested word automata (dNWAs$^\infty$). These have the advantage to run naturally in streaming mode, while being able to pass information top-down, bottom-up, and left to right. In contrast, dSHAs cannot pass any information top-down.

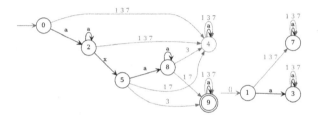

Fig. 1. A dSHA for the monadic query on hedges with letters in $\Sigma = \{a\}$ that selects the positions $1, \ldots, n - 1$ on hedges of the form $a^n \cdot \langle h \rangle \cdot h'$ if h does not start with letter "a" and position n otherwise.

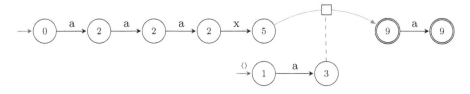

Fig. 2. A successful run of the dSHA A in Fig. 1 on $aaax\langle a \rangle a$.

Definition 2. *A dNWA$^\infty$ is a tuple $B = (\Omega, \mathcal{Q}, \Gamma, \delta, q_{init}, F)$, where Ω, Γ and $\mathcal{Q}$ are sets, $q_{init} \in \mathcal{Q}$, $F \subseteq \mathcal{Q}$, and $\delta = ((a^\delta)_{a \in \Omega}, \langle^\delta, \rangle^\delta)$ contains partial functions $a^\delta : \mathcal{Q} \hookrightarrow \mathcal{Q}$, $\langle^\delta : \mathcal{Q} \hookrightarrow \mathcal{Q} \times \Gamma$, and $\rangle^\delta : \mathcal{Q} \times \Gamma \hookrightarrow \mathcal{Q}$. A dNWA is a dNWA$^\infty$ whose sets $\mathcal{Q}$, Ω, and Γ are finite.*

The elements of Γ are called stack symbols. The transition rules in δ have three forms: letter rules $q \xrightarrow{a} q'$, opening rules $q \xrightarrow{\langle \downarrow \gamma} q'$ for pushing a stack symbol if $\langle^\delta(q) = (q', \gamma)$ and closing rules $q \xrightarrow{\rangle \uparrow \gamma} q'$ popping a stack symbol if $\rangle^\delta(q, \gamma) = q'$. Any dNWA$^\infty$ defines a pushdown machine that runs on words with parentheses. A configuration of this machine is a pair in $\mathcal{K} = \mathcal{Q} \times \Gamma^*$ of a state and a stack. For any word $v \in \hat{\Omega}^*$, we define a streaming transition $[\![v]\!]^\delta_{str} = [\![v]\!]_{str} : \mathcal{K} \hookrightarrow \mathcal{K}$ such that for all $q \in \mathcal{Q}$, stacks $s \in \Gamma^*$ and $\gamma \in \Gamma$:

$$[\![a]\!]_{str}(q, s) = (a^\delta(q), s) \qquad [\![\rangle]\!]_{str}(q, s \cdot \gamma) = (\rangle^\delta(q, \gamma), s)$$
$$[\![\varepsilon]\!]_{str}(q, s) = (q, s) \qquad\quad [\![\langle]\!]_{str}(q, s) = (q', s \cdot \gamma) \text{ where } (q', \gamma) = \langle^\delta(q)$$
$$[\![v \cdot v']\!]_{str}(q, s) = [\![v']\!]_{str}(q', s') \text{ where } (q', s') = [\![v]\!]_{str}(q, s)$$

The language of a dNWA$^\infty$ is the set of nested words that it accepts: $\mathcal{L}(B) = \{v \in \hat{\Omega}^* \mid [\![v]\!]_{str}(q_{init}, \varepsilon) \in F \times \{\varepsilon\}\}$. Since the initial and final stack are required to be empty it follows that any word $\mathcal{L}(B)$ is well-nested.

For any dSHA $A = (\Omega, \mathcal{Q}, \delta, q_{init}, F)$, we define the dNWA $A^{nwa} = (\Omega, \mathcal{Q}, \Gamma, \delta^{nwa}, q_{init}, F)$ with $\Gamma = \mathcal{Q}$ such that δ^{nwa} contains for all $a \in \Omega$ and $q, p \in \mathcal{Q}$ the rules $q \xrightarrow{a} a^\delta(q)$, $q \xrightarrow{\langle \downarrow q} \langle \rangle^\delta$, and $q \xrightarrow{\rangle \uparrow p} p@^\delta q$.

Lemma 3. $\mathcal{L}(A^{nwa}) = nw(\mathcal{L}(A))$.

5   Earliest Membership

A late streaming evaluator of a dSHA A on hedges $h \in \mathcal{H}_\Omega$ can be obtained by evaluating the dNWA A^{nwa} in streaming mode on the nested word of h, i.e., by testing $[\![nw(h)]\!]_{str}^{\delta^{nwa}}(q_{init}, \varepsilon) \in F \times \{\varepsilon\}$. In this manner, h is never fully loaded into the memory, but rather treated event by event . Only a state and a stack are stored at any event, i.e. at any prefix of $nw(h)$. The memory cost thus only depends on the depth of the hedge.

The decision of whether membership holds, however, is taken at the very end of the stream. Instead, we want to decide language membership at the earliest event when it becomes certain. We consider Σ-certain membership to languages $L \subseteq \mathcal{H}_\Omega$ where $\Omega \supseteq \Sigma$ as needed for certain query answering later on.

Definition 4. Let $\Sigma \subseteq \Omega$ and $L \subseteq \mathcal{H}_\Omega$. A nested word prefix v with letters in Ω satisfies cert-mem$_\Sigma^L(v)$ if $\forall h \in \mathcal{H}_\Omega$. $(\exists w \in \hat{\Sigma}^*. v \cdot w = nw(h)) \rightarrow h \in L$.

In other words, a nested word prefix v is Σ-certain for membership in $L \subseteq \mathcal{H}_\Omega$, if any completion of v with letters from Σ to a hedge in $\mathcal{H}_\Omega$ belongs to L. For instance, if $\Sigma = \{a\}$ then the prefix $v = aaax\langle a$ is Σ-certain for the language of the dSHA A with signature $\Omega = \{a, x\}$ in Fig. 1, since any completion of v without further x'es will be accepted by A.

Since certain membership is a universality property, we need to consider universal automata states. Given a state $q \in \mathcal{Q}$ let $A[q_{init}/q] = (\Omega, \mathcal{Q}, \delta, q, F)$ be obtained from A by replacing its initial state by q. We define:

$$q \in universal_\Sigma^A \Leftrightarrow \mathcal{H}_\Sigma \subseteq L(A[q_{init}/q])$$

In order to characterize universality by accessibility, we define for all $Q \subseteq \mathcal{Q}$:

$$safe^\delta(Q) = \{q \in \mathcal{Q} \mid acc^\delta(q) \subseteq Q\}$$

If δ is complete then $safe^\delta(Q) = \mathcal{Q} \setminus invacc^\delta(\mathcal{Q} \setminus Q)$, so it can be computed in $O(m)$. For any $\Sigma \subseteq \Omega$, let $\delta|_\Sigma$ be the restriction of δ to the letters of Σ, i.e., $\delta|_\Sigma = ((a^\delta)_{a \in \Sigma}, \langle\rangle^\delta, @^\delta)$.

Lemma 5. Let $A = (\Omega, \mathcal{Q}, \delta, q_{init}, F)$ be a dSHA and $\Sigma \subseteq \Omega$ such that $\delta|_\Sigma$ is complete, and $q \in \mathcal{Q}$. Then: $q \in universal_\Sigma^A \Leftrightarrow q \in safe^{\delta|_\Sigma}(F)$.

Safety can be used to detect certain language membership. For this, we define for any $Q \subseteq \mathcal{Q}$ and $q \in \mathcal{Q}$ such that $q@^\delta p$ is well-defined for some $p \in \mathcal{Q}$:

$$d^\delta(q, Q) = safe^\delta(dn_{@^\delta}(q, Q)) \quad where \ dn_{@^\delta}(q, Q) = \{p \in \mathcal{Q} \mid q@^\delta p \in Q\}.$$

Note that if $q@^\delta p$ is undefined for all p then $d^\delta(q, Q)$ remains undefined too. We define the dNWA A_Σ^c for testing certain Σ-membership to $L(A)$ as follows:

$$\mathcal{Q}_\Sigma^c = \mathcal{Q} \times 2^\mathcal{Q} = \Gamma_\Sigma^c, \qquad {q_{init}}_\Sigma^c = (q_{init}, safe^{\delta|_\Sigma}(F)).$$

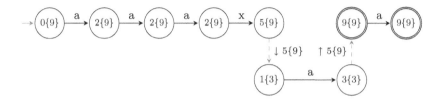

Fig. 3. A successful run of the dNwa A_Σ^c on $aaax\langle a\rangle a$.

The transition rules in δ_Σ^c allow for all $S \subseteq \mathcal{Q}$, $q, p \in \mathcal{Q}$, and $a \in \Omega$:

$$(q, S) \xrightarrow{\langle\downarrow(q,S)} (\langle\rangle^\delta, d^{\delta|_\Sigma}(q, S)), \quad p \xrightarrow{\rangle\uparrow(q,S)} (q@^\delta p, S), \quad (q, S) \xrightarrow{a} (a^\delta(q), S).$$

Finally, let $A_\Sigma^c = (\Omega, \mathcal{Q}_\Sigma^c, \Gamma_\Sigma^c, \delta_\Sigma^c, q_{init}{}_\Sigma^c, F_\Sigma^c)$ where $F_\Sigma^c = F \times 2^\mathcal{Q}$. In the first component A_Σ^c behaves like A^{nwa}, while in the second component it computes safety information. Therefore, $L(A) = L(A^{nwa}) = L(A_\Sigma^c)$. We next show that the streaming evaluator of A_Σ^c detects certain membership at any time.

Proposition 6. *Let $A = (\Omega, \mathcal{Q}, \delta, q_{init}, \_)$ be a dSHA, $v \in nwprefs_\Omega$ a nested word prefix, and $\Sigma \subseteq \Omega$ such that $\delta|_\Sigma$ is complete. If $q \in \mathcal{Q}$ and $S \subseteq \mathcal{Q}$ such that $((q, S), \_) = [\![v]\!]_{str}^{\delta_\Sigma^c}(q_{init}{}_\Sigma^c, \varepsilon)$ then: $cert\text{-}mem_\Sigma^{\mathcal{L}(A)}(v) \Leftrightarrow q \in S$.*

We illustrate Proposition 6 in Fig. 3 at the dSHA A from Fig. 1. Recall that it has signature $\Omega = \Sigma^x$ where $\Sigma = \{a\}$. Given that $\delta|_\Sigma$ is complete, Σ-certain membership of $aaax\langle a\rangle a$ to $\mathcal{L}(A)$ can be detected at the earliest event $aaax\langle a$, by running the streaming evaluator of the earliest automaton A_Σ^c. Note that the earliest automaton is a dNwa passing safety information top-down (while dSHAs cannot pass any information top-down). We have $safe^{\delta|_\Sigma}(\{9\}) = \{9\}$ and $d^{\delta|_\Sigma}(5, \{9\}) = \{3\}$. Hence $[\![aaax\langle]\!]_{str}^{\delta_\Sigma^c}(q_{init}{}_\Sigma^c) = ((1, \{3\}), s)$ where the stack is $s = (5, \{9\})$. Since $1 \notin \{3\}$, membership is not yet Σ-certain. Indeed, the Σ-completion $aaax\langle\rangle$ is not accepted. After reading the next letter a, we have $[\![aaax\langle a]\!]_{str}^{\delta_\Sigma^c} = ((3, \{3\}), s)$. Since the current state 3 belongs to the current set of safe states $\{3\}$, membership is Σ-certain, i.e., membership of all completions without further x'es.

6 Late Monadic Query Answering

We now move to the problem of how to answer monadic queries on hedges in streaming mode, while selecting query answers lately at the end of the stream.

Our algorithm will generate candidates $[x/\pi]$ binding the selection variable x to positions π of the input hedge. We want to formulate the streaming algorithm without fixing the input hedge a priori, thus we consider the infinite set of candidates $Cands = \{\alpha \mid \alpha : \{x\} \hookrightarrow \mathbb{N}\}$. Given a dSHA A with signature Σ^x and a hedge $h \in \mathcal{H}_\Sigma$, our algorithm computes the answer set $qry_{\mathcal{L}(A)}(h)$ in streaming mode. For this, we compile A to the late $dNWA^\infty$ A^l and run the streaming

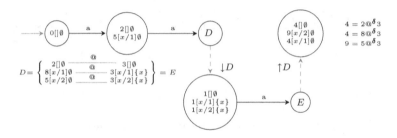

Fig. 4. The run of the late dNWA A^l for the dSHA A in Fig. 1 on $aa\langle a\rangle$.

evaluator of A^l on the nested word $\widetilde{nw(h)}$. The alphabet of A^l is $\Sigma \cup \mathbb{N}$. It has the state set $\mathcal{Q}^l = 2^{\mathcal{Q} \times Cands \times 2^{\{x\}}} = \Gamma^l$ and the initial state $q^l_{init} = \{(q_{init}, [], \emptyset)\}$. If the evaluator goes into some state $D \in \mathcal{Q}^l$, then $(q, \alpha, V) \in D$ means that the candidate α is in state q while the variables in $V \subseteq dom(\alpha)$ were bound in the context, i.e., before the last dangling opening parenthesis (so a preceding node that is not a preceding sibling). The set δ^l contains the following transition rules for all $D, E \in \mathcal{Q}^l$, $a \in \Sigma$, $V \subseteq \{x\}$, and $\pi \in \mathbb{N}$:

$$D \xrightarrow{a} \{(a^\delta(q), \alpha, V) \mid (q, \alpha, V) \in D\}$$
$$D \xrightarrow{\pi} \{(x^\delta(q), [x/\pi], \emptyset) \mid (q, [], \emptyset) \in D\} \cup D$$
$$D \xrightarrow{\langle \downarrow D} \{(\langle\rangle^\delta, \alpha, dom(\alpha)) \mid (q, \alpha, V) \in D\}$$
$$E \xrightarrow{\rangle \uparrow D} \{(q@^\delta p, \alpha', V) \mid (q, \alpha, V) \in D, (p, \alpha', dom(\alpha)) \in E, \alpha \in \{[], \alpha'\}\}$$

When reading a position $\pi \in \mathbb{N}$ in a state D that contains a triple with the empty candidate $(q, [], \emptyset)$, a new candidate $[x/\pi]$ is created, and the triple $(x^\delta(q), [x/\pi], \emptyset)$ is added to D. At opening events, the current state D of A^l is pushed onto the stack. It is also updated for continuation: if D contains a triple with candidate α, then the next state of A^l contains $(\langle\rangle^\delta, \alpha, dom(\alpha))$. At closing events, the state D of the parent hedge is popped from the stack. Let E be the current state. Any $(q, \alpha, V) \in D$ must be matched with some $(p, \alpha', dom(\alpha)) \in E$, so that A^l can continue in $q@^\delta p$. Matching here means that either $\alpha = \alpha'$ or, $\alpha' = [x/\pi]$ and $\alpha = []$. This is expressed by the condition $\alpha \in \{[], \alpha'\}$. Note that if $\alpha = []$ matches $\alpha' = [x/\pi]$ then $dom(\alpha) = \emptyset$ so that π was not bound in the context. This is where the knowledge of the context is needed.

An example run of A^l on the hedge $aa\langle a\rangle$ is given in Fig. 4, where A is the dSHA A from Fig. 1. The tuples are written there without commas and parentheses. The run of A^l first consumes aa and goes into state $D = \{(2, [], \emptyset), (8, [x/1], \emptyset), (5, [x/2], \emptyset)\}$. It contains the candidates $[x/1]$ and $[x/2]$ for the two leading a positions, plus the empty candidate $[]$. After the following open parenthesis $\langle$, the run goes into the set $\{(1, [], \emptyset), (1, [x/1], \{x\}), (1, [x/2], \{x\})\}$. The state of each of the candidates got set to $\langle\rangle^\delta = 1$. Furthermore, the set memoizes that the candidates $[x/1]$ and $[x/2]$ were bound in the context. It then consumes the letter a and reaches $E = \{(3, [], \emptyset), (3, [x/1], \{x\}), (3, [x/2], \{x\})\}$. When reading the closing parenthesis D is popped from the stack, its tuples in state q are matched with tuples in state p from E as illustrated in the figure, so that one can apply the apply rules $q@^\delta p$ of A. The tuple in state 5 of D, for instance, matches the tuple in state 3 of E, so A^l continues the candidate $[x/2]$ in state $9 = 5@^\delta 3$. Since $9 \in F^A$, position 2 is selected, i.e. $2 \in qry_{\mathcal{L}(A)}(aa\langle a\rangle)$.

Proposition 7 (Correctness of the late streaming evaluator). *If* $(D, \varepsilon) = [\![nw(\tilde{h})]\!]^{\delta^l}_{str}(q^l_{init}, \varepsilon)$ *then* $qry_{\mathcal{L}(A)}(h) = \{\pi \mid (q, [x/\pi], \emptyset) \in D, \ q \in F\}$.

7   Certain Answers and Earliest Query Answering

In order to justify early selection, we need the concept of certain answers. Let $\mathbf{Q}$ be a monadic query on $\mathcal{H}_\Sigma$ and $v \in nwprefs_\Sigma$ a nested word prefix.

Definition 8. *A position* $\pi \in \mathbb{N}$ *is a certain answer of* $\mathbf{Q}$ *at prefix* v – *written* $\pi \in CA^{\mathbf{Q}}(v)$ – *if* $\pi \in pos(v) \ \wedge \ \forall h \in \mathcal{H}_\Sigma. \ v \in prefs(nw(h)) \to \pi \in \mathbf{Q}(h)$.

A position π is thus a certain answer of query $\mathbf{Q}$ at prefix v of the stream if it answers the query for all completions h of v. Certain answers can be safely selected however the stream continues. For instance, position 3 is a certain answer on the prefix $aaa\langle a$ for the query defined by the dSHA in Fig. 1.

In analogy, we can define that π is certainly a nonanswer of $\mathbf{Q}$ at v, and denote this by $\pi \in CNA^{\mathbf{Q}}(v)$. Once π becomes a certain nonanswer then it can be safely rejected. The positions $1, ..., n-1$, for instance, are certain nonanswers on our example query on $a^n\langle a$. We call a position π alive for $\mathbf{Q}$ at v it is neither a certain answer nor a certain nonanswer of $\mathbf{Q}$ at v. The *concurrency* c of $\mathbf{Q}$ at v is its number of alive candidates. For the shorter prefix $aaa\langle$, for instance, all n positions $1, \ldots, n$ are alive, so the concurrency is n.

We next want to link certain answers to certain Σ-membership. For this, we need to annotate nested word prefixes with positions and variables, similarly as for hedges. Given a word $v \in \hat{\Sigma}$, the set of positions of v then is $pos(v) = \{1, \ldots, \#_{\Sigma \cup \{\langle\}}(v)\}$. We can define the annotation of v with its positions as a word $\tilde{v} \in (\hat{\Sigma} \cup pos(v))^*$. For any variable assignment $\alpha : \{x\} \hookrightarrow pos(v)$ we define an annotated word $v * \alpha \in \hat{\Sigma^x}^*$ in analogy as for hedges.

Lemma 9. *For any prefix* $v \in prefs(nw(\mathcal{H}_\Sigma))$, *language* $L \subseteq \mathcal{H}_{\Sigma^x}$ *and candidate* $\alpha = [x/\pi]$ *with* $\pi \in pos(v)$: *cert-mem*$^L_\Sigma(v * \alpha) \Leftrightarrow \pi \in CA^{qry_L}(v)$.

Proposition 10 (Corollary of Proposition 6 and Lemma 9). *Let* $A = (\Sigma^x, \_, \delta, q_{init}, \_)$ *be a dSHA such that* $\delta|_\Sigma$ *is complete. For any* $v \in nwprefs_\Sigma$, $\pi \in pos(v)$, *and* $((q, S), \_) = [\![v * [x/\pi]]\!]^{\delta^c_\Sigma}_{str}(q_{init}{}^c_\Sigma, \varepsilon)$: $\pi \in CA^{qry_{\mathcal{L}(A)}}(v) \Leftrightarrow q \in S$.

For any dSHA A over Σ^x, we construct the earliest $dNWA^\infty$ A^e with alphabet $\Sigma \cup \mathbb{N}$, testing for all candidates $[x/\pi]$ on prefixes $\tilde{v}$, whether π is certain for selection. For this A^e simulates the runs of A^c_Σ on all $v * [x/\pi]$. It has the states $\mathcal{Q}^e = \mathcal{Q} \times Cands \times 2^{\{x\}} \times 2^{\mathcal{Q}} = \Gamma^e$ and $q^e_{init} = \{(q_{init}, [], \emptyset, safe^{\delta|_\Sigma}(F))\}$. Initially,

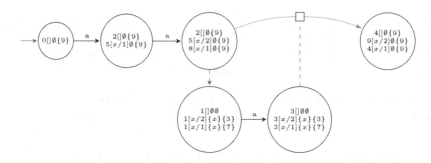

Fig. 5. A run of the earliest automaton A^e for the dSHA A in Fig. 1.

all states in $safe^{\delta|_{\Sigma}}(F)$ are safe for selection. Let δ^e contain the following rules for all $M, N \in \mathcal{Q}^e$, $a \in \Sigma$, $\pi \in \mathbb{N}$, $S \subseteq \mathcal{Q}$ and $q \in \mathcal{Q}$:

$$M \xrightarrow{a} \{(a^{\delta}(q), \alpha, V, S) \mid (q, \alpha, V, S) \in M\}$$
$$M \xrightarrow{\pi} \{(x^{\delta}(q), [x/\pi], \emptyset, S) \mid (q, [], \emptyset, S) \in M\} \cup M$$
$$M \xrightarrow{\langle\downarrow M} \{(\langle\rangle^{\delta}, \alpha, dom(\alpha), d^{\delta|_{\Sigma}}(q, S)) \mid (q, \alpha, V, S) \in M\}$$
$$N \xrightarrow{\rangle\uparrow M} \{(q@^{\delta}p, \alpha', V, S) \mid (q, \alpha, V, S) \in M, (p, \alpha', dom(\alpha), S') \in N, \alpha \in \{[], \alpha'\}\}$$

For the dSHA A in Fig. 1, for instance, a run of A^e is given in Fig. 5. It satisfies $[\![a1a2\langle 3a4]\!]^{\delta^e}_{str}(q^e_{init}, \varepsilon) = (\{(3, [x/2], \{x\}, \{3, 4\}), (3, [x/1], \{x\}, \{4, 9\})\}, \_)$. The certain answer $[x/2]$ is in a safe state now, while the other candidate $[x/1]$ can be seen to be safe for rejection so it could be removed.

Proposition 11. *Let A be a dSHA with signature Σ^x such that $\delta|_{\Sigma}$ is complete. For any nested word prefix $v \in nwprefs_{\Sigma}$ with $[\![\tilde{v}]\!]^{\delta^e}_{str}(q^e_{init}, \varepsilon) = (M, \_)$:*

$$CA^{qry_{\mathcal{L}(A)}}(v) = \{\alpha(x) \mid q \in S \wedge (q, \alpha, \emptyset, S) \in M \wedge dom(\alpha) = \{x\}\}$$

We can thus obtain an *EQA* algorithm by running the streaming evaluator of the earliest automaton A^e. Without removing candidates that are certainly nonanswers, however, it would maintain and update many candidates that are no more alive, leading to quadratic time in $O(m^2)$ even for bounded concurrency.

Theorem 1. *EQA for monadic dSHA queries can be done in time $O(c\,m)$ per event, where c is the concurrency of the query at the event.*

This complexity for dSHAs improves on Gauwin et al. [7] for dNWAs, which required time $O(c\,n^2)$ per event after $O(n^3)$ preprocessing time. Note that *EQA* for monadic queries can also be used to detect certain membership for language of dSHAs. In this case, we have $c = 1$ so the time per event is reduced to $O(m)$. Moreover, for monadic queries where c is bounded for all events and input hedges, the complexity per event is also reduced to $O(m)$. It is worth noting that as preprocessing, we only have to load the automaton in a linear time. On the other hand, m is equal to n^2 at the worst case for a fixed signature Σ. For SHAs where $O(m) = O(n)$ our complexity is better than that of Gauwin. This holds for all dSHAs in our experiments in particular.

8   Experimentation

We present experimental results of our *EQA* algorithm that we implemented in the AStream tool, with current version 1.01. The objective is to relate the theoretical complexity to practical efficiency. We note that we implemented AStream in Scala while using Java's abc-datalog for safety computation.

First, we consider a collection of 77 regular XPATH queries [2] that was selected from the larger collection of XPath queries harvested by Lick and Schmitz [11] from real-world XSLT programs . These queries represent the regular part of the benchmark that can be directly compiled to automata(4500 queries up to removing duplicates, renaming of XML namespace prefixes and local names and syntactical detail). A single matching XML document per XSLT program of size less than 2 MB was provided too. We used the dSHAs for these XPATH queries from [2] as inputs (so the automaton construction time is ignored here). We could correctly answer all 77 queries, yielding the same answer set as with Saxon. The overall time for computing the 77 answer sets was 110000 ms on a Macbook pro Apple M1 laptop with 16 GB of RAM. With Saxon in-memory evaluation it required 45000 ms. The low running time of AStream reflects the low concurrency of all the queries on all these documents according to Theorem 1. There are 12 queries with concurrency 1, 47 with concurrency 2, 6 with concurrency 3, and 12 with concurrency 4. Our efficiency results for AStream thus show for the first time, that *EQA* is indeed feasible in practical scenarios with queries of low concurrency.

Second, we compare AStream to existing streaming tools for regular XPATH queries with large coverage. We focus on the most efficient streaming evaluator called QuiXPath [5]. A detailed comparison to the many alternative tools is given there too. We note that QuiXPath is not always earliest, but still earliest in most cases. As done there, we consider the queries A1-A8 of the XPathMark collection [6] which also provides a generator for scalable XML documents. The other queries are either not regular or contain backward axis that our compiler to SHAS does not support. We also added the queries O1 and O2 from [5], in order to illustrate difficulties of non-earliest query answering and high concurrency.

We run AStream on XML documents of increasing size up to 1.2 GB, but can also stream much larger documents >100 GB. Up to 1 GB, we verified the correctness of the answer sets by comparison to Saxon's in-memory evaluator (which is limited to 1 GB).

The times grow linearly for all these queries given that their concurrency is bounded to 2, except for O1 where it grows quadratically since its concurrency grows linearly with the size of the document. The quadratic growth can be observed on smaller documents scaling from 27KB to 5MB. On average, for A1-A8, AStream 1.01 is by a factor of 60 slower than QuiXPath, so requiring minutes instead of seconds. The main reason is the lack of streaming projection algorithms for dSHAS. In contrast, QuiXPath uses streaming projection for queries defined by dNWAs with selection states [19]. On the one hand side, QuiXPath cannot stream O2 on large documents, since not being *earliest*. While the concurrency of O2 is 1, linearly many candidates are buffered by QuiXPath, until the buffer overflows for documents larger than 5GB. O2 poses no problem for our tool since it has a low concurrency. On the other hand, QuiXPath can stream queries with high concurrency(O1), whereas AStream 1.01 runs out of time for documents of 1MB. This is due to QuiXPath's state sharing, i.e. the sharing of the computations of all concurrent candidates in the same state.

Conclusion and Future Work. We introduced an *EQA* algorithm for regular monadic queries represented by dSHAs with a time complexity of $O(c\ m)$ per event. Its implementation in the AStream tool has demonstrated its efficiency on queries in practical scenarios with low concurrency. However, in order to compete with the current best non-earliest streaming algorithms, we need to develop streaming projection for dSHAs (as done previously for NWAs [19]), and to add factorization for candidates in the same state [5]. Additionally, we plan to extend our streaming algorithm to hyperstreaming, which involves handling multiple streams with references and holes [18].

References

1. Al Serhali, A., Niehren, J.: Earliest query answering for deterministic stepwise hedge automata. http://nria.hal.science/hal-04106420
2. Al Serhali, A., Niehren, J.: A Benchmark Collection of Deterministic Automata for XPath Queries. In: XML Prague 2022. Prague, Czech Republic (2022). https://inria.hal.science/hal-03527888
3. Alur, R.: Marrying words and trees. In: 26th ACM SIGMOD-SIGACT-SIGART Symposium on Principles of Database Systems, pp. 233–242. ACM-Press (2007). https://dl.acm.org/doi/10.1145/1265530.1265564
4. Benedikt, M., Jeffrey, A., Ley-Wild, R.: Stream firewalling of XML constraints. In: ACM SIGMOD International Conference on Management of Data, pp. 487–498. ACM-Press (2008)
5. Debarbieux, D., Gauwin, O., Niehren, J., Sebastian, T., Zergaoui, M.: Early nested word automata for Xpath query answering on XML streams. Theor. Comput. Sci. **578**, 100–125 (2015). https://doi.org/10.1016/j.tcs.2015.01.017
6. Franceschet, M.: Xpathmark performance test. http://users.dimi.uniud.it/~massimo.franceschet/xpathmark/PTbench.html. Accessed 13 June 2023
7. Gauwin, O., Niehren, J., Tison, S.: Earliest query answering for deterministic nested word automata. In: Kutylowski, M., Charatonik, W., Gebala, M. (eds.) FCT 2009. LNCS, vol. 5699, pp. 121–132. Springer, Heidelberg (2009). https://doi.org/10.1007/978-3-642-03409-1_12
8. Grez, A., Riveros, C., Ugarte, M.: A formal framework for complex event processing, pp. 1–18 (2019). https://doi.org/10.4230/LIPIcs.ICDT.2019.5
9. Kay, M.: Streaming in the Saxon XSLT processor. In: XML Prague, pp. 81–101 (2014)
10. Libkin, L.: How to define certain answers. In: Yang, Q., Wooldridge, M.J. (eds.) Proceedings of the Twenty-Fourth International Joint Conference on Artificial Intelligence, IJCAI 2015, Buenos Aires, Argentina, 25–31 July 2015, pp. 4282–4288. AAAI Press (2015). https://www.ijcai.org/Abstract/15/609
11. Lick, A., Sylvain, S.: XPath Benchmark. Accessed 13 Apr 2022. https://archive.softwareheritage.org/browse/directory/1ea68cf5bb3f9f3f2fe8c7995f1802ebadf17fb5/

12. Mozafari, B., Zeng, K., Zaniolo, C.: High-performance complex event processing over XML streams. In: Candan, K.S., et al. (eds.) SIGMOD Conference, pp. 253–264. ACM (2012). https://doi.org/10.1145/2213836.2213866

13. Muñoz, M., Riveros, C.: Streaming enumeration on nested documents. In: Olteanu, D., Vortmeier, N. (eds.) 25th International Conference on Database Theory, ICDT 2022, 29 March to 1 April 2022, Edinburgh, UK (Virtual Conference). LIPIcs, vol. 220, pp. 1–18. Schloss Dagstuhl - Leibniz-Zentrum für Informatik (2022). https://doi.org/10.4230/LIPIcs.ICDT.2022.19

14. Niehren, J., Sakho, M.: Determinization and minimization of automata for nested words revisited. Algorithms **14**(3), 68 (2021). https://doi.org/10.3390/a14030068

15. Niehren, J., Sakho, M., Serhali, A.A.: Schema-based automata determinization. In: Ganty, P., Monica, D.D. (eds.) Proceedings of the 13th International Symposium on Games, Automata, Logics and Formal Verification, GandALF 2022, Madrid, Spain, 21–23 September 2022. EPTCS, vol. 370, pp. 49–65 (2022). https://doi.org/10.4204/EPTCS.370.4

16. Okhotin, A., Salomaa, K.: Complexity of input-driven pushdown automata. SIGACT News **45**(2), 47–67 (2014). https://doi.org/10.1145/2636805.2636821

17. Olteanu, D.: SPEX: streamed and progressive evaluation of XPath. IEEE Trans. Know. Data Eng. **19**(7), 934–949 (2007). https://doi.org/10.1109/TKDE.2007.1063

18. Sakho, M.: Certain Query Answering on Hyperstreams. Ph. D thesis, Université de Lille (2020). https://theses.hal.science/tel-03028074

19. Sebastian, T., Niehren, J.: Projection for nested word automata speeds up Xpath evaluation on XML streams. In: Freivalds, R.M., Engels, G., Catania, B. (eds.) SOFSEM 2016. LNCS, vol. 9587, pp. 602–614. Springer, Heidelberg (2016). https://doi.org/10.1007/978-3-662-49192-8_49

Constrained Multi-Tildes

Samira Attou[1], Ludovic Mignot[2(✉)], Clément Miklarz[2], and Florent Nicart[2]

[1] LIGM, Université Gustave Eiffel, 5 Boulevard Descartes — Champs s/ Marne, Marne-la-Vallée Cedex 2, 77454 Champs-sur-Marne, France
`samira.attou@univ-eiffel.fr`
[2] GR2IF, Université de Rouen Normandie, Avenue de l'Université, 76801 Saint-Étienne-du-Rouvray, France
`{ludovic.mignot,clement.miklarz1,florent.nicart}@univ-rouen.fr`

Abstract. Multi-tildes are regular operators that were introduced to enhance the factorization power of regular expressions, allowing us to add the empty word in several factors of a catenation product of languages. In addition to multi-bars, which dually remove the empty word, they allow representing any acyclic automaton by a linear-sized expression, whereas the lower bound is exponential in the classic case.

In this paper, we extend multi-tildes from disjunctive combinations to any Boolean combination, allowing us to exponentially enhance the factorization power of tildes expressions. Moreover, we show how to convert these expressions into finite automata and give a Haskell implementation of them using advanced techniques of functional programming.

Keywords: Regular expressions · Partial derivatives · Boolean formulae · Multi-tildes operators

1 Introduction

Regular expressions are widely used inductively defined objects that allow us to easily represent (potentially infinite) set of words. In order to solve efficiently the membership test, they can be turned into finite automata [1], where the number of states is linear w.r.t. the number of symbols of the expressions. Numerous operators where added in order to enhance their representation powers, such as Boolean operators. However, the number of states after the conversion is not necessarily linear anymore [5].

Another class of operators, the multi-tildes [2], was introduced in order to allow a constrained adjunction of the empty word in some factors of the catenation product of regular languages. In combination with multi-bars [3], multi-tildes allow to improve the factorization power of regular languages: as an example, it is shown that any acyclic automaton can be turned into a linear-sized equivalent multi-tildes-bars expression, whereas the lower bound is exponential in the classical case [7]. However, they can be applied only across continuous positions.

In this paper, we extend the idea behind the conception of (disjunctive) multi-tildes to any Boolean combination of them. These Boolean combinations allow us to extend the specification power of expressions, by *e.g.*, applying tildes across

B. Nagy (Ed.): CIAA 2023, LNCS 14151, pp. 66–78, 2023.
https://doi.org/10.1007/978-3-031-40247-0_4

non-continuous intervals of positions. We show that their actions over languages preserve regularity, that they may lead to exponentially smaller expressions and how to solve the membership test by defining a finite automaton.

This is the first step of a more general plan: we aim to develop a characterization of the produced automaton in order to inverse the computation, *i.e.*, the conversion of an automaton into a short constrained tildes expression.

The paper is organized as follows. Section 2 contains general preliminaries. Then, we recall in Sect. 3 classical definitions and constructions for Boolean formulae. These latter allow us to define constrained tildes in Sect. 4. We study their factorization power in Sect. 5, and show how to convert these expressions into finite automata in Sect. 6. Finally, in Sect. 7, we present an Haskell implementation of these objects.

2   Preliminaries

Throughout this paper, we use the following notations:

- $\mathbb{B}$ is the Boolean set $\{0, 1\}$,
- $S \rightarrow S'$ is the set of functions from a set S to a set S',
- for a Boolean b and a set S, $S \mid b$ is the set S if b, $\emptyset$ otherwise,
- $\subset$ is to be understood as not necessarily strict subset.

A *regular expression* E over an alphabet Σ is inductively defined by

$$E = a, \qquad E = \emptyset, \qquad E = \varepsilon, \qquad E = F \cdot G, \qquad E = F + G, \qquad E = F^*,$$

where a is a symbol in Σ and F and G two regular expressions over Σ. Classical priority rules hold: $^* > \cdot > +$. The language *denoted by* E is the set $L(E)$ inductively defined by

$$L(a) = \{a\}, \qquad L(\emptyset) = \emptyset, \qquad L(\varepsilon) = \{\varepsilon\},$$
$$L(F \cdot G) = L(F) \cdot L(G), \qquad L(F + G) = L(F) \cup L(G), \qquad L(F^*) = L(F)^*,$$

where a is a symbol in Σ and F and G two regular expressions over Σ. A (non-deterministic) *automaton* A over an alphabet Σ is a 5-tuple $(\Sigma, Q, I, F, \delta)$ where

- Q is a finite set of *states*,
- $I \subset Q$ is the set of *initial states*,
- $F \subset Q$ is the set of *final states*,
- δ is a function in $\Sigma \times Q \rightarrow 2^Q$.

The function δ is extended to $\Sigma \times 2^Q \rightarrow 2^Q$ by $\delta(a, P) = \bigcup_{p \in P} \delta(a, p)$ and to $\Sigma^* \times 2^Q \rightarrow 2^Q$ by $\delta(\varepsilon, P) = P$ and $\delta(aw, P) = \delta(w, \delta(a, P))$. The language *denoted* by A is the set $L(A) = \{w \in \Sigma^* \mid \delta(w, I) \cap F \neq \emptyset\}$.

Any regular expression with n symbols can be turned into an equivalent automaton with at most $(n + 1)$ states, for example by computing the derived term automaton [1]. The *partial derivative* of E w.r.t. a symbol a in Σ is the set of expressions $\delta_a(E)$ inductively defined as follows:

$$\delta_a(b) = (\{\varepsilon\} \mid a = b), \qquad\qquad \delta_a(\emptyset) = \emptyset,$$

$$\delta_a(\varepsilon) = \emptyset, \qquad\qquad\qquad \delta_a(F + G) = \delta_a(F) \cup \delta_a(G),$$
$$\delta_a(F \cdot G) = \delta_a(F) \odot G \cup (\delta_a(G) \mid \text{Null}(F)), \qquad \delta_a(F^*) = \delta_a(F) \odot F^*,$$

where b is a symbol in Σ, F and G two regular expressions over Σ, $\text{Null}(F) = \varepsilon \in L(F)$ and $\mathcal{E} \odot G = \bigcup_{E \in \mathcal{E}} \{E \cdot G\}$, where $\odot$ has priority over $\cup$. The partial derivative of E w.r.t. a word w is defined by $\delta_\varepsilon(E) = \{E\}$ and $\delta_{aw}(E) = \bigcup_{E' \in \delta_a(E)} \delta_w(E')$. The *derived term automaton* of E is the automaton $(\Sigma, Q, \{E\}, F, \delta)$ where

$$Q = \bigcup_{w \in \Sigma^*} \delta_w(E), \qquad F = \{E' \in Q \mid \text{Null}(E')\}, \qquad \delta(a, E') = \delta_a(E').$$

The derived term automaton of E, with n symbols, is a finite automaton with at most $(n + 1)$ states that recognizes $L(E)$.

A *multi-tilde* is an n-ary operator which is parameterized by a set S of couples (i, j) (called *tildes*) in $\{1, \ldots, n\}^2$ with $i \leq j$. Such an expression is denoted by $\text{MT}_S(E_1, \ldots, E_n)$ while it is applied over n expressions $(E_1, \ldots, E_n)$. Two tildes (i, j) and (i', j') are overlapping if $\{i, \ldots, j\} \cap \{i', \ldots, j'\} \neq \emptyset$. A *free subset* of S is a subset where no tildes overlap each other. As far as $S = (i_k, j_k)_{k \leq m}$ is free, the action of a tilde is to add the empty word in the catenation of the languages denoted by the expression it overlaps in the catenation of all the denoted languages, *i.e.*

$$L(\text{MT}_S(E_1, \ldots, E_n)) = L(E_1) \cdots L(E_{i_1-1}) \cdot (L(E_{i_1}) \cdots L(E_{j_1}) \cup \{\varepsilon\}) \cdot L(E_{j_1+1}) \cdots$$
$$\cdots L(E_{i_m-1}) \cdot (L(E_{i_m}) \cdots L(E_{j_m}) \cup \{\varepsilon\}) \cdot L(E_{j_m+1}) \cdots L(E_n).$$

Inductively extended with multi-tildes operators, regular expressions with n symbols can be turned into equivalent automata with at most n states, using the position automaton [4] or the partial derivation one [6].

In the following, we show how to extend the notion of tildes from unions of free subsets to any Boolean combinations of tildes.

3   Boolean Formulae and Satisfiability

A *Boolean formula* ϕ over an alphabet Γ is inductively defined by

$$\phi = a, \qquad\qquad\qquad \phi = o(\phi_1, \ldots, \phi_n),$$

where a is an *atom* in Γ, o is an *operator* associated with an n-ary function o_f from $\mathbb{B}^n$ to $\mathbb{B}$, and $\phi_1, \ldots, \phi_n$ are n Boolean formulae over Γ.

As an example, $\neg$ is the operator associated with the Boolean negation, $\wedge$ with the Boolean conjunction and $\vee$ with the Boolean disjunction. We denote by $\bot$ the constant (0-ary function) 0 and by $\top$ the constant 1.

Let ϕ be a Boolean formula over an alphabet Γ. A function i from Γ to $\mathbb{B}$ is said to be an *interpretation* (of Γ). The *evaluation* of ϕ with respect to i is the Boolean $\text{eval}_i(\phi)$ inductively defined by

$$\text{eval}_i(a) = i(a), \qquad \text{eval}_i(o(\phi_1, \ldots, \phi_n)) = o_f(\text{eval}_i(\phi_1), \ldots, \text{eval}_i(\phi_n)),$$

where a an atom in Γ, o is an operator associated with an n-ary function o_f from $\mathbb{B}^n$ to $\mathbb{B}$, and $\phi_1, \ldots, \phi_n$ are n Boolean formulae over Γ. Non-classical Boolean functions can also be considered, like in the following example.

Example 1. The operator Mirror_n is associated with the $(2 \times n)$-ary Boolean function f defined for any $(2 \times n)$ Boolean $(b_1, \ldots, b_{2n})$ by

$$f(b_1, \ldots, b_{2n}) \Leftrightarrow (b_1, \ldots, b_n) = (b_{2n}, \ldots, b_{n+1}) \Leftrightarrow (b_1 = b_{2n}) \wedge \cdots \wedge (b_n = b_{n+1})$$
$$\Leftrightarrow (b_1 \wedge b_{2n} \vee \neg b_1 \wedge \neg b_{2n}) \wedge \cdots \wedge (b_n \wedge b_{n+1} \vee \neg b_n \wedge \neg b_{n+1}).$$

A Boolean formula is said to be: *satisfiable* if there exists an interpretation leading to a positive evaluation; a *tautology* if every interpretation leads to a positive evaluation; a *contradiction* if it is not satisfiable.

Even if it is an NP-Hard problem [9], checking the satisfiability of a Boolean formula can be performed by using incremental algorithms [10,11,15]. The following method can be performed: If there is no atom in the formula, it can be reduced to either $\bot$ or $\top$, and it is either a tautology or a contradiction; Otherwise, choose an atom a, replace it with $\bot$ (denoted by $a := \bot$), reduce and recursively reapply the method; If it is not satisfiable, replace a with $\top$ (denoted by $a := \top$), reduce and recursively reapply the method. The reduction step can be performed by recursively simplifying the subformulae of the form $o(\phi_1, \ldots, \phi_n)$ such that there exists $k \leq n$ satisfying $F_k \in \{\bot, \top\}$. As an example, the satisfiability of $\neg(a \wedge b) \wedge (a \wedge c)$ can be checked as shown in Fig. 1.

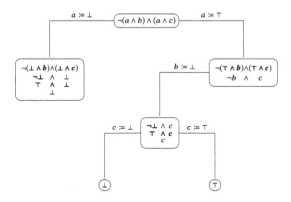

Fig. 1. The formula $\neg(a \wedge b) \wedge (a \wedge c)$ is satisfiable.

Two Boolean formulae ϕ and ϕ' are *equivalent*, denoted by $\phi \sim \phi'$, if for any interpretations i, $\mathrm{eval}_i(\phi) = \mathrm{eval}_i(\phi')$.

Example 2. Let us consider the operator Mirror_n defined in Example 1. It can be shown, following the equation in Example 1, that for any $(2n - 1)$ Boolean formulae $(\phi_1, \ldots, \phi_{2n-1})$,

$$\text{Mirror}_n(\bot, \phi_1, \ldots, \phi_{2n-1}) \sim \text{Mirror}_{n-1}(\phi_1, \ldots, \phi_{2n-2}) \wedge \neg\phi_{2n-1},$$
$$\text{Mirror}_n(\top, \phi_1, \ldots, \phi_{2n-1}) \sim \text{Mirror}_{n-1}(\phi_1, \ldots, \phi_{n-2}) \wedge \phi_{2n-1},$$
$$\text{Mirror}_n() \sim \top.$$

For any two Boolean formulae ϕ and ϕ' and for any atom a, we denote by $\phi_{a:=\phi'}$ the formula obtained by replacing any occurrence of a in ϕ with ϕ'. For any two sequences $(\phi'_1, \ldots, \phi'_n)$ of Boolean formulae and $(a_1, \ldots, a_n)$ of distinct atoms, we denote by $\phi_{a_1:=\phi'_1, \ldots, a_n:=\phi'_n}$ the formula obtained by replacing any occurrence of a_k in ϕ with ϕ'_k for any $1 \le k \le n$.

It is well known that for any Boolean formula ϕ and for any atom a in ϕ,

$$\phi \sim \neg a \wedge \phi_{a:=\bot} \vee a \wedge \phi_{a:=\top}. \tag{1}$$

4   Constrained Multi-Tildes

Multi-Tildes operators define languages by computing free sublists of tildes from a set of couples. This can be viewed as a particular disjunctive combination of these tildes, since sublists of a free list ℓ define languages that are included in the one ℓ defines. This disjunctive interpretation can be extended to any Boolean combination. One may choose to apply conjunctive sequences of not contiguous tildes, or may choose to exclude some combinations of free tildes. In this section, we show how to model this interpretation using Boolean formulae.

The action of a tilde is to add the empty word in the catenation of the languages it overhangs. If the tilde is considered as an interval of contiguous positions $(p_1, p_2, \ldots, p_k)$, its action can be seen as the conjunction of the substitution of each language at position p_1, position p_2, etc. with $\{\varepsilon\}$.

In fact, each position can be considered as an atom of a Boolean formula ϕ. For any interpretation i leading to a positive evaluation of ϕ, we can use $i(k)$ to determine whether the language L_k can be replaced by $\{\varepsilon\}$ in $L_1 \cdots L_n$. Let us formalize these thoughts as follows.

Let i be an interpretation over $\{1, \ldots, n\}$. Let $L_1, \ldots, L_n$ be n languages. We denote by $i(L_1, \ldots, L_n)$ the language $L'_1 \cdots \cdots L'_n$ where $L'_k = \begin{cases} \{\varepsilon\} & \text{if } i(k), \\ L_k & \text{otherwise.} \end{cases}$

Let ϕ be a Boolean formula over the alphabet $\{1, \ldots, n\}$ and $L_1, \ldots, L_n$ be n languages. We denote by $\phi(L_1, \ldots, L_n)$ the language

$$\bigcup_{i \,|\, \text{eval}_i(\phi)} i(L_1, \ldots, L_n). \tag{2}$$

Example 3. Let us consider the operator Mirror_n defined in Example 1 and the two alphabets $\Gamma_n = \{1, \ldots, 2n\}$ and $\Sigma_n = \{a_1, \ldots, a_{2n}\}$. Then:

$$\text{Mirror}_n(1, \ldots, 2n)(\{a_1\}, \ldots, \{a_{2n}\})$$
$$= \{w_1 \cdots w_{2n} \mid \forall k \le 2n, w_k \in \{\varepsilon, a_k\} \wedge (w_k = \varepsilon \Leftrightarrow w_{2n-k+1} = \varepsilon)\}$$
$$= \{a_1 \cdots a_{2n}, a_1 a_3 a_4 \cdots a_{2n-3} a_{2n-2} a_{2n}, \ldots, a_1 a_{2n}, \ldots, a_n a_{n+1}, \varepsilon\}.$$

First, we remark that the action of constrained tildes preserves regularity, since it is a finite union of catenations of regular languages, following Equation (2).

Theorem 1. *Let ϕ be a Boolean formula over the alphabet $\{1,\ldots,n\}$ and $L_1,\ldots,L_n$ be n regular languages. Then $\phi(L_1,\ldots,L_n)$ is regular.*

Moreover, this definition also allows us to explicit some remarkable identities. As an example, considering n languages $(L_1,\ldots,L_n)$ and a Boolean formula ϕ over the alphabet $\{1,\ldots,n\}$, it can be shown that the two following identities hold:

1. if ϕ is a contradiction, then $\phi(L_1,\ldots,L_n) = \emptyset$;
2. if ϕ is a tautology, then $\phi(L_1,\ldots,L_n) = (L_1 \cup \{\varepsilon\}) \cdots (L_n \cup \{\varepsilon\})$.

Some properties of Boolean formulae can also be transferred while acting over language sequences, as direct consequences of Eq. (2).

Lemma 1. *Let ϕ_1 and ϕ_2 be two equivalent Boolean formulae over the alphabet $\{1,\ldots,n\}$ and $L_1,\ldots,L_n$ be n languages. Then $\phi_1(L_1,\ldots,L_n) = \phi_2(L_1,\ldots,L_n)$.*

Lemma 2. *Let ϕ_1 and ϕ_2 be two Boolean formulae over $\{1,\ldots,n\}$ and $L_1,\ldots,L_n$ be n languages. Then $(\phi_1 \vee \phi_2)(L_1,\ldots,L_n) = \phi_1(L_1,\ldots,L_n) \cup \phi_2(L_1,\ldots,L_n)$.*

Lemma 3. *Let ϕ be a Boolean formula over the alphabet $\{2,\ldots,n\}$ and $L_1,\ldots,L_n$ be n languages. Then*

$$(1 \wedge \phi)(L_1,\ldots,L_n) = \phi_{2:=1,\ldots,n:=n-1}(L_2,\ldots,L_n),$$
$$(\neg 1 \wedge \phi)(L_1,\ldots,L_n) = L_1 \cdot \phi_{2:=1,\ldots,n:=n-1}(L_2,\ldots,L_n).$$

As a consequence of Eq. (1), Lemma 1, Lemma 2 and Lemma 3, it holds:

Proposition 1. *Let ϕ be a Boolean formula over the alphabet $\{1,\ldots,n\}$ and $L_1,\ldots,L_n$ be n languages. Then*

$$\phi(L_1,\ldots,L_n) = L_1 \cdot \phi'(L_2,\ldots,L_n) \cup \phi''(L_2,\ldots,L_n),$$
$$where \quad \phi' = \phi_{1:=\bot,2:=1,\ldots,n:=n-1} \quad and \quad \phi'' = \phi_{1:=\top,2:=1,\ldots,n:=n-1}.$$

Example 4. Let us consider the language $L_n = \mathrm{Mirror}_n(1,\ldots,2n)(\{a_1\},\ldots,\{a_{2n}\})$ of Example 3. Following Proposition 1 and rules in Example 2, it holds:

$$L_n = \{a_1\} \cdot \mathrm{Mirror}_{n-1}(1,\ldots,2n-2)(\{a_2\},\ldots,\{a_{2n-1}\}) \cdot \{a_{2n}\}$$
$$\cup \mathrm{Mirror}_{n-1}(1,\ldots,2n-2)(\{a_2\},\ldots,\{a_{2n-1}\}).$$

This first proposition allows us to show how to easily determine whether the empty word belongs to the action of a Boolean formula over a language sequence and how to compute the quotient of such a computation w.r.t. a symbol.

Corollary 1. *Let ϕ be a Boolean formula over $\{1, \ldots, n\}$ and $L_1, \ldots, L_n$ be n languages. Then:*

$$\varepsilon \in \phi(L_1, \ldots, L_n) \Leftrightarrow \varepsilon \in L_1 \wedge \varepsilon \in \phi'(L_2, \ldots, L_n) \vee \varepsilon \in \phi''(L_2, \ldots, L_n)$$

$$\text{where} \quad \phi' = \phi_{1:=\perp, 2:=1, \ldots, n:=n-1} \quad \text{and} \quad \phi'' = \phi_{1:=\top, 2:=1, \ldots, n:=n-1}.$$

Corollary 2. *Let ϕ be a Boolean formula over the alphabet $\{1, \ldots, n\}$, $L_1, \ldots, L_n$ be n languages and a be a symbol. Then:*

$$a^{-1}(\phi(L_1, \ldots, L_n)) = a^{-1}(L_1) \cdot \phi'(L_2, \ldots, L_n) \cup a^{-1}(\phi'(L_2, \ldots, L_n)) \mid \varepsilon \in L_1$$

$$\cup\, a^{-1}(\phi''(L_2, \ldots, L_n)),$$

$$\text{where} \quad \phi' = \phi_{1:=\perp, 2:=1, \ldots, n:=n-1} \quad \text{and} \quad \phi'' = \phi_{1:=\top, 2:=1, \ldots, n:=n-1}.$$

Let us now extend classical regular expressions with the action of a Boolean formula considered as a constrained Multi-Tildes.

An *Extended to Constrained Multi-Tildes Expression* E over an alphabet Σ (*extended expression* in the following) is inductively defined by

$$E = a, \quad E = \emptyset, \quad E = \varepsilon, \quad E = E_1 + E_2, \quad E = E_1 \cdot E_2, \quad E = E_1^*,$$
$$E = \phi(E_1, \ldots, E_n),$$

where a is a symbol in Σ, ϕ is a Boolean formula over the alphabet $\{1, \ldots, n\}$ and $E_1, \ldots, E_n$ are n extended expressions over Σ. The *language denoted* by an extended expression E is the language $L(E)$ inductively defined by

$$L(a) = \{a\}, \qquad\qquad L(\emptyset) = \emptyset, \qquad\qquad L(\varepsilon) = \{\varepsilon\},$$
$$L(E_1 + E_2) = L(E_1) \cup L(E_2), \quad L(E_1 \cdot E_2) = L(E_1) \cdot L(E_2), \quad L(E_1^*) = L(E_1)^*,$$
$$L(\phi(E_1, \ldots, E_n)) = \phi(L(E_1), \ldots, L(E_n)),$$

where a is a symbol in Σ, ϕ is a Boolean formula over the alphabet $\{1, \ldots, n\}$ and $E_1, \ldots, E_n$ are n extended expressions over Σ.

Since the Boolean satisfiability is an NP-hard problem [9], so is the emptiness problem for extended expressions, as a direct consequence of Eq. (2) and of denoted language definition.

Proposition 2. *Let $\Sigma_k = \{a_1, \ldots, a_k\}$ be an alphabet and ϕ be a Boolean formula over the alphabet $\{1, \ldots, k\}$. Then $L(\phi(a_1, \ldots, a_k)) \neq \emptyset \iff \phi$ is satisfiable.*

Corollary 3. *Determining whether the language denoted by an extended expression is empty is NP-hard.*

5   Factorization Power

In this section, we exhibit a parameterized family of expressions E_n such that the smallest NFA recognizing $L(E_n)$ admits a number of states exponentially larger than the sum of the number of symbols, the number of atoms and the number of operators of E_n. Let us consider the alphabet $\Sigma_{2n} = \{a_1, \ldots, a_{2n}\}$ and the expression $E_n = \text{Mirror}_n(1, \ldots, 2n)(a_1, \ldots, a_{2n})$. The expression E_n contains $2n$ atoms, $2n$ symbols and 1 operator. Using classical Boolean operators, like $\wedge$, $\vee$ and $\neg$, the Boolean formula $\text{Mirror}_n(1, \ldots, 2n)$ can be turned into the equivalent one $(1 \wedge 2n \vee \neg 1 \wedge \neg 2n) \wedge \cdots \wedge (n \wedge (n+1) \vee \neg n \wedge \neg(n+1))$ following equation in Example 1, that contains $4n$ atoms and $(6n-1)$ operators, which is a linearly larger Boolean formula. In order to exhibit a lower bound of the number of states of any NFA recognizing $L(E_n)$, let us consider the following property [12]:

Theorem 2 (*[12]*). *Let $L \subset \Sigma^*$ be a regular language, and suppose there exists a set of pairs $P = \{(x_i, w_i) : 1 \leq i \leq n\}$ such that $x_i w_i \in L$ for $1 \leq i \leq n$ and $x_i w_j \notin L$ for $1 \leq i, j \leq n$ and $i \neq j$. Then any NFA accepting L has at least n states.*

For any sequences of n Booleans $bs = (b_1, \ldots, b_n)$, let us consider the words $v_{bs} = w_1 \cdots w_n$ and $v'_{bs} = w'_1 \cdots w'_n$ where

$$w_k = \begin{cases} a_k & \text{if } \neg b_k, \\ \varepsilon & \text{otherwise,} \end{cases} \qquad w'_k = \begin{cases} a_{n+k} & \text{if } \neg b_k, \\ \varepsilon & \text{otherwise.} \end{cases} \tag{3}$$

Denoting by $\text{rev}(b_1, \ldots, b_n)$ the sequence $(b_n, \ldots, b_1)$, since the only words in $L(E_n)$ are the words $v_{bs} \cdot v'_{\text{rev}(bs)}$, and since the words $v_{bs} \cdot v'_{bs'}$ for any $bs' \neq \text{rev}(bs)$ are not in $L(E_n)$, it holds according Theorem 2 that

Proposition 3. *There is at least 2^n states in any automaton recognizing $L(E_n)$.*

Theorem 3. *There exist extended regular expressions exponentially smaller than any automaton recognizing their denoted languages.*

6   Partial Derivatives and Automaton Computation

Let us now show how to extend the Antimirov method in order to syntactically solve the membership test and to compute a finite automaton recognizing the language denoted by an extended expression. First, we define the partial derivative of an expression w.r.t. a symbol, where the derivation formula for the action of a Boolean combination is obtained by considering the fact that the empty word may appear at the first position for two reasons, if the first operand is nullable, or because the empty word is inserted by the multi-tilde.

Definition 1. *Let E be an extended expression and a be a symbol. The partial derivative of E w.r.t. a is the set $\delta_a(E)$ of extended expressions inductively defined*

as follows:

$$\delta_a(b) = \begin{cases} \{\varepsilon\} & \text{if } b = a, \\ \emptyset & \text{otherwise,} \end{cases} \qquad\qquad \delta_a(\varepsilon) = \emptyset,$$

$$\delta_a(\emptyset) = \emptyset, \qquad\qquad \delta_a(E_1 + E_2) = \delta_a(E_1) \cup \delta_a(E_2),$$

$$\delta_a(E_1 \cdot E_2) = \delta_a(E_1) \odot E_2 \cup \delta_a(E_2) \mid \varepsilon \in L(E_1), \qquad \delta_a(E_1^*) = \delta_a(E_1) \odot E_1^*,$$

$$\delta_a(\phi(E_1, \ldots, E_n)) = \delta_a(E_1) \odot \phi'(E_2, \ldots, E_n)$$
$$\cup \, \delta_a(\phi'(E_2, \ldots, E_n)) \mid \varepsilon \in L(E_1)$$
$$\cup \, \delta_a(\phi''(E_2, \ldots, E_n)),$$

where b is a symbol in Σ, ϕ is a Boolean formula over the alphabet $\{1, \ldots, n\}$, $E_1, \ldots, E_n$ are n extended expressions over Σ and

$$\phi' = \phi_{1:=\bot, 2:=1, \ldots, n:=n-1}, \qquad\qquad \phi'' = \phi_{1:=\top, 2:=1, \ldots, n:=n-1}.$$

In the following, to shorten the expressions in the next examples, we consider the trivial quotients $E \cdot \varepsilon = \varepsilon \cdot E = E$ and $E \cdot \emptyset = \emptyset \cdot E = \emptyset$. Furthermore, when ϕ is a contradiction, we consider that $\phi(E_1, \ldots, E_n) = \emptyset$.

Example 5. Let us consider the alphabet $\Sigma = \{a, b\}$ and the expression $E = \text{Mirror}_2(1, 2, 3, 4)(a^+, b^+, a^+, b^+)$, where $x^+ = x \cdot x^*$. The derived terms of E w.r.t. the symbols in Σ are the following, where underlined computations equal $\emptyset$:

$$\delta_a(E) = \delta_a(a^+) \odot \text{Mirror}_2(\bot, 1, 2, 3)(b^+, a^+, b^+) \cup \underline{\delta_a(\text{Mirror}_2(\top, 1, 2, 3)(b^+, a^+, b^+))}$$
$$= \{a^*\} \odot (\text{Mirror}_1(1, 2) \wedge \neg 3)(b^+, a^+, b^+)$$
$$= \{a^* \cdot (\text{Mirror}_1(1, 2) \wedge \neg 3)(b^+, a^+, b^+)\},$$

$$\delta_b(E) = \underline{\delta_b(a^+) \odot \text{Mirror}_2(\bot, 1, 2, 3)(b^+, a^+, b^+)} \cup \delta_b(\text{Mirror}_2(\top, 1, 2, 3)(b^+, a^+, b^+))$$
$$= \delta_b((\text{Mirror}_1(1, 2) \wedge 3)(b^+, a^+, b^+))$$
$$= \delta_b(b^+) \odot (\text{Mirror}_1(\bot, 1) \wedge 2)(a^+, b^+) \cup \underline{\delta_b((\text{Mirror}_1(\top, 1) \wedge 2)(a^+, b^+))}$$
$$= \{b^*\} \odot (\neg 1 \wedge 2)(a^+, b^+) = \{b^* \cdot (\neg 1 \wedge 2)(a^+, b^+)\}.$$

As usual, the partial derivative is closely related to the computation of the quotient of the denoted language, as a direct consequence of Corollary 2, and by induction over the structure of E.

Proposition 4. *Let E be an extended expression and a be a symbol. Then*

$$\bigcup_{E' \in \delta_a(E)} L(E') = a^{-1}(L(E)).$$

The partial derivative can be classically extended from symbols to words by repeated applications. Let E be an extended expression, a be a symbol and w be a word. Then

$$\delta_\varepsilon(E) = \{E\}, \qquad\qquad \delta_{aw}(E) = \bigcup_{E' \in \delta_a(E)} \delta_w(E').$$

Example 6. Let us consider the expression E and its derived terms computed in Example 5. Then:

$$\delta_{aa}(E) = \delta_a(a^* \cdot (\text{Mirror}_1(1,2) \wedge \neg 3)(b^+, a^+, b^+)) = \{a^* \cdot (\text{Mirror}_1(1,2) \wedge \neg 3)(b^+, a^+, b^+)\},$$

$$\delta_{ab}(E) = \delta_b(a^* \cdot (\text{Mirror}_1(1,2) \wedge \neg 3)(b^+, a^+, b^+))$$

$$= \delta_b((\text{Mirror}_1(1,2) \wedge \neg 3)(b^+, a^+, b^+))$$

$$= \delta_b(b^+) \odot (\text{Mirror}_1(\perp, 1) \wedge \neg 2)(a^+, b^+) \cup \delta_b((\text{Mirror}_1(\top, 1) \wedge \neg 2)(a^+, b^+))$$

$$= \{b^*\} \odot (\neg 1 \wedge \neg 2)(a^+, b^+) \cup \delta_b((1 \wedge \neg 2)(a^+, b^+))$$

$$= \{b^* \cdot (\neg 1 \wedge \neg 2)(a^+, b^+)\} \cup \delta_b((\neg 1)(b^+))$$

$$= \{b^* \cdot (\neg 1 \wedge \neg 2)(a^+, b^+), b^*\}.$$

Once again, this operation is a syntactical representation of the quotient computation, as a direct consequence of Proposition 4, and by induction over the structure of words.

Proposition 5. *Let E be an extended expression and w be a word. Then*

$$\bigcup_{E' \in \delta_w(E)} L(E') = w^{-1}(L(E)).$$

As a direct consequence, the membership test is solved for extended expressions. Indeed, determining whether a word w belongs to the language denoted by an extended expression E can be performed by computing the partial derivative of E w.r.t. w and then by testing whether it contains a nullable expression, *i.e.* an expression whose denoted language contains the empty word.

Let us now show that the partial derivative automaton of an extended expression E is a finite one that recognizes $L(E)$.

In the following, we denote by $\mathcal{D}_E$ the set of derived terms of an expression E, *i.e.*, the set of expressions $\bigcup_{w \in \Sigma^*} \delta_w(E)$.

Moreover, given an expression $\phi(E_1, \ldots, E_n)$, an integer $1 \leq k \leq n-1$ and an interpretation i in $\{1, \ldots, k\} \rightarrow \mathbb{B}$, we denote by $\mathcal{D}_{E,k,i}$ the set $\mathcal{D}_{E_k} \odot \phi'(E_{k+1}, \ldots, E_n)$, where

$$\phi' = \phi_{1:=\begin{cases} \top & \text{if } i(1), \\ \perp & \text{otherwise,} \end{cases} \ldots, k:=\begin{cases} \top & \text{if } i(k), \\ \perp & \text{otherwise,} \end{cases} k+1:=1,\ldots,n:=n-k}.$$

First, the union of these sets includes the partial derivatives and is stable w.r.t. derivation by a symbol, by induction over the structures of extended expressions, of words and over the integers.

Proposition 6. *Let E be an extended expression and a be a symbol. Then the two following conditions hold:*

1. $\delta_a(E) \subset \displaystyle\bigcup_{\substack{1 \leq k \leq n, \\ i \in \{1,\dots,k\} \to \mathbb{B}}} \mathcal{D}_{E,k,i}$,

2. $\displaystyle\bigcup_{E' \in \mathcal{D}_{E,k,i}} \delta_a(E') \subset \displaystyle\bigcup_{\substack{k \leq k' \leq n, \\ i' \in \{1,\dots,k'\} \to \mathbb{B}}} \mathcal{D}_{E,k',i'}$.

As a direct consequence, the set of the derived terms of an extended expression is included in the union of the $\mathcal{D}_{E,k,i}$ sets.

Corollary 4. *Let E be an extended expression. Then*

$$\mathcal{D}_E \subset \bigcup_{\substack{1 \leq k \leq n, \\ i \in \{1,\dots,k\} \to \mathbb{B}}} \mathcal{D}_{E,k,i}.$$

According to a trivial inductive reasoning, one can show that such a set is finite.

Corollary 5. *Let E be an extended expression and w be a word. Then*

$$\bigcup_{w \in \Sigma^*} \delta_w(E) \text{ is a finite set.}$$

As a direct consequence, the derived term automaton of an extended expression, defined as usual with derived terms as states and transitions computed from partial derivation, fulfils finiteness and correction.

Theorem 4. *Let E be an extended expression and a be a symbol. The partial derivative automaton of E is a finite automaton recognizing $L(E)$.*

Example 7. Let us consider the expression E defined in Example 5. The derived term automaton of E is given in Fig. 2.

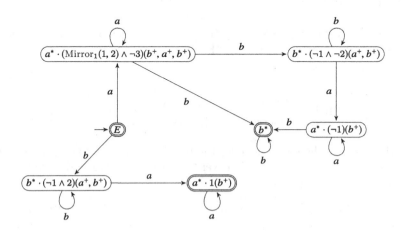

Fig. 2. The derived term automaton of E.

7   Haskell Implementation

The computation of partial derivatives and derived term automaton has been implemented in Haskell and is publicly available on GitHub [14]. Constrained tildes are implemented using dependently typed programming: a Boolean formula encoding a constrained tildes uses an alphabet the size of which cannot be greater than the length of the list of expressions the formula is applied on. Derived term automaton can be graphically represented using Dot and Graphviz, and converted in PNG. A parser from string declaration is also included.

8   Conclusion and Perspectives

In this paper, we have extended (disjunctive) multi-tildes operators to any Boolean combinations of these tildes, the constrained multi-tildes, and defined their denoted languages. We have shown that the action of these operators preserves regularity, that they may lead to exponentially smaller expressions and how to solve the membership test by defining the partial derivatives and the (finite) derived term automaton.

The next step of our plan is to study the conversion of an automaton into an equivalent expression, by first characterizing the structure of derived term automaton like it was previously done in the classical regular case [8,13].

References

1. Antimirov, V.M.: Partial derivatives of regular expressions and finite automaton constructions. Theor. Comput. Sci. **155**(2), 291–319 (1996)
2. Caron, P., Champarnaud, J.-M., Mignot, L.: Multi-tilde operators and their Glushkov automata. In: Dediu, A.H., Ionescu, A.M., Martín-Vide, C. (eds.) LATA 2009. LNCS, vol. 5457, pp. 290–301. Springer, Heidelberg (2009). https://doi.org/10.1007/978-3-642-00982-2_25
3. Caron, P., Champarnaud, J.-M., Mignot, L.: A new family of regular operators fitting with the position automaton computation. In: Nielsen, M., Kučera, A., Miltersen, P.B., Palamidessi, C., Tůma, P., Valencia, F. (eds.) SOFSEM 2009. LNCS, vol. 5404, pp. 645–655. Springer, Heidelberg (2009). https://doi.org/10.1007/978-3-540-95891-8_57
4. Caron, P., Champarnaud, J., Mignot, L.: Multi-bar and multi-tilde regular operators. J. Autom. Lang. Comb. **16**(1), 11–36 (2011)
5. Caron, P., Champarnaud, J.-M., Mignot, L.: Partial derivatives of an extended regular expression. In: Dediu, A.-H., Inenaga, S., Martín-Vide, C. (eds.) LATA 2011. LNCS, vol. 6638, pp. 179–191. Springer, Heidelberg (2011). https://doi.org/10.1007/978-3-642-21254-3_13
6. Caron, P., Champarnaud, J.-M., Mignot, L.: Multi-tilde-bar derivatives. In: Moreira, N., Reis, R. (eds.) CIAA 2012. LNCS, vol. 7381, pp. 321–328. Springer, Heidelberg (2012). https://doi.org/10.1007/978-3-642-31606-7_28
7. Caron, P., Champarnaud, J., Mignot, L.: Multi-tilde-bar expressions and their automata. Acta Inform. **49**(6), 413–436 (2012)

8. Caron, P., Ziadi, D.: Characterization of Glushkov automata. Theor. Comput. Sci. **233**(1–2), 75–90 (2000)

9. Cook, S.A.: The complexity of theorem-proving procedures. In: STOC, pp. 151–158. ACM (1971)

10. Davis, M., Logemann, G., Loveland, D.W.: A machine program for theorem-proving. Commun. ACM **5**(7), 394–397 (1962)

11. Davis, M., Putnam, H.: A computing procedure for quantification theory. J. ACM **7**(3), 201–215 (1960)

12. Glaister, I., Shallit, J.O.: A lower bound technique for the size of nondeterministic finite automata. Inf. Process. Lett. **59**(2), 75–77 (1996)

13. Lombardy, S., Sakarovitch, J.: How expressions can code for automata. RAIRO Theor. Informatics Appl. **39**(1), 217–237 (2005)

14. Mignot, L.: ConstrainedTildesHaskell. https://github.com/LudovicMignot/ConstrainedTildesHaskell (2023)

15. Quine, W.V.O.: Methods of logic. Harvard University Press (1982)

On the Smallest Synchronizing Terms of Finite Tree Automata

Václav Blažej[1] , Jan Janoušek[2] , and Štěpán Plachý[2(✉)]

[1] University of Warwick, Coventry, UK
vaclav.blazej@warwick.ac.uk
[2] Faculty of Information Technology, Czech Technical University in Prague,
Prague, Czech Republic
{Jan.Janousek,Stepan.Plachy}@fit.cvut.cz

Abstract. This paper deals with properties of synchronizing terms for finite tree automata, which is a generalization of the synchronization principle of deterministic finite string automata (DFA) and such terms correspond to a connected subgraph, where a state in the root is always the same regardless of states of subtrees attached to it. We ask, what is the maximum height of the smallest synchronizing term of a deterministic bottom-up tree automaton (DFTA) with n states, which naturally leads to two types of synchronizing terms, called weak and strong, that depend on whether a variable, i.e., a placeholder for a subtree, must be present in at least one leaf or all of them. We prove that the maximum height in the case of weak synchronization has a theoretical upper bound $\mathrm{sl}(n) + n - 1$, where $\mathrm{sl}(n)$ is the maximum length of the shortest synchronizing string of an n-state DFAs. For strong synchronization, we prove exponential bounds. We provide a theoretical upper bound of $2^n - n - 1$ for the height and two constructions of automata approaching it. One achieves the height of $\Theta(2^{n-\sqrt{n}})$ with an alphabet of linear size, and the other achieves $2^{n-1} - 1$ with an alphabet of quadratic size.

Keywords: Tree automata · synchronizing string · synchronizing term · Černý conjecture · automata theory

1 Introduction

Synchronization is a widely studied property of deterministic finite string automata (DFA). A string is called synchronizing (or sometimes called a reset word) if it reaches the same state regardless of what state we start in.

One problem of interest in this field is to determine what is the maximum length of the shortest synchronizing string for synchronizing DFAs with n states.

The authors acknowledge the support of the OP VVV MEYS funded project CZ.02.1.01/0.0/0.0/16_019/0000765 "Research Center for Informatics" and the Grant Agency of the Czech Technical University in Prague, grant No. SGS20/208/OHK3/3T/18. V. Blažej was supported by the Engineering and Physical Sciences Research Council [grant number EP/V044621/1].

© The Author(s), under exclusive license to Springer Nature Switzerland AG 2023
B. Nagy (Ed.): CIAA 2023, LNCS 14151, pp. 79–90, 2023.
https://doi.org/10.1007/978-3-031-40247-0_5

We denote this value by $sl(n)$. State-of-the-art bounds on this value differ based on whether we consider complete or partial automata.

For complete automata, we know that the function is somewhere between $(n-1)^2$ and $(n^3 - n)/6$. The lower bound was shown by Černý [11,12] and the upper bound by Pin [4], where the factor $1/6 = 0.1666\ldots$ was then further slightly improved couple of times, most recently by Shitov [10] down to 0.1654. Černý conjectured his lower bound to be tight. For decades now Černý's conjecture has not been resolved and is one of the most longstanding open problems in automata theory. The conjecture had been shown to hold for certain classes of automata. For a detailed survey of results in the theory of complete synchronizing automata refer to the Handbook of Automata Theory [5, Chapter 15].

In the case of partial automata, we can identify two types of synchronizing strings, called careful and exact, depending on how we handle missing transitions. Reading a carefully synchronizing string ends in a single state regardless of the starting state. For exactly synchronizing strings, the automaton either ends in a single state or fails (as the transition function is incomplete). The exact synchronization shares many properties of complete automata synchronization, including polynomial bounds for the length of the shortest exactly synchronizing string [9]. In the case of careful synchronization, it gets more complicated. An exponential upper bound of $\mathcal{O}(3^{(n+\varepsilon)/3})$ has been shown [8] together with several constructions of automata approaching this bound with alphabets of various sizes, from $3^{n/(6 \log_2 n)}$ for binary and up to $\Theta(3^{n/3})$ for linear size alphabet [2,3,7].

Finite tree automaton (FTA) is a standard model of computation for the class of regular tree languages [1] and is a very straightforward extension of string automata. The class is concerned with ranked tree structures, i.e., the label of each node, which is from a finite alphabet, determines the number of its children (also called arity), therefore the maximum degree is bounded by a constant. One can look at ranked trees as branching strings and adapt the language theory to them. Many concepts and properties of string automata have their analogous principle in tree automata as well, including synchronization.

A synchronizing term of a deterministic bottom-up tree automaton (DFTA) corresponds to a connected subgraph such that the state within the root is the same regardless of states obtained from subtrees attached to it. Places, where a subtree can be attached are marked with a variable in such term. This concept was first touched on by Janoušek and Plachý [6] when investigating tree automata analog for string k-locality (i.e. all strings of length k are synchronizing), which allows work-optimal parallelization.

In this paper, we deal with size bounds of synchronizing terms of tree automata. The aim is to determine the maximum height of the smallest (minimum height) synchronizing term in a complete n-state DFTA. We improve upon the definition of synchronizing terms from [6] to handle automata with unreachable states properly. We also identify two types of synchronizing terms – weakly and strongly synchronizing – which depend on whether a variable must be in at least one leaf or every leaf. These two types of synchronizing terms exhibit wildly different bounds. We prove, that the upper bound for weak synchronization is polynomial and closely tied to the upper bound for string synchronization. For strong synchronization, we prove that the bounds are super-polynomial. We prove an exponential upper bound and provide two constructions of n-state automata approaching this bound. One with an alphabet of linear size that reaches the height of $\Theta(2^{n-\sqrt{n}})$ and the other with a quadratic alphabet reaching $2^{n-1} - 1$.

2 Notation

We use notions from the theory of tree languages similarly as defined in Comon, et al. [1].

An *alphabet* Σ is a finite set of *symbols*. A *string* w over an alphabet Σ is a sequence $w = a_1 a_2 \ldots a_n$ such that $a_i \in \Sigma$ for all $1 \leq i \leq n$. The length of w is denoted $|w|$, ε denotes the string of length 0, and Σ^* denotes the set of all string over Σ.

A *ranked alphabet* $\mathcal{F}$ is a non-empty alphabet, where each symbol has a non-negative integer arity (or rank). The *arity* of a symbol f is denoted by $\mathrm{arity}(f)$ and the set of symbols of arity r is denoted $\mathcal{F}_r$. The symbols of $\mathcal{F}_0$ are called *constant*.

Let *terms* $T(\mathcal{F}, X)$ over a ranked alphabet $\mathcal{F}$ and a set of variables X, $X \cap \mathcal{F} = \emptyset$, be an infinite set that is inductively defined as follows. As a base, the sets X and $\mathcal{F}_0$ belong to $T(\mathcal{F}, X)$. As an inductive step, for any $f \in \mathcal{F}_r$ and $t_1, \ldots, t_r$ that all already belong to $T(\mathcal{F}, X)$ the term $(f, t_1, \ldots, t_r)$ also belongs to $T(\mathcal{F}, X)$. See Fig. 1 for an example of many notions explained in this section.

Terms where $X = \emptyset$ are called *ground terms* $T(\mathcal{F}) = T(\mathcal{F}, \emptyset)$. A *tree language* is a set of ground terms. Terms, where each variable occurs at most once, are called *linear* terms. In this paper, we assume all terms to be linear.

Let $\sigma_g \colon X \to T(\mathcal{F})$ be *ground substitution* of a set of variables X over a ranked alphabet $\mathcal{F}$. Ground substitution can be extended to $\sigma \colon T(\mathcal{F}, X) \to T(\mathcal{F})$ in such a way that the non-variable values remain unchanged, i.e.,

$$\sigma(t) = \begin{cases} t & \text{if } t \in \mathcal{F}_0, \\ \sigma_g(t) & \text{if } t \in X, \\ (f, \sigma(t_1), \ldots, \sigma(t_r)) & t = (f, t_1, \ldots, t_r), \text{otherwise.} \end{cases}$$

A *height* of a term $t \in T(\mathcal{F}, X)$ denoted $\mathrm{height}(t)$ is a function defined as 0 for all $t \in \mathcal{F}_0 \cup X$ and as $1 + \max_{i=1}^{r} \mathrm{height}(t_i)$ where $t = (f, t_1, \ldots, t_r)$ otherwise. Each term contains within itself many other terms, we call them *subterms* of a term. To get a specific subterm, we introduce its position through the child terms that we have to traverse to it. A *position* is a sequence of numbers (i.e. a string over $\mathbb{N}$) $p = (p_1, \ldots, p_d) \in \mathbb{N}^d$ where $d \in \mathbb{N}$ is its *depth*, also denoted as $\mathrm{depth}(p)$. A subterm $t|_p$ of a term $t \in T(\mathcal{F}, X)$ on a position p is defined as t if $|p| = 0$ and $t_i|_{p'}$ where $t = (f, t_1, \ldots, t_r)$, $p = ip'$ for $i \leq r$ and $p' \in \mathbb{N}^*$, otherwise. The set of all subterms of t is denoted by $\mathrm{Subterms}(t)$ and the set

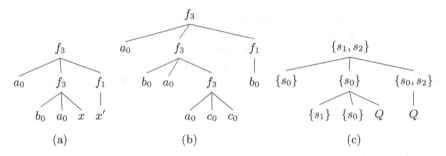

Fig. 1. An example term and the ways it can be processed. **(a):** A term t over the ranked alphabet $\mathcal{F} = \{a_0, b_0, f_1, f_3\}$ with variables $X = \{x, x'\}$. The subterm $t|_{21} = b_0$ is on position $(2, 1)$. We denote t as (f_3, t_1, t_2, t_3), however, with fully expanded subterms we get $(f_3, (a_0), (f_3, (b_0), (a_0), (x)), (f_1, (x')))$. **(b):** Application of the ground substitution $\sigma(t)$. We see that $\sigma(x) = (f_3, (a_0), (c_0), (c_0))$ and $\sigma(x') = (b_0)$. **(c):** Evaluation of $\widetilde{\Delta}_Q(t)$ of a DFTA. Labels of all subterms t' are substituted with $\widetilde{\Delta}_Q(t')$ in the image. We have $Q = \{s_0, s_1, s_2\}$ and $\widetilde{\Delta}_Q(a_0) = \{s_0\}$, $\widetilde{\Delta}_Q(b_0) = \{s_1\}$, $\widetilde{\Delta}_Q(f_3, \{s_1\}, \{s_0\}, Q) = s_0$, $\widetilde{\Delta}_Q(f_1, Q) = \{s_0, s_2\}$, and $\widetilde{\Delta}_Q(f_3, \{s_0\}, \{s_0\}, \{s_0, s_2\}) = \{s_1, s_2\}$. As $|\widetilde{\Delta}_Q(t)| > 1$ the term is not synchronizing for the DFTA.

of all subterm positions of t is denoted by $\mathrm{SubtPos}(t)$. For t we call the empty position *root* and a position of a constant or a variable a *leaf*.

For us, a *deterministic (bottom-up) finite tree automaton* (DFTA) over a ranked alphabet $\mathcal{F}$ is a 3-tuple $A = (Q, \mathcal{F}, \Delta)$, where Q is a finite set of states, and $\Delta \colon \mathcal{F}_r \times Q^r \to Q$ is a transition function; we omit the set of final states $Q_f \subseteq Q$ from the standard definition of A as it is irrelevant to our paper.

We will use several extensions of the transition function starting with sets of states as arguments $\Delta(f, S_1, \ldots, S_r) = \{\Delta(f, q_1, \ldots, q_r) \mid \forall i : q_i \in S_i\}$ for $f \in F_r$ and $S_1, \ldots, S_r \subseteq Q$. We further extend the transition function to terms provided a mapping assigning sets of states to variables is given. This function corresponds to the evaluation of a term by the DFTA. For a DFTA $A = (Q, \mathcal{F}, \Delta)$ and a set of variables X let $\widetilde{\Delta}$ be a mapping $\widetilde{\Delta} \colon T(\mathcal{F}, X) \times (X \to 2^Q) \to 2^Q$ such that

$$
\widetilde{\Delta}(t, \widetilde{X}) = \begin{cases} \{\Delta(t)\} & \text{if } t \in \mathcal{F}_0, \\ \{\widetilde{X}(t)\} & \text{if } t \in X, \\ \Delta\big(f, \widetilde{\Delta}(t_1, \widetilde{X}), \ldots, \widetilde{\Delta}(t_p, \widetilde{X})\big) & \text{where } t = (f, t_1, \ldots, t_r), \text{otherwise.} \end{cases}
$$

We also denote $\widetilde{\Delta}_Q(t) = \widetilde{\Delta}(t, \widetilde{X})$ when every variable evaluates to the set of all states, i.e., $\widetilde{X}(x) = Q$ for each $x \in X$. When t is a ground term we also denote $\widetilde{\Delta}(t) = q$ where $\widetilde{\Delta}_Q(t) = \{q\}$ since ground terms don't contain any variable and they always evaluate to a single state because of determinism.

Definition 1 (Synchronizing term). *For a DFTA $A = (Q, \mathcal{F}, \Delta)$ a term $t \in T(\mathcal{F}, X)$ is called* synchronizing *if $|\widetilde{\Delta}_Q(t)| = 1$.*

3 Tree Automata Synchronization

In the case of finite string automata, the automaton is called synchronizing if there exists a synchronizing string for it, i.e., a string that, when read by the automaton, reaches the same state regardless of what state the automaton started in. However, for tree automata defining a synchronizing DFTA by the existence of a synchronizing term is not as simple since every ground term is synchronizing because the DFTA is total deterministic.

From the point of view of tree automata, the synchronizing string can be viewed as a term with symbols of arity 1 and a variable in its single leaf. In the case of DFTA, however, a synchronizing term can also contain constants in leaves, not just variables. Constants do not have analogy in synchronizing strings since in the DFTA they play an analogous role to the initial state in string automata and the initial state is not relevant for string synchronization.

For tree automata, we can therefore distinguish three types of synchronizing terms based on the existence of constants and variables.

The first type are ground terms, i.e., terms with constants and without variables, which are always synchronizing. The second type are terms that contain at least one variable. We will call these terms weakly synchronizing and show polynomial worst-case bounds for the height of the smallest such terms.

The third type, are terms without constants, i.e., terms containing variables in all leaves. We will call these terms strongly synchronizing and show that the bounds on the worst case for this type of synchronization are exponential. Constants and their transitions in the automaton are irrelevant for this type of synchronization and so without loss of generality, we will be omitting them.

Definition 2 (Strong and weak synchronization). *Let us have a DFTA* $A = (Q, \mathcal{F}, \Delta)$. *A synchronizing term* $t \in T(\mathcal{F}, X)$ *for* A *is called*

– *weakly synchronizing if* $t \in T(\mathcal{F}, X) \setminus T(\mathcal{F})$, *i.e., it has at least one variable.*
– *strongly synchronizing if* $t \in T(\mathcal{F} \setminus \mathcal{F}_0, X)$, *i.e., it has no constant subterms.*

A DFTA is called strongly *(resp.* weakly*) synchronizing if there exists a strongly (resp. weakly) synchronizing term for it. A set* $ST_{weak}(A)$ *is a set of all weakly synchronizing terms of* A *and* $ST_{strong}(A)$ *is a set of all strongly synchronizing terms of* A.

Observe that strong synchronization implies weak synchronization and both synchronization types are equivalent either when the alphabet of the DFTA does not contain constants or when the maximum arity is 1.

We shall focus on the analysis of the following functions that describe the value we wish to bound.

Definition 3 (wsh & ssh). *Let* $\mathcal{A}_s$ *be the set of all strongly synchronizing DFTAs where* $\forall A \in \mathcal{A}_s$ *and* $A = (Q, \mathcal{F}, \Delta)$ *we have* $|Q| = n$, $|\mathcal{F} \setminus \mathcal{F}_0| = m$ *and* $\forall f \in \mathcal{F} : \mathrm{arity}(f) \leq r$, *then*

$$\mathrm{ssh}(n, m, r) = \max_{A \in \mathcal{A}_s} \left(\min_{t \in ST_{strong}(A)} \mathrm{height}(t) \right).$$

Similarly, let $\mathcal{A}_w$ be the set of all weakly synchronizing DFTAs, where additionally $\mathcal{F}_0 \neq \emptyset$, then $\mathrm{wsh}(n, m, r) = \max_{A \in \mathcal{A}_w} \left(\min_{t \in ST_{weak}(A)} \mathrm{height}(t) \right).$

The reason why the function wsh requires DFTAs with constants is that in automata without constants all weakly synchronizing terms are strongly synchronizing and it is not hard to see that such automata would be the worst case and therefore wsh would equal ssh. The existence of constants in the DFTAs, however, is a crucial assumption that allows polynomial bounds of wsh.

A DFTA with symbols of arity 1 is equivalent to a string automaton (omitting initial and final states), and so they share synchronization bounds.

Remark 1. $\forall n \geq 1, m \geq 2 : \mathrm{wsh}(n, m, 1) = \mathrm{ssh}(n, m, 1) = \mathrm{sl}(n)$

4   Weak Synchronization

In this section, we show the upper bound for the height of the smallest weakly synchronizing term which is similar to the string case. Although the term can contain any number of variables, the worst case can actually be reduced to a single variable and such terms can be encoded into a string for which we can create a synchronizing DFA. However, we are not sure whether the lower bound is better or worse than the string case. We leave this as an open question.

Lemma 1. *If for a DFTA $A = (Q, \mathcal{F}, \Delta)$ exists a ground term $t \in T(\mathcal{F})$ such that $\widetilde{\Delta}(t) = q$ for some $q \in Q$, then there exists a ground term t' such that $\widetilde{\Delta}(t') = q$ and $\mathrm{height}(t') < n$ where $n = |Q|.$*

Proof (sketch). In the evaluation of $\widetilde{\Delta}_Q(t)$, each state can occur only once on every path from the root to a constant. Otherwise, the term can be reduced while the state in the root will not change.   □

Theorem 1 (Bounds for smallest weakly synchronizing term height).
For every $n \geq 1$, $m \geq 2$ and $r \geq 2$ applies $\mathrm{wsh}(n, m, r) < \mathrm{sl}(n) + n$

Proof (sketch). In a synchronizing term, substituting a variable with arbitrary constant preserves synchronization, therefore without loss of generality, we can assume the smallest weakly synchronizing term contains just one variable.

The term is therefore a path from the root to the variable with ground terms attached to it. Since each ground term has a well-defined state in its root, we can encode the path into a string, such that each symbol contains information about the ranked symbol, the index of the child on the path to the variable, and states of ground terms in remaining children. The root is the last symbol in the string.

From the DFTA we can then construct a string automaton for such strings, that ends in the same states as the DFTA would in the root of the original term and therefore it is synchronizing iff the DFTA is weakly synchronizing. From a synchronizing word, we can then construct a weakly synchronizing term by substituting states of ground terms with some actual ground term. The depth of the variable is therefore at most $\mathrm{sl}(n)$ and based on Lemma 1 a ground term can further increase the height by less than n.   □

5 Strong Synchronization

In this section, we study the bounds of strong synchronization. We will first show an exponential upper bound for the height of the smallest strongly synchronizing term, then we will show, that, unlike the weak synchronization, the lower bound will be super-polynomial. We will show that by constructing a DFTA and proving the height of its smallest strongly synchronizing term. We provide two such constructions, one with an alphabet of linear size with a smaller bound, and the other with an alphabet of quadratic size, that is closer to the upper bound.

First, let us show a general upper bound for the height of the smallest synchronizing term.

Theorem 2. *For a strongly synchronizing DFTA $A = (Q, \mathcal{F}, \Delta)$ with $n = |Q|$ there exists a strongly synchronizing term t such that $\mathrm{height}(t) \leq 2^n - n - 1$. Therefore $\mathrm{ssh}(n, m, r) \leq 2^n - n - 1$ for $n \geq 1$.*

Proof (sketch). The evaluation $\widetilde{\Delta}_Q$ of the sequence of subterms starting from a variable and going up to t cannot repeat any set of states. Otherwise, we could exchange the bigger subterm with the smaller one with the same evaluation and make the whole synchronizing term smaller.   □

5.1 Lower Bound with Linear Size Alphabet

In this section, we create an n-state DFTA with linear size alphabet, where the height of the smallest strongly synchronizing term is roughly $2^{n-\sqrt{n}}$. The main idea is to simulate the behavior of a binary subtractor. We imagine a subset of states as a binary number where a bit is set to 1 if a specific state is present in the subset. Evaluation starts in variables, where the binary number has all bits set to 1. In each transition, we decrease the value of the number by at most 1. The children of the term have to contain identical terms and a correct symbol must be used to decrease the binary number. The synchronization is achieved when we reach the value of 0. The transition function is designed in such a way that this sequence of transitions is forced. When an incorrect symbol is used the binary number may not decrease which delays synchronization.

Recall that subtracting 1 from a binary number sets the least significant bit containing 1 to 0 and all less significant bits to 1 (e.g. 10$\underline{1}$00 → 10$\underline{0}$11). Alphabet symbols will be used to represent the position of the least significant one. To set many zeroes to ones, the automaton will use roughly $\sqrt{n}$ auxiliary states, which will be kept until the very last transition. One additional special state $\top$ will be present at all times and the automaton will synchronize to this state. The remaining approximately $n - \sqrt{n}$ states will represent the binary number.

Definition 4 (Binary subtractor DFTA). *For $n \geq 1$ let us have integers $\alpha \geq 0$ and $\beta \geq 0$ such that $n = 1 + \beta + \alpha$ and if $n > 1$ then $\beta \geq 1$ and $(\beta - 1)^2 \leq \alpha \leq \beta^2$. We define Binary subtractor DFTA as $A_n^- = (Q, \mathcal{F}, \Delta)$ where $Q = \{\top\} \cup P_\beta \cup Q_\alpha$, $P_\beta = \{p_{\beta-1}, \ldots, p_0\}$, $Q_\alpha = \{q_{\alpha-1}, \ldots, q_0\}$, $\mathcal{F} = \{a_2^0, \ldots, a_2^{\alpha-1}, b_2\}$, and Δ is described in Table 1.*

Table 1. Transition function Δ of A_n^-. Underlined transitions are not supposed to be used in the smallest strongly synchronizing term. Gray transitions are redundant for the result and default to $\top$.

$\Delta(a_2^i, \downarrow, \rightarrow)$	$\top$	p_j	$q_j, j > i$	$q_j, j = i$	$q_j, j < i$
$\top$	$\top$	p_j	q_j	$\top$	q_j
p_k	p_k	$q_{(j\beta+k) \bmod i}$	$\top$	$\top$	q_i
$q_k, k > i$	q_k	$\top$	$\top$	$\top$	$\top$
$q_k, k = i$	$\top$	$\top$	$\top$	$\top$	$\top$
$q_k, k < i$	q_k	q_i	$\top$	$\top$	$\top$

$\Delta(b_2, \downarrow, \rightarrow)$	$\top$	p_j	q_j
$\top$	$\top$	$\top$	q_j
p_k	$\top$	$\top$	p_k
q_k	q_k	p_j	$\top$

For a set of states $S \subseteq Q$ we denote $B_\alpha(S)$ a binary integer and $B_{\alpha i}(S)$ is i-th bit of $B_\alpha(S)$ (starting from the least significant bit), such that $B_{\alpha i}(S) = 1 \iff q_i \in S$ for all $0 \le i < \alpha$. We also denote $\mathrm{LSO}(k) = \log_2(k \oplus (k-1))$ function giving the position of least significant one in the binary representation of integer $k > 0$, where $\oplus$ is bit-wise xor operator (e.g. $\mathrm{LSO}(100\underline{1}00_2) = 2$).

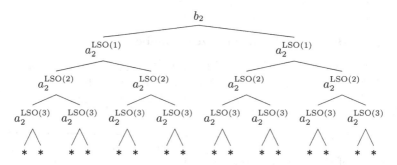

Fig. 2. The smallest strongly synchronizing term (nodes denoted by $*$ contain some distinct variable), for DFTA $A_5^- = (Q, \mathcal{F}, \Delta)$, with $Q = \{\top, p_0, p_1, q_1, q_0\}$ and $\mathcal{F} = \{a_2^0, a_2^1, b_2\}$. For such automaton therefore $\alpha = 2$ and $B_\alpha(Q) = 2^\alpha - 1 = 3$. LSO gives the position of the least significant one in a binary number, therefore $\mathrm{LSO}(3) = 0, \mathrm{LSO}(2) = 1$, and $\mathrm{LSO}(1) = 0$.

Figure 2 shows an example of the smallest synchronizing term of A_n^- with $n = 5$.

Notice, that both $\Delta(a_2^i, \top, \top) = \top$ and $\Delta(b_2, \top, \top) = \top$ so the $\top$ state is ever-present. States P_β are also in all sets except after transitioning on symbol b_2 when no state from Q_α is present in states of children, which produces a set containing only $\top$ and so achieves synchronization.

Next, we prove that the transition function of A_n^- either decreases the value of B_α by 1 or otherwise leaves the value unchanged in the best case. The idea is that all bits of B_α of children are merged, except for the i-th bit when transitioning on symbol a_2^i when no bit after i is set in both children. In such case, i-th bit is removed and all less significant bits are set to 1 using auxiliary states P_β, thus the value decreases exactly by 1. Therefore to decrease the value, both children must have identical sets of states and transition must be done with an appropriate symbol corresponding to the position of the least significant one.

Lemma 2. *For all $n \geq 1$ and A_n^- let $S = \{\top\} \cup P_\beta \cup Q'_\alpha$ and $S' = \{\top\} \cup P_\beta \cup Q''_\alpha$ where $Q'_\alpha, Q''_\alpha \subseteq Q_\alpha$. Then $B_\alpha(\Delta(a_2^i, S, S')) = \max(B_\alpha(S), B_\alpha(S')) - 1$ if $LSO(B_\alpha(S)) = LSO(B_\alpha(S')) = i$.*
  Otherwise $B_\alpha(\Delta(f, S, S')) \geq \max(B_\alpha(S), B_\alpha(S'))$, when $f \in \mathcal{F} \setminus \{a_2^i\}$.

Proof. (sketch). In the case of transition on a_2^i the transition function creates the binary number $B_\alpha(\Delta(a_2^i, S, S'))$, such that, using state $\top$, all bits set in $B_\alpha(S)$ and $B_\alpha(S')$ on position higher than i remain also set, and every bit after i is set using pairs of auxiliary states P_β (which remain active till the last transition). Except for the position i the new number always has all bits from $B_\alpha(S)$ and $B_\alpha(S')$ set and maybe some more. The bit on position i is set only when $B_\alpha(S)$ or $B_\alpha(S')$ has a bit set on a position smaller than i and in such case, the value can always be at least the maximum of $B_\alpha(S)$ and $B_\alpha(S')$. The decrease in value, therefore, happens when $LSO(LSO(B_\alpha(S)) = LSO(B_\alpha(S')) = i$ and both children don't have disjoint bits set. In the case of transition on b_2, all states are preserved when $B_\alpha(S)$ and $B_\alpha(S')$ are not zero. □

The height of the smallest strongly synchronizing term will be shown to be 2^α. However, based on how α is defined, an exact formula for the height with given n would be very complicated and not very interesting. We will show asymptotic bounds for it instead.

Lemma 3. *For $n \geq 1$ and A_n^- applies $2^\alpha = \Theta(2^{n-\sqrt{n}})$.*

Proof (sketch). Using $(\beta - 1)^2 \leq \alpha \leq \beta^2$ we lower and upper bound the "worst case" and solve for 2^α. In both cases substituting α with n results in a value that falls into $\Theta(2^{n-\sqrt{n}})$. □

Finally, we prove the height of the smallest synchronizing term in A_n^-.

Theorem 3. *For $n \geq 1$ the height of the smallest strongly synchronizing term in automaton A_n^- is $2^\alpha = \Theta(2^{n-\sqrt{n}})$ and therefore $ssh(n, m, r) = \Omega(2^{n-\sqrt{n}})$ for alphabet size $m \geq f(n)$ where $f(n) \in \Omega(n - \sqrt{n})$ and maximum arity $r \geq 2$.*

Proof. Any strongly synchronizing term of A_n^- must have in its root symbol b_2 and for sets of states of children S and S' must apply that $B_\alpha(S) = B_\alpha(S') = 0$. This is the only way to get rid of states P_β and since state $\top$ is always present, every synchronizing term must synchronize to this state. Since $B_\alpha(Q) = 2^\alpha - 1$ and based on Lemma 2 any transition can decrease the binary number by at most 1, every variable in the synchronizing term has to be at a position in depth at least 2^α. To achieve a decrease of the binary number at every transition (except for the root), the smallest strongly synchronizing term t of A_n^-, therefore, has at every position $p \in SubtPos(t)$ symbol $a_2^{LSO(depth(p))}$ when $depth(p) > 0$, and symbol b_2 when $depth(p) = 0$ to achieve final synchronization. Therefore $height(t) = 2^\alpha$ and based on Lemma 3 $height(t) = \Theta(2^{n-\sqrt{n}})$. □

5.2   Lower Bound with Quadratic Size Alphabet

The main idea is to represent combinations $\binom{n-1}{k}$ on $n-1$ states and force the synchronizing term to go through every combination starting from $\binom{n-1}{n-1}$, then $\binom{n-1}{n-2}, \binom{n-1}{n-3}, \ldots, \binom{n-1}{1}$, and last $\binom{n-1}{0}$. The last state denoted $\top$ is always on, hence, $\binom{n-1}{0}$ is the only combination that together with $\top$ has only 1 state active and it is the state in which evaluation of any strongly synchronizing term needs to end in.

Now we recall an operation that iterates through all combinations $\binom{n-1}{k}$ for a fixed k. We denote it NEXT(B) and it reorders bits in a binary number B, shown in Fig. 3, such that it takes the maximum block of least significant consecutive ones and moves the most significant one within the block to a bit higher by one and it moves all the remaining ones to the least significant bits. The number representing the first combination has all ones aligned in the least significant bits.

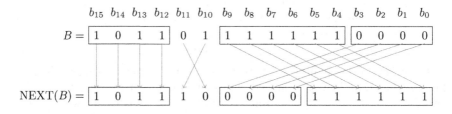

Fig. 3. An example application of the NEXT operation on a binary number.

The last binary number we can create with the NEXT operation has all ones aligned in the most significant bits. For such a number then the operation is undefined. We denote NEXT*(B) a set of binary numbers that is a closure of $\{B\}$ on the NEXT operation, i.e., all numbers following from B onward.

Remark 2. Let B and B' be binary numbers with the same number of bits set to 1. If $B' \geq B$ then $B' \in$ NEXT*(B).

To lower k we shall use a FLIP operation that assumes that all 1 bits are all in the block of most significant bits and results in a number that has one less bits set to 1 and all of them are in a block of least significant bits.

Definition 5 (Combination DFTA). *For $n \geq 1$ let* combination DFTA *denoted as $A_n^c = (Q, \mathcal{F}, \Delta)$ be such that $Q = \{\top, s_{n-2}, \ldots, s_0\}$, $\mathcal{F} = \{f_2\} \cup \{a_2^{i,j} \mid n-2 > i \geq j \geq 0\}$ all with arity 2, and let Δ be defined as in Table 2.*

Let $B(S)$ be a binary representation of states $S = s_{n-2}, \ldots, s_0$ after evaluation of $\widetilde{\Delta}_Q(t_1)$. We will see that transitions $\widetilde{\Delta}_Q(a_2^{i,j}, t_1, t_2)$ for appropriate i and j represent operation NEXT$(B(S))$, $\widetilde{\Delta}_Q(f_2, t_1, t_2)$ represents FLIP$(B(S))$,

Table 2. Transition function Δ of A_n^c. Underlined transitions are not supposed to be used in the smallest strongly synchronizing term and underdashed transitions are redundant for such term. Gray transitions are redundant for the result and default to $\top$.

$\Delta(a_2^{i,j}, \rightarrow, \downarrow)$	$\top$	s_k $k>i+1$	s_{i+1} $j=0$	s_{i+1} $j>0$	s_i $j=0$	s_i $j>0$	s_k $i>k\geq j$	s_{j-1}	s_k $j-1>k$
$\top$	$\top$	s_k	s_i	s_i	s_{i+1}	s_{i+1}	s_{k-j}	s_{i-1}	s_{i-j+k}
rest	$\top$	$\top$	s_{i+1}	s_{i-j}	$\top$	s_0	s_{k-j+1}	s_{i+1}	$s_{i-j+k+1}$

$\Delta(f_2, \rightarrow, \downarrow)$	$\top$	s_{n-2}	$s_k, k<n-2$
$\top$	$\top$	$\top$	s_{n-3-k}
rest	$\top$	$\top$	$s_{n-3-(k+1 \bmod n-2)}$

and that "wrong" use of these symbols does not happen in the smallest strongly synchronizing term.

We can see that the transition function with $a_2^{i,j}$ and $\top$ on the right side maps every state on the left side as in Fig. 3, resulting in the NEXT operation when i and j are correct. Transitions with other states than $\top$ on the right side work with two blocks of states, one from s_i to s_j (where we expect all states to be present) and the other consisting of s_{i+1} and states before s_j (which we don't expect to be present), and each state is cyclically mapped to the next state within the block. If each block is not full or empty, then this operation increases the number of states in the result, and if the first block is empty and the second is full, then the binary number representing the result will be smaller than the left side. A similar thing happens in transitions with f_2.

Now we prove the height of the smallest strongly synchronizing term in automaton A_n^c. Note that the size of the alphabet is $|\mathcal{F}| = 1 + (\binom{n-2}{2} + (n-2)) = \frac{1}{2}(n^2 - 3n + 4)$.

Theorem 4. *For $n \geq 1$ the height of the smallest strongly synchronizing term in automaton A_n^c is $2^{n-1} - 1$ and therefore $\mathrm{ssh}(n, m, r) \geq 2^{n-1}$ for alphabet size $m \geq \frac{1}{2}(n^2 - 3n + 4)$ and maximum arity $r \geq 2$.*

Proof (sketch). The transitions essentially depend only on the left child of every term. The number of ones in B is decreased only if all ones occupy the most significant bits by using FLIP. As NEXT moves to the next combination when used correctly and goes to a state that was already visited when used incorrectly. Hence, we iterate through all 2^{n-1} combinations on $n-1$ states. Term height is one less than the number of states so it is $2^{n-1} - 1$. □

6   Conclusion

We defined weakly and strongly synchronizing terms of a DFTA and proved possible bounds of functions corresponding to the maximum height of the smallest synchronizing term of the respective type.

For weak synchronization, we proved the upper bound $\mathrm{wsh}(n, m, r) < \mathrm{sl}(n) + n$ for $n \geq 1, m \geq 2$ and $r \geq 2$, i.e., it is slightly bigger than the length of the shortest synchronizing string, which is polynomial. We are not sure whether the actual lower bound will be smaller or bigger than the string case, therefore we leave it as an open problem.

For strong synchronization, we proved the upper bound $\mathrm{ssh}(n, m, r) \leq 2^n - n - 1$ for $n \geq 1$, and lower bounds with $r \geq 2$ of $\Omega(2^{n-\sqrt{n}})$ for $m \geq f(n), f(n) \in \Omega(n - \sqrt{n})$, and $2^{n-1} - 1$ for $m \geq \frac{1}{2}(n^2 - 3n + 4)$. A lower bound for constant size alphabet as well as closing the gap in general are topics for further research.

References

1. Comon, H., et al.: Tree automata techniques and applications (Nov 2008). https://jacquema.gitlabpages.inria.fr/files/tata.pdf
2. Martyugin, P.: A lower bound for the length of the shortest carefully synchronizing words. Russian Math. **54**, 46–54 (2010). https://doi.org/10.3103/S1066369X10010056
3. Martyugin, P.V.: Careful synchronization of partial automata with restricted alphabets. In: Bulatov, A.A., Shur, A.M. (eds.) CSR 2013. LNCS, vol. 7913, pp. 76–87. Springer, Heidelberg (2013). https://doi.org/10.1007/978-3-642-38536-0_7
4. Pin, J.E.: On two combinatorial problems arising from automata theory. In: Combinatorial Mathematics, North-Holland Mathematics Studies, vol. 75, pp. 535–548. North-Holland (1983). https://doi.org/10.1016/S0304-0208(08)73432-7
5. Pin, J. (ed.): Handbook of Automata Theory. European Mathematical Society Publishing House, Zürich, Switzerland (2021). https://doi.org/10.4171/Automata
6. Plachý, Š, Janoušek, J.: On synchronizing tree automata and their work–optimal parallel run, usable for parallel tree pattern matching. In: Chatzigeorgiou, A., et al. (eds.) SOFSEM 2020. LNCS, vol. 12011, pp. 576–586. Springer, Cham (2020). https://doi.org/10.1007/978-3-030-38919-2_47
7. Ruszil, J.: Some results concerning careful synchronization of partial automata and subset synchronization of DFA's. In: Implementation and Application of Automata, pp. 106–115. Springer International Publishing, Cham (2022). https://doi.org/10.1007/978-3-031-07469-1_8
8. Rystsov, I.: Asymptotic estimate of the length of a diagnostic word for a finite automaton. Cybern. Syst. Anal. **16**, 194–198 (1980). https://doi.org/10.1007/BF01069104
9. Shabana, H.: Exact synchronization in partial deterministic automata. J. Phys.: Conf. Ser. **1352**(1) (Oct 2019). https://doi.org/10.1088/1742-6596/1352/1/012047
10. Shitov, Y.: An improvement to a recent upper bound for synchronizing words of finite automata. J. Autom. Lang. Comb. **24**(2–4), 367–373 (2019). https://doi.org/10.25596/jalc-2019-367
11. Černý, J.: Poznámka k homogénnym experimentom s konečnými automatmi. Matematicko-fyzikálny časopis **14**(3), 208–216 (1964). https://eudml.org/doc/29801
12. Černý, J., Pirická, A., Rosenauerová, B.: On directable automata. Kybernetika **07**(4), 289–298 (1971). https://eudml.org/doc/28384

Universal First-Order Quantification
over Automata

Bernard Boigelot[(✉)] [iD], Pascal Fontaine [iD], and Baptiste Vergain [iD]

Montefiore Institute, B28, University of Liège, Liège, Belgium
`bernard.boigelot@uliege.be`

Abstract. Deciding formulas that mix arithmetic and uninterpreted predicates is of practical interest, notably for applications in verification. Some decision procedures consist in building by structural induction an automaton that recognizes the set of models of the formula under analysis, and then testing whether this automaton accepts a non-empty language. A drawback is that universal quantification is usually handled by a reduction to existential quantification and complementation. For logical formalisms in which models are encoded as infinite words, this hinders the practical use of this method due to the difficulty of complementing infinite-word automata. The contribution of this paper is to introduce an algorithm for directly computing the effect of universal first-order quantifiers on automata recognizing sets of models, for formulas involving natural numbers encoded in unary notation. This paves the way to implementable decision procedures for various arithmetic theories.

Keywords: Infinite-word automata · first-order logic · quantifier elimination · satisfiability

1 Introduction

Automated reasoning with arithmetic theories is of primary importance, notably for verification, where Satisfiability Modulo Theories (SMT) solvers are regularly used to discharge proof obligations. It is well known however that mixing arithmetic with uninterpreted symbols quickly leads to undecidable languages. For instance, extending Presburger arithmetic, i.e., the first-order additive theory of integer numbers, with just one uninterpreted unary predicate makes it undecidable [7,8,15]. There exist decidable fragments mixing arithmetic and uninterpreted symbols that are expressive enough to be interesting, for instance, the monadic second-order theory of $\mathbb{N}$ under one successor (S1S).

The decidability of S1S has been established thanks to the concept of infinite-word automaton [6]. In order to decide whether a formula φ is satisfiable, the

---

Research reported in this paper was supported in part by an Amazon Research Award, Fall 2022 CFP. Any opinions, findings, and conclusions or recommendations expressed in this material are those of the authors and do not reflect the views of Amazon.

B. Nagy (Ed.): CIAA 2023, LNCS 14151, pp. 91–102, 2023.
https://doi.org/10.1007/978-3-031-40247-0_6

approach consists in building an automaton that recognizes the set of its models, encoded in a suitable way, and then checking that this automaton accepts a non-empty language. Such an automaton has one separate input tape for each first-order and second-order free variable of φ. It is constructed by starting from elementary automata representing the atoms of φ, and then translating the effect of Boolean connectives and quantifiers into corresponding operations over automata. For instance, applying an existential quantifier simply amounts to removing from the automaton the input tape associated to the quantified variable. Universal quantification reduces to existential quantification thanks to the equivalence $\forall x\, \varphi \equiv \neg \exists x\, \neg \varphi$.

Even though this approach has originally been introduced as a purely theoretical tool, it is applied in practice to obtain usable decision procedures for various logics. In particular, the tool MONA [9] uses this method to decide a restricted version of S1S, and tools such as LASH [3] and Shasta [14] use a similar technique to decide Presburger arithmetic. The LASH tool also generalizes this result by providing an implemented decision procedure for the first-order additive theory of mixed integer and real variables [2].

A major issue in practice is that the elimination of universal quantifiers relies on complementation, which is an operation that is not easily implemented for infinite-word automata [13,17]. Actual implementations of automata-based decision procedures elude this problem by restricting the language of interest or the class of automata that need to be manipulated. For instance, the tool MONA only handles *Weak* S1S (WS1S) which is, schematically, a restriction of S1S to finite subsets of natural numbers [5]. The tool LASH handles the mixed integer and real additive arithmetic by working with *weak deterministic automata*, which are a restricted form of infinite-word automata admitting an easy complementation algorithm [2].

The contribution of this paper is to introduce a direct algorithm for computing the effect of universal first-order quantification over infinite-word automata. This is an essential step towards practical decision procedures for more expressive fragments mixing arithmetic with uninterpreted symbols. The considered automata are those that recognize models of formulas over natural numbers encoded in unary notation. This algorithm does not rely on complementation, and can be implemented straightforwardly on unrestricted infinite-word automata. As an example of its potential applications, this algorithm leads to a practically implementable decision procedure for the first-order theory of natural numbers with the order relation and uninterpreted unary predicates. It also paves the way to a decision procedure for SMT solvers for the UFIDL (Uninterpreted Functions and Integer Difference Logic) logic with only unary predicates.

2   Basic Notions

2.1   Logic

We address the problem of deciding satisfiability for formulas expressed in first-order structures of the form $(\mathbb{N}, R_1, R_2, \ldots)$, where $\mathbb{N}$ is the *domain* of natural numbers, and R_1, R_2, ... are (interpreted) *relations* over tuples of values in $\mathbb{N}$. More precisely, each R_i is defined as a relation $R_i \subseteq \mathbb{N}^{\alpha_i}$ for some $\alpha_i > 0$ called the *arity* of R_i.

The formulas in such a structure involve *first-order variables* x_1, x_2, ..., and *second-order variables* X_1, X_2, ... Formulas are recursively defined as

- $\top$, $\bot$, $x_i = x_j$, $X_i = X_j$, $X_i(x_j)$ or $R_i(x_{j_1}, \ldots, x_{j_{\alpha_i}})$, where $i, j, j_1, j_2, \ldots \in \mathbb{N}_{>0}$ (*atomic formulas*),
- $\varphi_1 \wedge \varphi_2$, $\varphi_1 \vee \varphi_2$ or $\neg\varphi$, where φ_1, φ_2 and φ are formulas, or
- $\exists x_i\, \varphi$ or $\forall x_i\, \varphi$, where φ is a formula.

We write $\varphi(x_1, \ldots, x_k, X_1, \ldots, X_\ell)$ to express that $x_1, \ldots, x_k, X_1, \ldots, X_\ell$ are the free variables of φ, i.e., that φ does not involve other unquantified variables.

An *interpretation* I for a formula $\varphi(x_1, \ldots, x_k, X_1, \ldots, X_\ell)$ is an assignment of values $I(x_i) \in \mathbb{N}$ for all $i \in [1, k]$ and $I(X_j) \subseteq \mathbb{N}$ for all $j \in [1, \ell]$ to its free variables. An interpretation I that makes φ true, which is denoted by $I \models \varphi$, is called a *model* of φ.

The semantics is defined in the usual way. One has

- $I \models \top$ and $I \not\models \bot$ for every I.
- $I \models x_i = x_j$ and $I \models X_i = X_j$ iff (respectively) $I(x_i) = I(x_j)$ and $I(X_i) = I(X_j)$.
- $I \models X_i(x_j)$ iff $I(x_j) \in I(X_i)$.
- $I \models R_i(x_{j_1}, \ldots, x_{j_{\alpha_i}})$ iff $(I(x_{j_1}), \ldots, I(x_{j_{\alpha_i}})) \in R_i$.
- $I \models \varphi_1 \wedge \varphi_2$, $I \models \varphi_1 \vee \varphi_2$ and $I \models \neg\varphi$ iff (respectively) $(I \models \varphi_1) \wedge (I \models \varphi_2)$, $(I \models \varphi_1) \vee (I \models \varphi_2)$, and $I \not\models \varphi$.
- $I \models \exists x_i\, \varphi(x_1, \ldots, x_k, X_1, \ldots, X_\ell)$ iff there exists $n \in \mathbb{N}$ such that $I[x_i = n] \models \varphi(x_1, \ldots, x_k, X_1, \ldots, X_\ell)$.
- $I \models \forall x_i\, \varphi(x_1, \ldots, x_k, X_1, \ldots, X_\ell)$ iff for every $n \in \mathbb{N}$, one has $I[x_i = n] \models \varphi(x_1, \ldots, x_k, X_1, \ldots, X_\ell)$.

In the two last rules, the notation $I[x_i = n]$, where $n \in \mathbb{N}$, stands for the extension of the interpretation I to one additional first-order variable x_i that takes the value n, i.e., the interpretation such that $I[x_i = n](x_j) = I(x_j)$ for all $j \in [1, k]$ such that $j \neq i$, $I[x_i = n](x_i) = n$, and $I[x_i = n](X_j) = I(X_j)$ for all $j \in [1, \ell]$.

A formula is said to be *satisfiable* if it admits a model.

2.2   Automata

A finite-word or infinite-word automaton is a tuple $\mathcal{A} = (\Sigma, Q, \Delta, Q_0, F)$ where Σ is a finite *alphabet*, Q is a finite set of *states*, $\Delta \subseteq Q \times (\Sigma \cup \{\varepsilon\}) \times Q$ is a *transition relation*, $Q_0 \subseteq Q$ is a set of *initial states*, and $F \subseteq Q$ is a set of *accepting states*.

A *path* of $\mathcal{A}$ from q_0 to q_m, with $q_0, q_m \in Q$ and $m \geq 0$, is a finite sequence $\pi = (q_0, a_0, q_1); (q_1, a_1, q_2); \ldots; (q_{m-1}, a_{m-1}, q_m)$ of transitions from Δ. The finite word $w \in \Sigma^*$ *read* by π is $w = a_0 a_1 \ldots a_{m-1}$; the existence of such a path is denoted by $q_0 \xrightarrow{w} q_m$. A *cycle* is a non-empty path from a state to itself. If $\mathcal{A}$ is a finite-word automaton, then a path π from q_0 to q_m is *accepting* if $q_m \in F$. A word $w \in \Sigma^*$ is *accepted* from the state q_0 if there exists an accepting path originating from q_0 that reads w.

For infinite-word automata, we use a Büchi acceptance condition for the sake of simplicity, but the results of this paper straightforwardly generalize to other types of infinite-word automata. If $\mathcal{A}$ is an infinite-word automaton, then a *run* of $\mathcal{A}$ from a state $q_0 \in Q$ is an infinite sequence $\sigma = (q_0, a_0, q_1); (q_1, a_1, q_2); \ldots$ of transitions from Δ. This run reads the infinite word $w = a_0 a_1 \ldots \in \Sigma^\omega$. The run σ is *accepting* if the set $inf(\sigma)$ formed by the states q_i that occur infinitely many times in σ is such that $inf(\sigma) \cap F \neq \emptyset$, i.e., there exists a state in F that is visited infinitely often by σ. A word $w \in \Sigma^\omega$ is *accepted* from the state $q_0 \in Q$ if there exists an accepting run from q_0 that reads w.

For both finite-word and infinite-word automata, a word w is *accepted* by $\mathcal{A}$ if it is accepted from an initial state $q_0 \in Q_0$. The set of all words accepted from a state $q \in Q$ (resp. by $\mathcal{A}$) forms the *language* $L(\mathcal{A}, q)$ accepted from q (resp. $L(\mathcal{A})$ accepted by $\mathcal{A}$). An automaton accepting $L(\mathcal{A}, q)$ can be derived from $\mathcal{A}$ by setting Q_0 equal to $\{q\}$. The language of finite-words w read by paths from q_1 to q_2, with $q_1, q_2 \in Q$, is denoted by $L(\mathcal{A}, q_1, q_2)$; a finite-word automaton accepting this language can be obtained from $\mathcal{A}$ by setting Q_0 equal to $\{q_1\}$ and F equal to $\{q_2\}$. A language is said to be *regular* (resp. *ω-regular*) if it can be accepted by a finite-word (resp. an infinite-word) automaton.

3   Deciding Satisfiability

3.1   Encoding Interpretations

In order to decide whether a formula $\varphi(x_1, \ldots, x_k, X_1, \ldots, X_\ell)$ is satisfiable, Büchi introduced the idea of building an automaton that accepts the set of all models of φ, encoded in a suitable way, and then checking whether it accepts a non-empty language [5, 6].

A simple encoding scheme consists in representing the value of first-order variables x_i in *unary notation*: A number $n \in \mathbb{N}$ is encoded by the infinite word $0^n 1 0^\omega$ over the alphabet $\{0, 1\}$, i.e., by a word in which the symbol 1 occurs only once, at the position given by n. This leads to a compatible encoding scheme for the values of second-order variables X_j: a predicate $P \subseteq \mathbb{N}$ is encoded by the infinite word $a_0 a_1 a_2 \ldots$ such that for every $n \in \mathbb{N}$, $a_n \in \{0, 1\}$ satisfies $a_n = 1$ iff $n \in P$, i.e., if $P(n)$ holds.

Encodings for the values of first-order variables $x_1, \ldots, x_k$ and second-order variables $X_1, \ldots, X_\ell$ can be combined into a single word over the alphabet $\Sigma = \{0, 1\}^{k+\ell}$: A word $w \in \Sigma^\omega$ encodes an interpretation I for those variables iff $w = (a_{0,1}, \ldots, a_{0,k+\ell})(a_{1,1}, \ldots, a_{1,k+\ell}) \ldots$, where for each $i \in [1, k]$, $a_{0,i} a_{1,i} \ldots$ encodes $I(x_i)$, and for each $j \in [1, \ell]$, $a_{0,k+j} a_{1,k+j} \ldots$ encodes $I(X_j)$. Note that not all infinite words over Σ form valid encodings: For each first-order variable x_i, an encoding must contain exactly one occurrence of the symbol 1 for the i-th

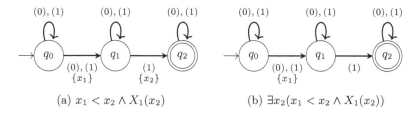

(a) $x_1 < x_2 \wedge X_1(x_2)$ (b) $\exists x_2(x_1 < x_2 \wedge X_1(x_2))$

Fig. 1. Automata recognizing sets of models.

component of its tuple symbols. Assuming that the set of variables is clear from the context, we write $e(I)$ for the encoding of I with respect to those variables.

3.2   Automata Recognizing Sets of Models

Let S be a set of interpretations for k first-order and ℓ second-order variables. The set of encodings of the elements of S forms a language L over the alphabet $\{0,1\}^{k+\ell}$. If this language is ω-regular, then we say that an automaton $\mathcal{A}$ that accepts L *recognizes*, or *represents*, the set S. Such an automaton can be viewed as having $k + \ell$ input tapes reading symbols from $\{0,1\}$, each of these tapes being associated to a variable. Equivalently, we can write the label of a transition $(q_1, (a_1, \ldots, a_{k+\ell}), q_2) \in \Delta$ as $V^{(a_{k+1},\ldots,a_{k+\ell})}$ where V is the set of the variables x_i, with $i \in [1, k]$, for which $a_i = 1$. In other words, each transition label distinct from ε specifies the set of first-order variables whose value corresponds to this transition, and provides one symbol for each second-order variable. For each $x_i \in V$, we then say that x_i is *associated* to the transition. Note that every transition for which $V \neq \emptyset$ can only be followed at most once in an accepting run. Any automaton recognizing a set of valid encodings can therefore easily be transformed into one in which such transitions do not appear in cycles, and that accepts the same language.

An example of an automaton recognizing the set of models of the formula $\varphi(x_1, x_2, X_1) = x_1 < x_2 \wedge X_1(x_2)$ is given in Fig. 1a. For the sake of clarity, labels of transitions sharing the same origin and destination are grouped together, and empty sets of variables are omitted.

3.3   Decision Procedure

For the automata-based approach to be applicable, it must be possible to construct elementary automata recognizing the models of atomic formulas. This is clearly the case for atoms of the form $x_i = x_j$, $X_i = X_j$ and $X_i(x_j)$, and this property must also hold for each relation R_i that belongs to the structure of interest; in other words, the atomic formula $R_i(x_1, x_2, \ldots, x_{\alpha_i})$ must admit a set of models whose encoding is ω-regular. With the positional encoding of natural numbers, this is the case in particular for the order relation $x_i < x_j$ and the successor relation $x_j = x_i + 1$. Note that one can easily add supplementary variables to an automaton, by inserting a new component in the tuples of its alphabet, and making this component read a symbol 1 at any single position of a run for first-order variables, and any symbol at any position for second-order ones. Reordering the variables is a similarly immediate operation.

After automata recognizing the models of atomic formulas have been obtained, the next step consists in combining them recursively by following the syntactic structure of the formula to be decided. Let us denote by L_φ the language of encodings of all the models of a formula φ, i.e., $L_\varphi = \{e(I) \mid I \models \varphi\}$.

For the Boolean operator $\wedge$, we have $L_{\varphi_1 \wedge \varphi_2} = L_{\varphi_1} \cap L_{\varphi_2}$, where φ_1 and φ_2 are formulas over the same free variables. Similarly, we have $L_{\varphi_1 \vee \varphi_2} = L_{\varphi_1} \cup L_{\varphi_2}$. The case of the complement operator $\neg$ is slightly more complicated, since the complement of a language of encodings systematically contains words that do not validly encode an interpretation. The set of models of a formula $\neg\varphi$ is encoded by the language $\overline{L_\varphi} \cap L_{valid}$, where L_{valid} is the language of all valid encodings consistent with the free variables of φ. It is easily seen that this language is ω-regular.

It remains to compute the effect of quantifiers. The language $L_{\exists x_i \varphi}$ can be derived from L_φ by removing the i-th component from each tuple symbol, i.e., by applying a mapping $\Pi_{\neq i} : \Sigma^{k+\ell} \to \Sigma^{k+\ell-1} : (a_1, \ldots, a_{k+\ell}) \mapsto (a_1, \ldots, a_{i-1}, a_{i+1}, \ldots, a_{k+\ell})$ to each symbol of the alphabet. Indeed, the models of $\exists x_i \, \varphi$ correspond exactly to the models of φ in which the variable x_i is removed. In the rest of this paper, we will use the notation $\Pi_{\neq i}(w)$, where w is a finite or infinite word, to express the result of applying $\Pi_{\neq i}$ to each symbol in w. If L is a language, then we write $\Pi_{\neq i}(L)$ for the language $\{\Pi_{\neq i}(w) \mid w \in L\}$.

Finally, universal quantification can be reduced to existential quantification: For computing $L_{\forall x_i \varphi}$, we use the equivalence $\forall x_i \, \varphi \equiv \neg \exists x_i \, \neg \varphi$ which yields $L_{\forall x_i \varphi} = \overline{L_{\exists x_i \neg \varphi}} \cap L_{valid}$.

3.4   Operations over Automata

We now discuss how the operations over languages mentioned in Sect. 3.3 can be computed over infinite-word automata. Given automata $\mathcal{A}_1$ and $\mathcal{A}_2$, automata $\mathcal{A}_1 \cap \mathcal{A}_2$ and $\mathcal{A}_1 \cup \mathcal{A}_2$ accepting respectively $L(\mathcal{A}_1) \cap L(\mathcal{A}_2)$ and $L(\mathcal{A}_1) \cup L(\mathcal{A}_2)$ can be obtained by the so-called product construction. The idea consists in building an automaton $\mathcal{A}$ that simulates the combined behavior of $\mathcal{A}_1$ and $\mathcal{A}_2$ on identical input words. The states of $\mathcal{A}$ need to store additional information about the accepting states that are visited in $\mathcal{A}_1$ and $\mathcal{A}_2$. For $\mathcal{A}_1 \cap \mathcal{A}_2$, one ensures that each accepting run of $\mathcal{A}$ correspond to an accepting run in both $\mathcal{A}_1$ and $\mathcal{A}_2$. For $\mathcal{A}_1 \cup \mathcal{A}_2$, the condition is that the run should be accepting in $\mathcal{A}_1$ or $\mathcal{A}_2$, or both. A complete description of the product construction for Büchi automata is given in [16].

Modifying the alphabet of an automaton in order to implement the effect of an existential quantification is a simple operation. As an example, Fig. 1b shows an automaton recognizing the set of models of $\exists x_2(x_1 < x_2 \wedge X_1(x_2))$, obtained by removing all occurrences of the variable x_2 from transition labels. Testing whether the language accepted by an automaton is not empty amounts to checking the existence of a reachable cycle that visits at least one accepting state, which is simple as well.

The only problematic operation is complementation, which consists in computing from an automaton $\mathcal{A}$ an automaton that accepts the language $\overline{L(\mathcal{A})}$. Although it preserves ω-regularity, this operation is difficult to perform on Büchi automata [13,17]. In the context of our decision procedure, it is only useful for applying universal quantifiers. Indeed, other instances of the negation operator in formulas can be pushed inwards until they are applied to atomic formulas, and it is easy to construct the complement of the elementary automata recognizing the models of those atomic formulas, provided that for each relation R_i in the structure of interest, an automaton recognizing $\{(x_1, \ldots, x_{\alpha_i}) \in \mathbb{N}^{\alpha_i} \mid \neg R_i(x_1, \ldots, x_{\alpha_i})\}$ is available. In order to eliminate the need for complementation, we develop in the next section a direct algorithm for computing the effect of universal quantifiers on automata recognizing sets of models.

4   Universal Quantification

4.1   Principles

Let $\varphi(x_1, \ldots, x_k, X_1, \ldots, X_\ell)$, with $k > 0$ and $\ell \geq 0$, be a formula. Our goal is to compute an automaton $\mathcal{A}'$ accepting $L_{\forall x_i \varphi}$, given an automaton $\mathcal{A}$ accepting L_φ and $i \in [1, k]$.

By definition of universal quantification, we have $I \models \forall x_i \, \varphi$ iff $I[x_i = n] \models \varphi$ holds for every $n \in \mathbb{N}$. In other words, $L_{\forall x_i \varphi}$ contains $e(I)$ iff L_φ contains $e(I[x_i = n])$ for every $n \in \mathbb{N}$. Conceptually, we can then obtain $L_{\forall x_i \varphi}$ by defining for each $n \in \mathbb{N}$ the language $S_n = \{e(I) \mid e(I[x_i = n]) \in L_\varphi\}$, which yields $L_{\forall x_i \varphi} = \bigcap_{n \in \mathbb{N}} S_n$.

An automaton $\mathcal{A}'$ accepting $L_{\forall x_i \varphi}$ can be obtained as follows. Each language S_n, with $n \in \mathbb{N}$, is accepted by an automaton $\mathcal{A}_n$ derived from $\mathcal{A}$ by restricting the transitions associated to x_i to be followed only after having read exactly n symbols. In other words, the accepting runs of $\mathcal{A}_n$ correspond to the accepting runs of $\mathcal{A}$ that satisfy this condition. After imposing this restriction, the variable x_i is removed from the set of variables managed by the automaton, i.e., the operator $\Pi_{\neq i}$ is applied to the language that this automaton accepts, so as to get $S_n = L(\mathcal{A}_n)$. The automaton $\mathcal{A}'$ then corresponds to the infinite intersection product of the automata $\mathcal{A}_n$ for all $n \in \mathbb{N}$, i.e., an automaton that accepts the infinite intersection $\bigcap_{n \in \mathbb{N}} S_n$. We show in the next section how to build $\mathcal{A}'$ by means of a finite computation.

4.2   Construction

The idea of the construction is to make $\mathcal{A}'$ simulate the join behavior of the automata $\mathcal{A}_n$, for all $n \in \mathbb{N}$, on the same input words. This can be done by making each state of $\mathcal{A}'$ correspond to one state q_n in each $\mathcal{A}_n$, i.e., to an infinite tuple $(q_0, q_1, \ldots)$. By definition of $\mathcal{A}_n$, there exists a mapping $\mu : Q_n \to Q$, where Q_n and Q are respectively the sets of states of $\mathcal{A}_n$ and $\mathcal{A}$, such that whenever a run of $\mathcal{A}_n$ visits q_n, the corresponding run of $\mathcal{A}$ on the same input word visits $\mu(q_n)$.

If two automata $\mathcal{A}_{n_1}$ and $\mathcal{A}_{n_2}$, with $n_1, n_2 \in \mathbb{N}$, are (respectively) in states q_{n_1} and q_{n_2} such that $\mu(q_{n_1}) = \mu(q_{n_2})$, then they share the same future behaviors, except for the requirement to follow a transition associated to x_i after having read (respectively) n_1 and n_2 symbols. It follows that the states of $\mathcal{A}'$ can be characterized by sets of states of $\mathcal{A}$: The infinite tuple $(q_0, q_1, \ldots)$ is described by the set $\{\mu(q_i) \mid i \in \mathbb{N}\}$. Each element of this set represents the current state of one or several automata among the $\mathcal{A}_n$. This means that the number of these automata that are in this current state is not counted. We will establish that this abstraction is precise and leads to a correct construction.

During a run of $\mathcal{A}'$, each transition with a label other than ε must correspond to a transition reading the same symbol in every automaton $\mathcal{A}_n$, which in turn can be mapped to a transition of $\mathcal{A}$. In the automaton $\mathcal{A}_n$ for which n is equal to the number of symbols already read during the run, this transition of $\mathcal{A}$ is necessarily associated to x_i, by definition of $\mathcal{A}_n$. It follows that every transition of $\mathcal{A}'$ with a non-empty label is characterized by a set of transitions of $\mathcal{A}$, among which one of them is associated to x_i.

We are now ready to describe formally the construction of $\mathcal{A}'$, leaving for the next section the problem of determining which of its runs should be accepting or not: From the automaton $\mathcal{A} = (\Sigma, Q, \Delta, Q_0, F)$, we construct $\mathcal{A}' = (\Sigma', Q', \Delta', Q'_0, F')$ such that

- $\Sigma' = \Pi_{\neq i}(\Sigma)$.
- $Q' = 2^Q \setminus \{\emptyset\}$.
- Δ' contains
  - the transitions $(q'_1, (a'_1, \ldots, a'_{k+\ell-1}), q'_2)$ for which there exists a set $T \subseteq \Delta$ that satisfies the following conditions:
    * $q'_1 = \{q_1 \mid (q_1, (a_1, \ldots, a_{k+\ell}), q_2) \in T\}$.
    * $q'_2 = \{q_2 \mid (q_1, (a_1, \ldots, a_{k+\ell}), q_2) \in T\}$.
    * For all $(q_1, (a_1, \ldots, a_{k+\ell}), q_2) \in T$, one has $a'_j = a_j$ for all $j \in [1, i-1]$, and $a'_j = a_{j+1}$ for all $j \in [i, k+\ell-1]$.
    * There exists exactly one $(q_1, (a_1, \ldots, a_{k+\ell}), q_2) \in T$ such that $a_i = 1$.
  - the transitions $(q'_1, \varepsilon, q'_2)$ for which there exists a transition $(q_1, \varepsilon, q_2) \in \Delta$ such that
    * $q_1 \in q'_1$.
    * $q'_2 = q'_1 \cup \{q_2\}$ or $q'_2 = (q'_1 \setminus \{q_1\}) \cup \{q_2\}$.
- $Q'_0 = 2^{Q_0} \setminus \{\emptyset\}$.
- $F' = Q'$ for now. The problem of characterizing more finely the accepting runs will be addressed in the next section.

The rule for the transitions $(q'_1, (a'_1, \ldots, a'_{k+\ell-1}), q'_2)$ ensures that for each $q_1 \in q'_1$, each automaton $\mathcal{A}_n$ that is simulated by $\mathcal{A}'$ has the choice of following any possible transition originating from q_1 that has a label consistent with $(a'_1, \ldots, a'_{k+\ell-1})$. One such automaton must nevertheless follow a transition

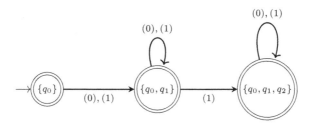

Fig. 2. First step of construction for $\forall x_1 \exists x_2 (x_1 < x_2 \wedge X_1(x_2))$.

associated to the quantified variable x_i. The rule for the transitions $(q'_1, \varepsilon, q'_2)$ expresses that one automaton $\mathcal{A}_n$, or any number of identical copies of this automaton, must follow a transition labeled by ε, while the other automata stay in their current state.

As an example, applying this construction to the automaton in Fig. 1b, as a first step of the computation of a representation of $\forall x_1 \exists x_2 (x_1 < x_2 \wedge X_1(x_2))$, yields the automaton given in Fig. 2. For the sake of clarity, unreachable states and states from which the accepted language is empty are not depicted.

4.3   A Criterion for Accepting Runs

The automaton $\mathcal{A}'$ defined in the previous section simulates an infinite combination of automata $\mathcal{A}_n$, for all $n \in \mathbb{N}$. By construction, every accepting run of this infinite combination corresponds to a run of $\mathcal{A}'$.

The reciprocal property is not true, in the sense that there may exist a run of $\mathcal{A}'$ that does not match an accepting run of the infinite combination of automata $\mathcal{A}_n$. Consider for instance a run of the automaton in Fig. 2 that ends up cycling in the state $\{q_0, q_1, q_2\}$, reading 0^ω from that state. Recall that for this example, the automaton $\mathcal{A}$ that undergoes the universal quantification operation is the one given in Fig. 1b. The run that we have considered can be followed in $\mathcal{A}'$, but cannot be accepting in every $\mathcal{A}_n$. Indeed, in this example, the transition of $\mathcal{A}_n$ reading the $(n + 1)$-th symbol of the run corresponds, by definition of this automaton, to the transition of $\mathcal{A}$ that is associated to the quantified variable x_1. By the structure of $\mathcal{A}$, this transition is necessarily followed later in any accepting run by one that reads the symbol 1, which implies that no word of the form $u \cdot 0^\omega$, with $u \in \{0, 1\}^*$, can be accepted by a run of $\mathcal{A}_n$ such that $n \geq |u|$. This represents the fact that the words accepted by all $\mathcal{A}_n$ correspond to the encodings of predicates that are true infinitely often.

One thus needs a criterion for characterizing the runs of $\mathcal{A}'$ that correspond to combinations of accepting runs in all automata $\mathcal{A}_n$.

It is known [11] that two ω-regular languages over the alphabet Σ are equal iff they share the same set of *ultimately periodic* words, i.e., words of the form $u \cdot v^\omega$ with $u \in \Sigma^*$ and $v \in \Sigma^+$. It follows that it is sufficient to characterize the accepting runs of $\mathcal{A}'$ that read ultimately periodic words. The automaton $\mathcal{A}'$ accepts a word $u \cdot v^\omega$ iff every $\mathcal{A}_n$, with $n \in \mathbb{N}$, admits an accepting run that reads this word. Note that such a run also matches a run of $\mathcal{A}$, and that this run of $\mathcal{A}$ always ends up following a cycle from an accepting state to itself.

Our solution takes the following form. For each state q of $\mathcal{A}$, we define a language $U_q \subseteq \Sigma^+$ of non-empty words u such that $\mathcal{A}$ accepts u^ω from q, after dismissing the input tape associated to the quantified variable x_i. The alphabet Σ is thus equal to $\{0,1\}^{k+\ell-1}$. Remember that each state q' of $\mathcal{A}'$ is defined as a subset of states of $\mathcal{A}$, corresponding to the current states in the combination of copies of $\mathcal{A}$ that are jointly simulated by $\mathcal{A}'$. In order for the word u^ω to be accepted by $\mathcal{A}'$ from q', it should therefore be accepted by $\mathcal{A}$ from each state $q \in q'$, i.e., u must belong to all the languages U_q such that $q \in q'$.

It must also be possible to read u^ω from the state q' of $\mathcal{A}'$. We impose a stronger condition, by requiring that there exists a cycle from q' to itself labeled by u. This condition leads to a correct acceptance criterion.

In summary, the language $U'_{q'} = L(\mathcal{A}', q', q') \cap \bigcap_{q \in q'} U_q$ characterizes the words u such that u^ω must be accepted from the state q' of $\mathcal{A}'$. Note that for this property to hold, it is not necessary for the language U_q to contain all words u such that $u^\omega \in \Pi_{\neq i}(L(\mathcal{A}, q))$, but only some number of copies u^p, where $p > 0$ is bounded, of each such u. In other words, the finite words u whose infinite repetition is accepted from q do not have to be the shortest possible ones.

Once the language $U'_{q'}$ has been obtained, we build a *widget*, in the form of an infinite-word automaton accepting $(U'_{q'})^\omega$, along the state q' of $\mathcal{A}'$, and add a transition labeled by ε from q' to the initial state of this widget. This ensures that every path that ends up in q' can be suitably extended into an accepting run. Such a widget does not have to be constructed for every state q' of $\mathcal{A}'$: Since the goal is to accept from q' words of the form u^ω, we can require that at least one state $q \in q'$ is accepting in $\mathcal{A}$. We then only build widgets for the states q' that satisfy this requirement.

4.4 Computation Steps

The procedure for modifying $\mathcal{A}'$ in order to make it accept the runs that match those of the infinite combination of automata $\mathcal{A}_n$, outlined in the previous section, can be carried out by representing the regular languages U_q and $U'_{q'}$ by finite-state automata. The construction proceeds as follows:

1. For each state $q \in Q$ of $\mathcal{A}$, build a finite-word automaton $\mathcal{A}_q$ that accepts all the non-empty words u for which there exists a path $q \xrightarrow{v} q$ of $\mathcal{A}$ that visits at least one accepting state $q_F \in F$, such that $u = \Pi_{\neq i}(v)$. This automaton can be constructed in a similar way as one accepting $\Pi_{\neq i}(L(\mathcal{A}, q, q))$ (cf. Sects 2.2 and 3.4), keeping one additional bit of information in its states for determining whether an accepting state has already been visited or not.
2. For each pair of states $q_1, q_2 \in Q$ of $\mathcal{A}$, build a finite-word automaton $\mathcal{A}_{q_1, q_2}$ accepting the language $\Pi_{\neq i}(L(\mathcal{A}, q_1, q_2))$ (cf. Sects 2.2 and 3.4).
3. For each state $q \in Q$ of $\mathcal{A}$, build an automaton $\mathcal{A}_{U_q} = \bigcup_{r \in Q} (\mathcal{A}_{q,r} \cap \mathcal{A}_r)$ accepting the finite-word language U_q.

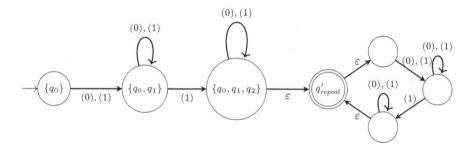

Fig. 3. Automaton recognizing the set of models of $\forall x_1 \exists x_2 (x_1 < x_2 \wedge X_1(x_2))$.

4. For each state q' of $\mathcal{A}'$ such that $q' \cap F \neq \emptyset$, where F is the set of accepting states of $\mathcal{A}$, build a finite-word automaton $\mathcal{A}'_{U'_{q'}} = \mathcal{A}'_{q'} \cap \bigcap_{q \in q'} \mathcal{A}_{U_q}$ accepting $U'_{q'}$, where $\mathcal{A}'_{q'}$ is an automaton accepting $L(\mathcal{A}', q', q')$ (cf. Sect. 2.2).

5. Then, turn each automaton $\mathcal{A}'_{U'_{q'}}$ into an infinite-word automaton $\mathcal{A}'_{(U'_{q'})^\omega}$ accepting $(U'_{q'})^\omega$:
   (a) Create a new state q'_{repeat}.
   (b) Add a transition $(q'_{repeat}, \varepsilon, q_0)$ for each initial state q_0, and a transition $(q_F, \varepsilon, q'_{repeat})$ for each accepting state q_F, of $\mathcal{A}'_{U'_{q'}}$.
   (c) Make q'_{repeat} the only initial and accepting state of $\mathcal{A}'_{(U'_{q'})^\omega}$.

6. For each state q' of $\mathcal{A}'$ considered at Step 4, add the widget $\mathcal{A}'_{U'_{q'}}$ alongside q', by incorporating its sets of states and transitions into those of $\mathcal{A}'$, and adding a transition $(q', \varepsilon, q'_{repeat})$. In the resulting automaton, mark as the only accepting states the states q'_{repeat} of all widgets.

This procedure constructs an automaton that accepts the language $L_{\forall x_i \varphi}$. Applied to the automaton $\mathcal{A}'$ in Fig. 2, it produces the result shown in Fig. 3. For the sake of clarity, the states from which the accepted language is empty have been removed. A detailed description of the computation steps for this example and the proof of correctness of the construction are given in [1].

5   Conclusions

This paper introduces a method for directly computing the effect of a first-order universal quantifier on an infinite-word automaton recognizing the set of models of a formula. It is applicable when the first-order variables range over the natural numbers and their values are encoded in unary notation. Among its potential applications, it provides a solution for deciding the first-order theory $\langle \mathbb{N}, < \rangle$ extended with uninterpreted unary predicates.

  The operation on regular languages that corresponds to the effect of a universal first-order quantifier has already been studied at the theoretical level [12]. Our contribution is to provide a practical algorithm for computing it, that does not require to complement infinite-word automata. This algorithm has an exponential worst-case time complexity, which is unavoidable since there exist automata for which universal quantification incurs an exponential blowup in their number of states (see [1] for details). The main advantage over the complementation-based approach is however that this exponential cost is not systematic, since only a fraction of the possible subsets of states typically need to be constructed.

Our solution is open to many possible improvements, one of them being to extend the algorithm so as to quantify over several first-order variables in a single operation. For future work, we plan to generalize this algorithm to automata over more expressive structures, such as the automata over linear orders defined in [4]. This would make it possible to obtain an implementable decision procedure for, e.g., the first-order theory $\langle \mathbb{R}, < \rangle$ with uninterpreted unary predicates [10].

References

1. Boigelot, B., Fontaine, P., Vergain, B.: Universal first-order quantification over automata. arXiv:2306.04210 [cs.LO] (2023)
2. Boigelot, B., Jodogne, S., Wolper, P.: An effective decision procedure for linear arithmetic over the integers and reals. ACM Tr. Comp. Logic **6**(3), 614–633 (2005)
3. Boigelot, B., Latour, L.: Counting the solutions of Presburger equations without enumerating them. Theo. Comp. Sc. **313**(1), 17–29 (2004)
4. Bruyère, V., Carton, O.: Automata on linear orderings. J. Comput. Syst. Sci. **74**(1), 1–24 (2007)
5. Büchi, J.R.: Weak second-order arithmetic and finite automata. Math. Logic Q. **6**(1–6), 66–92 (1960)
6. Büchi, J.R.: On a decision method in restricted second order arithmetic. In: Proceedings International Congress on Logic, Methodology and Philosophy of Science, pp. 1–12 (1962)
7. Downey, P.J.: Undecidability of Presburger arithmetic with a single monadic predicate letter. Harvard University, Technical Report (1972)
8. Halpern, J.Y.: Presburger arithmetic with unary predicates is Π_1^1 complete. Journal Symbolic Logic **56**(2), 637–642 (1991)
9. Klarlund, N.: Mona & Fido: The logic-automaton connection in practice. In: Nielsen, M., Thomas, W. (eds.) CSL 1997. LNCS, vol. 1414, pp. 311–326. Springer, Heidelberg (1998). https://doi.org/10.1007/BFb0028022
10. Läuchli, H., Leonard, J.: On the elementary theory of linear order. Fundamenta Mathematicae **59**(1), 109–116 (1966)
11. McNaughton, R.: Testing and generating infinite sequences by a finite automaton. Inf. Control **9**(5), 512–530 (1966)
12. Okhotin, A.: The dual of concatenation. Theo. Comp. Sc. **345**(2–3), 425–447 (2005)
13. Safra, S.: On the complexity of omega-automata. In: Proceedings of 29th FOCS, pp. 319–327. IEEE Computer Society (1988)
14. Shiple, T.R., Kukula, J.H., Ranjan, R.K.: A comparison of Presburger engines for EFSM reachability. In: Hu, A.J., Vardi, M.Y. (eds.) CAV 1998. LNCS, vol. 1427, pp. 280–292. Springer, Heidelberg (1998). https://doi.org/10.1007/BFb0028752
15. Speranski, S.O.: A note on definability in fragments of arithmetic with free unary predicates. Arch. Math. Logic **52**(5–6), 507–516 (2013)
16. Thomas, W.: Automata on infinite objects. In: Handbook of Theoretical Computer Science, Volume B, pp. 133–191. Elsevier and MIT Press (1990)
17. Vardi, M.Y.: The Büchi complementation saga. In: Thomas, W., Weil, P. (eds.) STACS 2007. LNCS, vol. 4393, pp. 12–22. Springer, Heidelberg (2007). https://doi.org/10.1007/978-3-540-70918-3_2

Average Complexity of Partial Derivatives for Synchronised Shuffle Expressions

Sabine Broda⬤, António Machiavelo⬤, Nelma Moreira$^{(\boxtimes)}$⬤, and Rogério Reis⬤

CMUP & DM-DCC, Faculdade de Ciências da Universidade do Porto,
Rua do Campo Alegre, 4169-007 Porto, Portugal
{sabine.broda,antonio.machiavelo,nelma.moreira,
rogerio.reis}@fc.up.pt

Abstract. Synchronised shuffle operators allow to specify symbols on which the operands must or can synchronise instead of interleave. Recently, partial derivative and position based automata for regular expressions with synchronised shuffle operators were introduced. In this paper, using the framework of analytic combinatorics, we study the asymptotic average state complexity of partial derivative automata for regular expressions with strongly and arbitrarily synchronised shuffles. The new results extend and improve the ones previously obtained for regular expressions with shuffle and intersection. For intersection, asymptotically the average state complexity of the partial derivative automaton is 3, which significantly improves the known exponential upper-bound.

1 Introduction

Synchronised shuffle operators allow to specify symbols on which the operands must or can synchronise instead of interleave. Intersection and shuffle can be seen as two extreme cases, corresponding to strict synchronisation and pure interleaving. Several variants were introduced and studied by ter Beek et al. [3], motivated by modelling synchronisation in concurrent systems or certain gene operations in molecular biology. Sulzmann and Thiemann [14] studied regular expressions with a general synchronised shuffling operator and extended the notions of derivatives and partial derivatives to these expressions. Broda et al. [7] defined a location based position automaton for regular expressions with strongly, arbitrarily, and weakly synchronised operators and showed that the partial derivative automaton (defined in [14]) is a quotient of the position automaton. For a standard regular expression α, the partial derivative automaton $\mathcal{A}_{PD}(\alpha)$, introduced by Antimirov [1], can also be obtained by solving a system of equations, whose solution is a support set $\pi(\alpha)$, due to Mirkin [12]. In fact, the set of states of

---

This work was partially supported by CMUP, which is financed by national funds through FCT under the project with reference UIDB/00144/2021.

B. Nagy (Ed.): CIAA 2023, LNCS 14151, pp. 103–115, 2023.
https://doi.org/10.1007/978-3-031-40247-0_7

$\mathcal{A}_{PD}(\alpha)$ is equal to $\pi(\alpha) \cup \{\alpha\}$. In Bastos et al. [2], the rules for computing $\pi(\alpha)$ were extended to regular expressions with intersection. However, in that case it was shown that both constructions were not identical, as Mirkin's construction could have some states that were not accessible. Still, the inductive definition of a support set is essential to obtain average complexity results using the framework of analytic combinatorics [2, 4–6]. In the worst-case, the number of states of a partial derivative automaton for expressions with shuffle or intersection can be exponential in the size of the expression. In the average-case, the known upper bounds are smaller, but still exponential [2,5]. In this paper, we define new rules for computing the support for a system of expression equations, where for each alphabet symbol the solution is independently computed. In the case of standard regular expressions, the support obtained by these rules is the same as the one defined by Mirkin. However, the new rules allow to consider synchronising and non-synchronising symbols separately, and thereby obtain a support for a regular expression with strongly and arbitrarily synchronised operators. Moreover, the new rules lead to a smaller support for the intersection operator [2], which corresponds to the strong synchronised operator when all alphabet symbols synchronise. Using the framework of analytic combinatorics, we give an upper bound for the asymptotic size of the support. We also estimate the asymptotic average number of partial derivatives by one symbol. In particular, for intersection we show that asymptotically, as the size of the alphabet grows, the size of the support set, thus the average state complexity of $\mathcal{A}_{PD}$, is 3. This is a surprising improvement with regard to the previous upper bound of $(1.056 + o(1))^n$, where n is the size of the expression [2]. However, experimental results in [2], suggested that the size of $\mathcal{A}_{PD}$ as the alphabet size grows could approach the constant 3.

2   Regular Expressions with Synchronised Shuffles

Let $\Sigma = \{\sigma_1, \ldots, \sigma_m\}$ be an alphabet and $\Sigma^\star$ be the set of words over Σ. A *language* is any subset of $\Sigma^\star$. The *empty word* is denoted by ε. The set of alphabet symbols that occur in a word $w \in \Sigma^\star$ is denoted by Σ_w. Given a set $\Gamma \subseteq \Sigma$, the *strongly synchronised shuffle* of two words w.r.t. Γ imposes synchronisation on all symbols of Γ. Formally, for $u, v \in \Sigma^\star$, $u\,{}^{s}\|_{\Gamma}\, v$ is defined inductively as follows [14]:

$$\varepsilon\,{}^{s}\|_{\Gamma}\, v = v\,{}^{s}\|_{\Gamma}\,\varepsilon = \begin{cases} \{v\}, & \text{if } \Sigma_v \cap \Gamma = \emptyset, \\ \emptyset, & \text{otherwise,} \end{cases}$$

$$\sigma u\,{}^{s}\|_{\Gamma}\, \tau v = \begin{cases} \{\, \sigma w \mid w \in u\,{}^{s}\|_{\Gamma}\, v \,\}, & \text{if } \sigma = \tau \wedge \sigma \in \Gamma, \\ \emptyset, & \text{if } \sigma \neq \tau \wedge \sigma, \tau \in \Gamma, \\ \{\, \sigma w \mid w \in u\,{}^{s}\|_{\Gamma}\, \tau v \,\}, & \text{if } \sigma \notin \Gamma \wedge \tau \in \Gamma, \\ \{\, \tau w \mid w \in \sigma u\,{}^{s}\|_{\Gamma}\, v \,\}, & \text{if } \sigma \in \Gamma \wedge \tau \notin \Gamma, \\ \{\, \sigma w \mid w \in u\,{}^{s}\|_{\Gamma}\, \tau v \,\} \\ \quad \cup \{\, \tau w \mid w \in \sigma u\,{}^{s}\|_{\Gamma}\, v \,\}, & \text{if } \sigma, \tau \notin \Gamma. \end{cases}$$

For $\Gamma = \emptyset$ the operator $^s\|_\emptyset$ coincides with the usual shuffle operator $\sqcup\!\sqcup$, given by $u\sqcup\!\sqcup\varepsilon = \varepsilon\sqcup\!\sqcup u = \{u\}$ and $\sigma u\sqcup\!\sqcup\tau v = \{\sigma w \mid w \in u\sqcup\!\sqcup\tau v\} \cup \{\tau w \mid w \in \sigma u\sqcup\!\sqcup v\}$, for $u, v \in \Sigma^\star$ and $\sigma, \tau \in \Sigma$. On the other hand, if $\Gamma = \Sigma$, the operator $^s\|_\Sigma$ corresponds to intersection ($\cap$). The *arbitrarily synchronised shuffle* of two words w.r.t. $\Gamma \subseteq \Sigma$ permits symbols in Γ to synchronise, but does not force their synchronisation. Formally, for $u, v \in \Sigma^\star$, $u\,^a\|_\Gamma v$, is defined as follows [3]:

$$\varepsilon\,^a\|_\Gamma v = v\,^a\|_\Gamma \varepsilon = \{v\},$$

$$\sigma u\,^a\|_\Gamma \tau v = \begin{cases} \{\sigma w \mid w \in u\,^a\|_\Gamma \tau v\} \cup \{\tau w \mid w \in \sigma u\,^a\|_\Gamma v\}, & \text{if } \sigma \neq \tau \vee \sigma \notin \Gamma, \\ \{\sigma w \mid w \in u\,^a\|_\Gamma \tau v\} \cup \{\tau w \mid w \in \sigma u\,^a\|_\Gamma v\} \\ \quad \cup \{\sigma w \mid w \in u\,^a\|_\Gamma v\}, & \text{if } \sigma = \tau \wedge \sigma \in \Gamma. \end{cases}$$

Example 1. We have $abca\,^s\|_{\{a\}}\,ada = \{abcda, abdca, adbca\}$, $ab\,^s\|_{\{a\}}\,da = \{dab\}$, and $ab\,^a\|_{\{a\}}\,da = \{abda, adba, adab, dab, daab, daba\}$.

Given two languages $L_1, L_2 \subseteq \Sigma^\star$ and $\circ \in \{^s\|_\Gamma, ^a\|_\Gamma\}$ one has, as usual, $L_1 \circ L_2 = \bigcup_{u\in L_1, v\in L_2} u \circ v$. If L_1 and L_2 are regular, $L_1 \circ L_2$ is regular [3]. The set of regular expressions with synchronised shuffles over the alphabet Σ, RE($\|$), contains $\emptyset$ and is generated by the following grammar

$$\alpha \to \varepsilon \mid \sigma \in \Sigma \mid (\alpha + \alpha) \mid (\alpha\alpha) \mid (\alpha^\star) \mid (\alpha\,^s\|_\Gamma \alpha) \mid (\alpha\,^a\|_\Gamma \alpha). \tag{1}$$

Let RE be the subset of RE($\|$) of standard regular expressions without the operators $^s\|_\Gamma$ and $^a\|_\Gamma$. The *language* associated with an expression $\alpha \in$ RE($\|$) is denoted by $\mathcal{L}(\alpha)$, which for $\alpha \in$ RE is defined as usual, and $\mathcal{L}(\alpha_1\circ\alpha_2) = \mathcal{L}(\alpha_1)\circ \mathcal{L}(\alpha_2)$ for $\circ \in \{^s\|_\Gamma, ^a\|_\Gamma\}$. Two regular expressions α and β are *equivalent*, and we write $\alpha \doteq \beta$, iff $\mathcal{L}(\alpha) = \mathcal{L}(\beta)$. We define $\varepsilon(\alpha)$ by $\varepsilon(\alpha) = \varepsilon$ if $\varepsilon \in \mathcal{L}(\alpha)$, and $\varepsilon(\alpha) = \emptyset$, otherwise. In the same way, given a language L one defines $\varepsilon(L)$. Given a set of expressions S, the *language* associated with S is $\mathcal{L}(S) = \bigcup_{\alpha\in S} \mathcal{L}(\alpha)$. If $\beta \in$ RE($\|$) $\setminus \{\varepsilon, \emptyset\}$, we define $S\beta = \{\alpha\beta \mid \alpha \in S \wedge \alpha \neq \varepsilon\} \cup (\varepsilon \in S)\{\beta\}$. We have $\varepsilon S = S\varepsilon = S$ and $\emptyset S = S\emptyset = \emptyset$. Moreover, for $S, T \subseteq$ RE($\|$) $\setminus \{\emptyset\}$ and $\circ \in \{^s\|_\Gamma, ^a\|_\Gamma\}$, we define $S \circ T = \{\alpha \circ \beta \mid \alpha \in S \wedge \beta \in T\}$. The size of $\alpha \in$ RE($\|$) is denoted by $|\alpha|$ and defined as the number of occurrences of symbols (parenthesis not counted) in α.

3   Automata and Systems of Equations

A *nondeterministic finite automaton* (NFA) is a quintuple $A = \langle Q, \Sigma, \delta, I, F\rangle$ where Q is a finite set of states, Σ is a finite alphabet, $I \subseteq Q$ is the set of initial states, $F \subseteq Q$ is the set of final states, and $\delta : Q \times \Sigma \to 2^Q$ is the transition function. The function δ can be naturally extended to sets of states and to words. In what follows we will take $Q = [1, n]$, for $|Q| = n$. The *language* of A is $\mathcal{L}(A) = \{w \in \Sigma^\star \mid \delta(I, w) \cap F \neq \emptyset\}$. The *right language* of a state q, denoted by $\mathcal{L}_q$, is the language accepted by A if we take $I = \{q\}$. It is well known that it is possible to associate to each n-state NFA A over $\Sigma = \{\sigma_1, \ldots, \sigma_m\}$, with right languages $\mathcal{L}_1, \ldots, \mathcal{L}_n$, a system of linear language equations

$$\mathcal{L}_i = \sigma_1 \mathcal{L}_{1i} \cup \cdots \cup \sigma_m \mathcal{L}_{mi} \cup \varepsilon(\mathcal{L}_i), \quad \text{for } i \in Q,$$

where $\mathcal{L}_{ji} = \bigcup_{h \in \delta(i, \sigma_j)} \mathcal{L}_h$ and $\mathcal{L}(A) = \bigcup_{i \in I} \mathcal{L}_i$. In the same way, it is possible to associate to each regular expression a system of equations. Given an expression α over Σ, a *support* for $\alpha = \alpha_0$ is a set of expressions $\{\alpha_1, \ldots, \alpha_n\}$ that satisfies a system of equations

$$\alpha_i \doteq \sigma_1 \alpha_{i,1} + \cdots + \sigma_m \alpha_{i,m} + \varepsilon(\alpha_i), \quad i \in [0, n] \tag{2}$$

where each of $\alpha_{i,1}, \ldots, \alpha_{i,m}$ is a (possibly empty) sum of elements in $\{\alpha_1, \ldots, \alpha_n\}$. Mirkin [8,12] showed that for a standard regular expression $\alpha \in \mathrm{RE}$, a support $\pi(\alpha)$ can be computed as follows:

$$\begin{aligned}
\pi(\emptyset) = \pi(\varepsilon) = \emptyset, && \pi(\alpha + \beta) = \pi(\alpha) \cup \pi(\beta), \\
\pi(\sigma) = \{\varepsilon\} \quad (\sigma \in \Sigma), && \pi(\alpha\beta) = \pi(\alpha)\beta \cup \pi(\beta), \\
\pi(\alpha^\star) = \pi(\alpha)\alpha^\star.
\end{aligned} \tag{3}$$

In the following, we show that the support of an expression α, $\pi(\alpha)$, can be written as the union of m sets, each set corresponding to a letter $\sigma_j \in \Sigma$ ($j \in [1, m]$) and denoted by $\pi^{\sigma_j}(\alpha)$, i.e., $\pi(\alpha) = \bigcup_{\sigma_j \in \Sigma} \pi^{\sigma_j}(\alpha)$. Moreover, each set $\pi^{\sigma_j}(\alpha)$ can be obtained independently, and contains exactly the components in the sums $\alpha_{0,j}, \ldots, \alpha_{n,j}$ in (2). For standard regular expressions this result is established in Lemma 1.

Lemma 1. *Consider a standard regular expression α over an alphabet Σ and $\pi(\alpha) = \{\alpha_1, \ldots, \alpha_n\}$ its support. Then, we have $\pi(\alpha) = \bigcup_{\sigma \in \Sigma} \pi^\sigma(\alpha)$, where:*

$$\begin{aligned}
\pi^\sigma(\emptyset) = \pi^\sigma(\varepsilon) = \emptyset, && \pi^\sigma(\alpha + \beta) = \pi^\sigma(\alpha) \cup \pi^\sigma(\beta), \\
\pi^\sigma(\sigma) = \{\varepsilon\}, && \pi^\sigma(\alpha\beta) = \pi^\sigma(\alpha)\beta \cup \pi^\sigma(\beta), \\
\pi^\sigma(\tau) = \emptyset \quad (\tau \neq \sigma). && \pi^\sigma(\alpha^\star) = \pi^\sigma(\alpha)\alpha^\star.
\end{aligned}$$

Furthermore, $\pi(\alpha)$ satisfies a system of equations of the form,

$$\alpha_i \doteq \sigma_1 \alpha_{i,1} + \cdots + \sigma_m \alpha_{i,m} + \varepsilon(\alpha_i), \quad i \in [0, n] \tag{4}$$

such that each $\alpha_{i,j}$ is a (possibly empty) sum of elements in $\pi^{\sigma_j}(\alpha)$, for $j \in [1, m]$.

We extend the notion of support to $\alpha \in \mathrm{RE}(\|)$, which in the case of $^s\|_\Sigma$, i.e. of intersection, represents a significant improvement with regard to the definition of a support given in [2]. For $\circ \in \{^s\|_\Gamma, {}^a\|_\Gamma\}$ we define

$$\pi(\alpha \circ \beta) = \bigcup_{\sigma \in \Sigma} \pi^\sigma(\alpha \circ \beta), \tag{5}$$

where

$$\pi^{\sigma}(\alpha\,{}^{\mathsf{s}}\|_{\Gamma}\,\beta) = \begin{cases} \pi^{\sigma}(\alpha)\,{}^{\mathsf{s}}\|_{\Gamma}\,\pi^{\sigma}(\beta), & \text{for } \sigma \in \Gamma; \\ \pi^{\sigma}(\alpha)\,{}^{\mathsf{s}}\|_{\Gamma}\,(\pi(\beta)\cup\{\beta\})\cup(\pi(\alpha)\cup\{\alpha\})\,{}^{\mathsf{s}}\|_{\Gamma}\,\pi^{\sigma}(\beta), & \text{otherwise;} \end{cases} \tag{6}$$

$$\pi^{\sigma}(\alpha\,{}^{\mathsf{a}}\|_{\Gamma}\,\beta) = \begin{cases} \pi^{\sigma}(\alpha)\,{}^{\mathsf{a}}\|_{\Gamma}\,(\pi(\beta)\cup\{\beta\}) \ \cup \ (\pi(\alpha)\cup\{\alpha\})\,{}^{\mathsf{a}}\|_{\Gamma}\,\pi^{\sigma}(\beta) \\ \qquad\qquad\qquad \cup\ \pi^{\sigma}(\alpha)\,{}^{\mathsf{a}}\|_{\Gamma}\,\pi^{\sigma}(\beta), & \text{for } \sigma \in \Gamma; \\ \pi^{\sigma}(\alpha)\,{}^{\mathsf{a}}\|_{\Gamma}\,(\pi(\beta)\cup\{\beta\}) \ \cup \ (\pi(\alpha)\cup\{\alpha\})\,{}^{\mathsf{a}}\|_{\Gamma}\,\pi^{\sigma}(\beta), & \text{otherwise.} \end{cases} \tag{7}$$

Using (5) one obtains a support for $\alpha \in \mathrm{RE}(\|)$.

Proposition 1. *Given $\alpha \in \mathrm{RE}(\|)$, $\pi(\alpha) = \{\alpha_1, \ldots, \alpha_n\}$ is a support for $\alpha = \alpha_0$, satisfying a system of equations of the form,*

$$\alpha_i \doteq \sigma_1\alpha_{i,1} + \cdots + \sigma_m\alpha_{i,m} + \varepsilon(\alpha_i), \quad i \in [0, n] \tag{8}$$

such that each $\alpha_{i,s}$ is a (possibly empty) sum of elements in $\pi^{\sigma_s}(\alpha)$, for $s \in [1, m]$.

Example 2. Let $\Sigma = \{a, b\}$ and consider $\alpha = (b + ab + aab + abab)\,{}^{\mathsf{s}}\|_{\{a,b\}}\,(ab)^{\star}$. We have $\pi^a(\alpha) = \{b\,{}^{\mathsf{s}}\|_{\{a,b\}}\,b(ab)^{\star}, ab\,{}^{\mathsf{s}}\|_{\{a,b\}}\,b(ab)^{\star}, bab\,{}^{\mathsf{s}}\|_{\{a,b\}}\,b(ab)^{\star}\}$, $\pi^b(\alpha) = \{\varepsilon\,{}^{\mathsf{s}}\|_{\{a,b\}}\,(ab)^{\star}, ab\,{}^{\mathsf{s}}\|_{\{a,b\}}\,(ab)^{\star}\}$, $\pi(\alpha) = \pi^a(\alpha) \cup \pi^b(\alpha)$, and $|\pi(\alpha)| = 5$. This is an improvement w.r.t. the definition of π for expressions with the intersection operator in [2, Example 12], for which $|\pi(\alpha)| = 8$. □

4   Partial Derivatives and Partial Derivative Automata

The notions of partial derivatives and partial derivative automata of standard regular expressions were introduced by Antimirov [1]. Champarnaud and Ziadi [8] showed that the partial derivative automaton and Mirkin's contruction are identical. Sulzmann and Thiemann [14] extended partial derivatives and the partial derivative automaton to regular expressions with synchronised shuffles. In this section, we recall those notions and relate the set of partial derivatives with the support defined in the previous section. The set of partial derivatives of an expression $\alpha \in \mathrm{RE}(\|)$ by a symbol $\sigma \in \Sigma$, denoted by $\partial_{\sigma}(\alpha)$, is defined inductively as follows.

$$\partial_{\sigma}(\emptyset) = \partial_{\sigma}(\varepsilon) = \emptyset, \quad \partial_{\sigma}(\alpha^{\star}) = \partial_{\sigma}(\alpha)\alpha^{\star}, \quad \partial_{\sigma}(\sigma') = \begin{cases} \{\varepsilon\} & \text{if } \sigma = \sigma', \\ \emptyset & \text{otherwise,} \end{cases}$$

$$\partial_{\sigma}(\alpha + \beta) = \partial_{\sigma}(\alpha) \cup \partial_{\sigma}(\beta), \quad \partial_{\sigma}(\alpha\beta) = \partial_{\sigma}(\alpha)\beta \cup \varepsilon(\alpha)\partial_{\sigma}(\beta),$$

$$\partial_{\sigma}(\alpha\,{}^{\mathsf{s}}\|_{\Gamma}\,\beta) = \begin{cases} \partial_{\sigma}(\alpha)\,{}^{\mathsf{s}}\|_{\Gamma}\,\partial_{\sigma}(\beta) \text{ if } \sigma \in \Gamma, \\ \partial_{\sigma}(\alpha)\,{}^{\mathsf{s}}\|_{\Gamma}\,\{\beta\} \cup \{\alpha\}\,{}^{\mathsf{s}}\|_{\Gamma}\,\partial_{\sigma}(\beta), \text{ otherwise,} \end{cases}$$

$$\partial_{\sigma}(\alpha\,{}^{\mathsf{a}}\|_{\Gamma}\,\beta) = \begin{cases} \partial_{\sigma}(\alpha)\,{}^{\mathsf{a}}\|_{\Gamma}\,\partial_{\sigma}(\beta) \cup \partial_{\sigma}(\alpha)\,{}^{\mathsf{a}}\|_{\Gamma}\,\{\beta\} \cup \{\alpha\}\,{}^{\mathsf{a}}\|_{\Gamma}\,\partial_{\sigma}(\beta) \text{ if } \sigma \in \Gamma, \\ \partial_{\sigma}(\alpha)\,{}^{\mathsf{a}}\|_{\Gamma}\,\{\beta\} \cup \{\alpha\}\,{}^{\mathsf{a}}\|_{\Gamma}\,\partial_{\sigma}(\beta), \text{ otherwise.} \end{cases} \tag{9}$$

As usual, the set of partial derivatives of $\alpha \in RE(\|)$ w.r.t. a word $w \in \Sigma^*$ is inductively defined by $\partial_\varepsilon(\alpha) = \{\alpha\}$ and $\partial_{w\sigma}(\alpha) = \partial_\sigma(\partial_w(\alpha))$, where, given a set $S \subseteq RE(\|)$, $\partial_\sigma(S) = \bigcup_{\alpha \in S} \partial_\sigma(\alpha)$. Moreover, $\mathcal{L}(\partial_w(\alpha)) = \{w_1 \mid ww_1 \in \mathcal{L}(\alpha)\}$. Let $\partial(\alpha) = \bigcup_{w \in \Sigma^*} \partial_w(\alpha)$, and $\partial^+(\alpha) = \bigcup_{w \in \Sigma^+} \partial_w(\alpha)$. The partial derivative automaton of $\alpha \in RE(\|)$ is $\mathcal{A}_{PD}(\alpha) = \langle \partial(\alpha), \Sigma, \delta_{PD}, \{\alpha\}, F_{PD} \rangle$, with $F_{PD} = \{\beta \in \partial(\alpha) \mid \varepsilon(\beta) = \varepsilon\}$ and $\delta_{PD}(\beta, \sigma) = \partial_\sigma(\beta)$, for $\beta \in \partial(\alpha)$, $\sigma \in \Sigma$.

We now relate the set of partial derivatives with the support from Proposition 1. The following lemma is essential to obtain Proposition 2.

Lemma 2. *For $\sigma, \tau \in \Sigma$ and $\alpha \in RE(\|)$, we have $\pi^\sigma(\pi^\tau(\alpha)) \subseteq \pi^\sigma(\alpha)$.*

Proposition 2. *For $w\sigma \in \Sigma^+$ and $\alpha \in RE(\|)$, we have $\partial_{w\sigma}(\alpha) \subseteq \pi^\sigma(\alpha)$.*

As an immediate consequence, we have the following proposition establishing that the support of an expression is a superset of the set of states of its $\mathcal{A}_{PD}$.

Proposition 3. *Given $\alpha \in RE(\|)$, $\partial^+(\alpha) \subseteq \pi(\alpha)$.*

Example 3. For α from Example 2, we have $\partial^+(\alpha) = \{bab^s\|_{\{a,b\}} b(ab)^\star, ab^s\|_{\{a,b\}} b(ab)^\star, b^s\|_{\{a,b\}} b(ab)^\star, ab^s\|_{\{a,b\}} (ab)^\star, \varepsilon^s\|_{\{a,b\}} (ab)^\star\}$. Thus, in this case $\partial^+(\alpha) = \pi^a(\alpha) \cup \pi^b(\alpha) = \pi(\alpha)$. However, in general one has $\partial^+(\alpha) \subsetneq \pi(\alpha)$. For instance, $\partial^+(ab^s\|_{\{a,b\}} b^\star a) = \{b^s\|_{\{a,b\}} \varepsilon\} \subsetneq \{b^s\|_{\{a,b\}} \varepsilon, \varepsilon^s\|_{\{a,b\}} b^\star a\} = \pi(\alpha)$.

5   Average Case Complexity

In this section, we will extensively use the Bachman-Landau notation, namely

$$f(n_1, \ldots, n_k) \underset{n_1, \ldots, n_k \to \infty}{\sim} g(n_1, \ldots, n_k) \text{ as } \lim_{n_k \to \infty} \cdots \lim_{n_1 \to \infty} \frac{f(n_1, \ldots, n_k)}{g(n_1, \ldots, n_k)} = 1.$$

Given some measure over the objects of a combinatorial class, $\mathcal{A}$, for each $n \in \mathbb{N}$, let a_n be the sum of the values of this measure for all objects of size n. Here we consider $\mathcal{A} = RE(\|)$, and the measure (cost function) is the number of partial derivatives for expressions of size n. Let $A(z) = \sum_n a_n z^n$ be the corresponding generating function. We will use the notation $[z^n]A(z)$ for a_n. Seeing this generating function $A(z)$ as a complex analytic function, if it has a unique dominant singularity ρ, the study of the behaviour of $A(z)$ around ρ gives us access to the asymptotic form of its coefficients. For more details see [10]. In this section, we will make ample use of the notions and the techniques expounded in [6,11]. Of particular relevance is [6, Theorem 3.2], which we here reproduce:

Theorem 1. *Let $G(z)$ be a generating function with non-negative integral coefficients, and $C(z, w) \in \mathbb{Q}[z, w]$ be such that $C(z, G(z)) = 0$. Assume that $G(z)$ has a unique dominant singularity, ρ. Then, if $\lim_{z \to \rho} G(z) = a \in \mathbb{R}$,*

$$[z^n]G(z) \underset{n \to \infty}{\sim} \frac{-b}{\Gamma(-\alpha)} \rho^{-n} n^{-\alpha-1},$$

where α is the smallest non-zero exponent of the Puiseux expansion of $G(\rho - \rho s)$ with respect to the variable s, and in the case $\alpha = \frac{1}{2}$ (which is true in all the cases considered in this paper), b is given by:

$$b = \sqrt{\frac{2\rho \frac{\partial C}{\partial z}}{\frac{\partial^2 C}{\partial w^2}}} \Bigg|_{\substack{z=\rho \\ w=a}}.$$

5.1   Average Number of Partial Derivatives by One Symbol

We start by estimating the asymptotic average number of partial derivatives by one symbol for $\alpha \in \mathrm{RE}(\|)$, which corresponds to the expected number of transitions from a state in $\mathcal{A}_{\mathrm{PD}}$. For standard regular expressions and for large alphabets that value is known to be the constant 6 [4,13]. The cases of the strong and arbitrarily synchronised shuffles are analysed individually, but the results will be essentially identical. For the strong synchronisation, we also consider the extreme cases of intersection and shuffle which were not considered before in the literature.

Strong Synchronisation. Let Σ be the alphabet, and let $\Gamma \subseteq \Sigma$ with $\ell = |\Gamma|$. We set $k = |\Sigma \setminus \Gamma|$, and $m = k + \ell = |\Sigma|$. We consider regular expressions with the strong synchronisation operator over the alphabet Σ generated by the grammar (1) without expressions with the operator $^{\mathrm{a}}\|_{\Gamma}$. Moreover we will consider all operators using the same Γ. We denote this set of expressions by $\mathrm{RE}(^{\mathrm{s}}\|_{\Gamma})$. The generating function, $R_m(z)$, whose coefficient $[z^n]R_m(z)$ is the cumulative number of those regular expressions of size n, satisfies

$$R_m(z) = (m+1)z + 3zR_m(z)^2 + zR_m(z).$$

Regular expressions that have ε in their language, denoted by α_ε, are unambiguously generated by the following grammar

$$\alpha_\varepsilon \to \varepsilon \mid (\alpha_\varepsilon + \alpha) \mid (\alpha_{\overline{\varepsilon}} + \alpha_\varepsilon) \mid (\alpha_\varepsilon \alpha_\varepsilon) \mid (\alpha^\star) \mid (\alpha_\varepsilon {}^{\mathrm{s}}\|_{\Gamma} \alpha_\varepsilon),$$

where $\alpha_{\overline{\varepsilon}}$ represents regular expressions that do not have ε in their language. Using $R_{\overline{\varepsilon},m}(z) = R_m(z) - R_{\varepsilon,m}(z)$, the generating function for α_ε satisfies

$$R_{\varepsilon,m}(z) = z + 2zR(z)R_{\varepsilon,m}(z) + zR_{\varepsilon,m}(z)^2 + zR_m(z).$$

Given $\tau \in \Sigma \setminus \Gamma$ and $\gamma \in \Gamma$, we denote by $\mathsf{t}(\alpha)$ and by $\mathsf{g}(\alpha)$ the cost functions for an upper bound of the cardinality of $\partial_\tau(\alpha)$ and the cardinality of $\partial_\gamma(\alpha)$, respectively. Using (9) we have

$$\begin{aligned}
\mathsf{t}(\varepsilon) = \mathsf{t}(\sigma) &= 0, \quad \sigma \neq \tau & \mathsf{t}(\tau) &= 1, \\
\mathsf{t}(\alpha + \beta) &= \mathsf{t}(\alpha) + \mathsf{t}(\beta), & \mathsf{t}(\alpha_\varepsilon \beta) &= \mathsf{t}(\alpha_\varepsilon) + \mathsf{t}(\beta), \\
\mathsf{t}(\alpha_{\overline{\varepsilon}} \beta) &= \mathsf{t}(\alpha_{\overline{\varepsilon}}), & \mathsf{t}(\alpha^\star) &= \mathsf{t}(\alpha), \\
\mathsf{t}(\alpha {}^{\mathrm{s}}\|_{\Gamma} \beta) &= \mathsf{t}(\alpha) + \mathsf{t}(\beta).
\end{aligned}$$

Note that, for $\alpha\,{}^s\|_\Gamma\,\beta$ the number of partial derivatives by τ equals the sum of the number of derivatives by τ of both α and β. Thus, $\mathsf{t}(\alpha\,{}^s\|_\Gamma\,\beta) = \mathsf{t}(\alpha) + \mathsf{t}(\beta)$. The definition of $\mathsf{g}(\alpha)$ is analogous, except that $\mathsf{g}(\sigma) = 0$ if $\sigma \neq \gamma$ and $\mathsf{g}(\gamma) = 1$. Also, since $\partial_\gamma(\alpha\,{}^s\|_\Gamma\,\beta) = \partial_\gamma(\alpha)\,{}^s\|_\Gamma\,\partial_\gamma(\beta)$, we have $\mathsf{g}(\alpha\,{}^s\|_\Gamma\,\beta) = \mathsf{g}(\alpha)\,\mathsf{g}(\beta)$. The corresponding generating functions $T_m(z) = \sum_\alpha \mathsf{t}(\alpha)z^{|\alpha|}$ and $G_m(z) = \sum_\alpha \mathsf{g}(\alpha)z^{|\alpha|}$ satisfy, respectively,

$$T_m(z) = z + 5zT_m(z)R_m(z) + zT_m(z)R_{\varepsilon,m}(z) + zT_m(z),$$
$$G_m(z) = z + 3zG_m(z)R_m(z) + zG_m(z)R_{\varepsilon,m}(z) + zG_m(z) + zG_m(z)^2.$$

The generating function for an upper bound of the cardinality of the set of partial derivatives by one symbol $\bigcup_{\sigma \in \Sigma} \partial_\sigma(\alpha)$ is given by

$$D_{k,\ell}(z) = kT_m(z) + \ell G_m(z).$$

To compute the asymptotic behaviour of the coefficients of $D_{k,\ell}(z)$, we proceed by dealing first with $G_m(z)$, and then with $T_m(z)$. Eliminating the auxiliary variables from the above equations [9], one obtains an algebraic curve given by $C_m(z,w) \in \mathbb{Q}[z,w]$ such that $C_m(z, G_m(z)) = 0$. The polynomial $C_m(z,w)$ has degree 8 in w, and degree 6 in z. Using the techniques expounded in [6,11], one finds that the minimal polynomial of the relevant singularity, let us call it $r_m(z)$, is a factor of the resultant of $C_m(z,w)$ and $\frac{\partial C_m}{\partial w}(z,w)$, which has 5 distinct irreducible factors. One of them is z, which of course cannot be $r_m(z)$, while another is $16\,m^2z^4 + 192\,mz^4 + 96\,mz^3 + 256z^4 + 8\,mz^2 + 128z^3 - 16z^2 - 8z + 1$, which is always positive for $z \in \mathbb{R}^+$, since $256z^4 + 128z^3 - 16z^2 - 8z + 1 = x^4 + 2x^3 - x^2 - 2x + 1 = (x^2 + x - 1)^2$, with $x = 4z$. One is left with three polynomials, and to find which is $r_m(z)$, one may proceed as follows. Looking at the (real part of the) graph of the curve $C_m(z,w)$, one realises that this curve has only one branch on the first quadrant, and that our generating function, being an increasing function, has a singularity that corresponds to the turning point with the lowest ordinate. In this way we found out that $r_m(z) = (12m+11)z^2 + 2z - 1$, and therefore the singularity is

$$\rho_m = \frac{1}{1 + 2\sqrt{3m+3}}, \tag{10}$$

while the minimal polynomial of $a_m = \lim_{z \to \rho_m} G_m(z)$ (see again [6,11]) has degree 8:

$$3w^8 + 6w^7 + (11 - 10m)w^6 + (10 - 14m)w^5 + (12 - 24m + 3m^2)w^4 +$$
$$+ (10 - 14m)w^3 + (11 - 10m)w^2 + 6w + 3. \tag{11}$$

With the help of a plotting program that can deal with functions given in an implicit form, one can identify the root a_m of this polynomial pertaining to ρ_m, and then one can use Puiseux expansions to obtain the expansion of the appropriate a_m, which is, as $m \to \infty$,

$$\sqrt{3}\,m^{-\frac{1}{2}} + \frac{1}{12}m^{-1} + \frac{307\sqrt{3}}{64}m^{-\frac{3}{2}} + o(m^{-\frac{3}{2}}).$$

Using the techniques described in [6,11], one gets:

$$[z^n]G_m(z) \underset{n,\,m\to\infty}{\sim} \frac{\sqrt{6}}{2\sqrt{\pi\,m}}\rho_m^{-n}n^{-\frac{3}{2}}. \tag{12}$$

With respect to T_m, the value of the singularity is the same as for G_m, i.e. ρ_m, whereas the minimal polynomial of a_m is:

$$3m^2w^4 - 6mw^3 - (10m+1)w^2 - 8w - 5, \tag{13}$$

and its Puiseux expansion is

$$\sqrt{3}\,m^{-\frac{1}{2}} + \frac{3}{4}m^{-1} + \frac{19\sqrt{3}}{64}m^{-\frac{3}{2}} + o(m^{-\frac{3}{2}}),$$

as $m \to \infty$. From this, one gets:

$$[z^n]T_m(z) \underset{n,\,m\to\infty}{\sim} \frac{5\sqrt{6}}{2\sqrt{\pi\,m}}\rho_m^{-n}n^{-\frac{3}{2}}. \tag{14}$$

Therefore,

$$[z^n]D_{k,\ell}(z) \underset{n,\,k,\,\ell\to\infty}{\sim} \frac{(k+5\ell)\sqrt{6}}{2\sqrt{\pi(k+\ell)}}\rho_m^{-n}n^{-\frac{3}{2}}. \tag{15}$$

Using the formulas in Sect. 5.1 of [2], with $s = m+1$, $u = 1$, $b = 3$, one obtains an estimate for the number of expressions, for large values of n and m,

$$[z^n]R_m(z) \underset{n\to\infty}{\sim} \frac{\sqrt{2-2\rho_m}}{12\rho_m\sqrt{\pi}}\rho_m^{-n}n^{-\frac{3}{2}} \underset{m\to\infty}{\sim} \sqrt{\frac{m}{6\pi}}\rho_m^{-n}n^{-\frac{3}{2}}. \tag{16}$$

Proposition 4. *The average of the upper bound (here considered) of the number of partial derivatives of an $\alpha \in \mathrm{RE}({}^s\|_\Gamma)$ of size n is*

$$\frac{[z^n]D_{k,\ell}(z)}{[z^n]R_m(z)} \underset{n,\,k,\,\ell\to\infty}{\sim} \frac{6\sqrt{3}(k+5\ell)\rho_m}{\sqrt{k+\ell}\sqrt{1-\rho_m}}. \tag{17}$$

In particular, when $\ell = 0$, and thus $m = k$, one has

$$\lim_{k\to\infty} \frac{[z^n]D_{k,0}(z)}{[z^n]R_k(z)} = 3, \tag{18}$$

whereas when $k = 0$, and thus $m = \ell$, one has

$$\lim_{\ell\to\infty} \frac{[z^n]D_{0,\ell}(z)}{[z^n]R_\ell(z)} = 15, \tag{19}$$

and when $k = \ell = \frac{m}{2}$

$$\lim_{m\to\infty} \frac{[z^n]D_{\frac{m}{2},\frac{m}{2}}(z)}{[z^n]R_m(z)} = 9. \tag{20}$$

We recall that if $\ell = 0$, ${}^s\|_\emptyset$ coincides with the shuffle operator, $\sqcup\!\sqcup$; and if $k = 0$ ${}^s\|_\Sigma$ coincides with intersection. The given results nicely relate with the estimated value for standard regular expressions mentioned above.

Arbitrary Synchronisation. Now consider the set of expressions only with the operator $^{\mathsf{a}}\|_\Gamma$ with Γ as above, i.e., $\mathrm{RE}(^{\mathsf{a}}\|_\Gamma)$ and k, ℓ, m as in the previous section. The generating functions $R_m(z)$, $R_{\varepsilon,m}(z)$, and $T_m(z)$ coincide with ones for $\mathrm{RE}(^{\mathsf{s}}\|_\Gamma)$. In the definition of $\mathbf{g}(\alpha)$ we have

$$\mathbf{g}(\alpha \,^{\mathsf{a}}\|_\Gamma \beta) = \mathbf{g}(\alpha)\,\mathbf{g}(\beta) + \mathbf{g}(\alpha) + \mathbf{g}(\beta),$$

and, thus,

$$G_m(z) = z + 4zG_m(z)R_m(z) + zG_m(z)R_{\varepsilon,m}(z) + zG_m(z) + zG_m(z)^2.$$

In this case, the minimal polynomial for the singularity, σ_m, is

$$\begin{cases} (64m^2 + 579m + 925)z^3 + (84m + 161)z^2 - 4(m+11)z - 6, & \text{when } m \le 6, \\ (12m + 11)z^2 + 2z - 1, & \text{otherwise,} \end{cases} \tag{21}$$

and, for $m \le 6$,

$$\sigma_m = \frac{1}{4}m^{-\frac{1}{2}} - \frac{3}{32}m^{-1} - \frac{77}{256}m^{-\frac{3}{2}} + o(m^{-\frac{3}{2}}), \text{ as } m \to \infty, \tag{22}$$

while $\sigma_m = \rho_m$ for $m \ge 7$. It turns out that $a_m = 1$, for $1 \le m \le 6$, while for $m \ge 7$ one has

$$a_m = \frac{\sqrt{3}}{2}m^{-\frac{1}{2}} + \frac{3}{16}m^{-1} + \frac{77\sqrt{3}}{256}m^{-\frac{3}{2}} + o(m^{-\frac{3}{2}}), \text{ as } m \to \infty. \tag{23}$$

Using this, one arrives at the same result as in (12). Thus, one will obtain the same asymptotic estimates for the average size of $\bigcup_{\sigma \in \Sigma} \partial_\sigma(\alpha)$ as for $\mathrm{RE}(^{\mathsf{s}}\|_\Gamma)$.

5.2 Average Size of the Support

In this section we consider $\alpha \in \mathrm{RE}(^{\mathsf{s}}\|_\Gamma)$ with Γ used as in Sect. 5.1, and we estimate an upper bound for the asymptotic average size of $\pi(\alpha)$, and thus of the average state complexity of $\mathcal{A}_{\mathrm{PD}}$. Let $\Gamma \subseteq \Sigma$ with $|\Gamma| = \ell$ and $|\Sigma \setminus \Gamma| = k$. Let $\mathsf{p}(\alpha)$ be the cost function for an upper bound of the size $\pi(\alpha) = \bigcup_{\gamma \in \Gamma} \pi^\gamma(\alpha) \cup \bigcup_{\tau \in \Sigma \setminus \Gamma} \pi^\tau(\alpha)$. For computing $\mathsf{p}(\alpha)$, let $\mathsf{s}(\alpha)$ be the cost function for an upper bound of the size of $\pi^\gamma(\alpha)$, where $\gamma \in \Gamma$. Using (6), we have:

$$\begin{array}{ll} \mathsf{s}(\varepsilon) = 0 = \mathsf{s}(\sigma), \text{ for } \sigma \ne \gamma, & \mathsf{s}(\alpha + \beta) = \mathsf{s}(\alpha) + \mathsf{s}(\beta), \\ \mathsf{s}(\gamma) = 1, & \mathsf{s}(\alpha\beta) = \mathsf{s}(\alpha) + \mathsf{s}(\beta), \\ \mathsf{s}(\alpha^\star) = \mathsf{s}(\alpha), & \mathsf{s}(\alpha \,^{\mathsf{s}}\|_\Gamma \beta) = \mathsf{s}(\alpha)\,\mathsf{s}(\beta). \end{array}$$

In the same way, let $\mathsf{u}(\alpha)$ be the cost function for an upper bound of the size of $\pi^\tau(\alpha)$, where $\tau \in \Sigma \setminus \Gamma$. We have:

$$\begin{array}{ll} \mathsf{u}(\varepsilon) = 0 = \mathsf{u}(\sigma), \text{ for } \sigma \ne \tau, & \mathsf{u}(\alpha + \beta) = \mathsf{u}(\alpha) + \mathsf{u}(\beta), \\ \mathsf{u}(\tau) = 1, & \mathsf{u}(\alpha\beta) = \mathsf{u}(\alpha) + \mathsf{u}(\beta), \\ \mathsf{u}(\alpha^\star) = \mathsf{u}(\alpha), & \mathsf{u}(\alpha \,^{\mathsf{s}}\|_\Gamma \beta) = \mathsf{p}(\alpha)\,\mathsf{u}(\beta) + \ell\,\mathsf{u}(\alpha)\mathsf{s}(\beta) + \mathsf{u}(\alpha) + \mathsf{u}(\beta), \end{array}$$

where in $u(\alpha^{\mathsf{s}}\|_{\Gamma}\beta)$ we avoid to count twice $u(\alpha)u(\beta)$. Then, $\mathsf{p}(\alpha) = \ell \cdot \mathsf{s}(\alpha) + k \cdot \mathsf{u}(\alpha)$. The generating functions for $\mathsf{s}(\alpha)$, $\mathsf{u}(\alpha)$, and $\mathsf{p}(\alpha)$, respectively $S_m(z)$, $U_m(z)$ and $P_m(z)$, satisfy the following equalities:

$$S_m(z) = z + 4zS_m(z)R_m(z) + zS_m(z) + zS_m(z)^2,$$
$$U_m(z) = z + 6zU_m(z)R_m(z) + zU_m(z)P_{k,\ell}(z) + \ell zS_m(z)U_m(z) + zU_m(z),$$
$$P_{k,\ell}(z) = \ell \cdot S_m(z) + k \cdot U_m(z).$$

Using the same procedure as described above, one obtains a polynomial $C_{k,\ell}(z,w) \in \mathbb{Q}[z,w]$, such that $C_{k,\ell}(z, P_{k,\ell}(z)) = 0$. This polynomial $C_{k,\ell}$ has degree 8 in w, and degree 6 in z.

For $k = \ell = \frac{m}{2}$, using graphical and numerical methods, as well as Puiseux expansions, one obtains that the behaviour of the relevant singularity, η_m, which is a root of a polynomial of degree 8, has the following asymptotic behaviour:

$$\eta_m \underset{m \to \infty}{\sim} \sqrt{2}\,\beta\, m^{-\frac{1}{2}}, \tag{24}$$

where $\beta \simeq 0.180866$ is the biggest root of the polynomial $1100z^4 + 8z^3 - 68z^2 + 1$.

Using the same techniques as above, one sees that:

$$[z^n]P_{\frac{m}{2},\frac{m}{2}}(z) \underset{n,\,m \to \infty}{\sim} \frac{\gamma\sqrt{m}}{2\sqrt{\pi}}\eta_m^{-n}n^{-\frac{3}{2}}, \tag{25}$$

where $\gamma \simeq 6.73978$, and therefore

$$\frac{[z^n]P_{\frac{m}{2},\frac{m}{2}}(z)}{[z^n]R_m(z)} \underset{n,\,m \to \infty}{\sim} \sqrt{\frac{3}{2}}\,\gamma\left(\frac{\rho_m}{\eta_m}\right)^n, \tag{26}$$

$$\lim_{m \to \infty} \frac{\rho_m}{\eta_m} = \frac{1}{2\sqrt{6}\,\beta} \simeq 1.12859. \tag{27}$$

Proposition 5. *For large values of m and n, and $k = \ell$, an upper bound for the average number of states of $\mathcal{A}_{\mathrm{PD}}(\alpha)$ for $\alpha \in \mathrm{RE}(\,^{\mathsf{s}}\|_{\Gamma}\,)$ is $(1.12859 + o(1))^n$.*

When $k = 0$, i.e., in case of intersection, one obtains a simpler polynomial $C_\ell(z,w)$ for the corresponding generating function, namely:

$$3z^2w^4 + 2\ell z(z-1)w^3 + \ell^2((16\ell+21)z^2 + 2z - 1)w^2 + 2\ell^3 z(z-1)w + 3\ell^4 z^2. \tag{28}$$

The singularity is the same ρ_ℓ as above, and $a_\ell = \frac{\ell\sqrt{2\ell-1-2\sqrt{\ell^2-\ell-2}}}{\sqrt{3}}$. This yields

$$[z^n]P_{0,\ell}(z) \underset{n,\,\ell \to \infty}{\sim} \sqrt{\frac{3}{2}\frac{\ell}{\pi}}\rho_\ell^{-n}n^{-\frac{3}{2}}. \tag{29}$$

Proposition 6. *With the notations introduced above, the average size of the number of states in the partial derivative automata for a (standard) regular expression with intersection is asymptotically*

$$\frac{[z^n]P_{0,\ell}(z)}{[z^n]R_\ell(z)} \underset{n,\,\ell \to \infty}{\sim} 3. \tag{30}$$

This result is a surprising improvement of the previous upper bound of $(1.056 + o(1))^n$ given in [2] but is compatible with experimental values. Moreover for $\ell = 1, n = 1000$, the ratio (30) is 710; for $\ell = 2, n = 1000$, we obtain 9.5; and for $\ell = 10, n = 300$, the value is 2.94445.

Finally, when $\ell = 0$, i.e., in the case of shuffle, a polynomial $C_k(z, w)$ for the corresponding generating function is:

$$z^2 w^4 + ((14k + 11)z^2 + 2z - 1)w^2 + k^2 z^2. \tag{31}$$

The singularity is now $\xi_k = \frac{1}{1+2\sqrt{4k+3}}$, and $a_k = \sqrt{k}$. This yields:

$$[z^n]P_{k,0}(z) \underset{n,\,k \to \infty}{\sim} \sqrt{\frac{2k}{\pi}}\xi_k^{-n}n^{-\frac{3}{2}}, \tag{32}$$

and therefore, we obtain the same result as in [5], but using different techniques.

Proposition 7. *With the notations introduced above, the average size of the number of states in the partial derivative automata for a (standard) regular expression with shuffle is asymptotically*

$$\frac{[z^n]P_{k,0}(z)}{[z^n]R_k(z)} \underset{n,\,k \to \infty}{\sim} 2\sqrt{3}\left(\frac{\rho_k}{\xi_k}\right)^n \underset{k \to \infty}{\sim} 2\sqrt{3}\left(\frac{4}{3}\right)^{\frac{n}{2}}. \tag{33}$$

References

1. Antimirov, V.M.: Partial derivatives of regular expressions and finite automaton constructions. Theoret. Comput. Sci. **155**(2), 291–319 (1996)
2. Bastos, R., Broda, S., Machiavelo, A., Moreira, N., Reis, R.: On the average complexity of partial derivative automata for semi-extended expressions. J. Autom. Lang. Comb. **22**(1–3), 5–28 (2017)
3. ter Beek, M.H., Martín-Vide, C., Mitrana, V.: Synchronized shuffles. Theoret. Comput. Sci. **341**, 263–275 (2005)
4. Broda, S., Machiavelo, A., Moreira, N., Reis, R.: On the average size of Glushkov and partial derivative automata. Int. J. Found. Comput. S. **23**(5), 969–984 (2012)
5. Broda, S., Machiavelo, A., Moreira, N., Reis, R.: Automata for regular expressions with shuffle. Inf. Comput. **259**(2), 162–173 (2018)
6. Broda, S., Machiavelo, A., Moreira, N., Reis, R.: Analytic combinatorics and descriptional complexity of regular languages on average. ACM SIGACT News **51**(1), 38–56 (2020)
7. Broda, S., Machiavelo, A., Moreira, N., Reis, R.: Location automata for synchronised shuffle expressions. J. Log. Algebr. Methods Program. **133**, 100847 (2023)
8. Champarnaud, J.M., Ziadi, D.: From Mirkin's prebases to Antimirov's word partial derivatives. Fundam. Inform. **45**(3), 195–205 (2001)
9. Cox, D.A., John Little, D.O.: Ideals, Varieties, and Algorithms. Springer, 4th edn. (2018)
10. Flajolet, P., Sedgewick, R.: Analytic combinatorics. CUP (2008)
11. Konstantinidis, S., Machiavelo, A., Moreira, N., Reis, R.: On the average state complexity of partial derivative transducers. In: Chatzigeorgiou, A., et al. (eds.) SOFSEM 2020. LNCS, vol. 12011, pp. 174–186. Springer, Cham (2020). https://doi.org/10.1007/978-3-030-38919-2_15

12. Mirkin, B.G.: An algorithm for constructing a base in a language of regular expressions. Eng. Cybern. **5**, 51–57 (1966)
13. Nicaud, C.: Average state complexity of operations on unary automata. In: Kutyłowski, M., Pacholski, L., Wierzbicki, T. (eds.) MFCS 1999. LNCS, vol. 1672, pp. 231–240. Springer, Heidelberg (1999). https://doi.org/10.1007/3-540-48340-3_21
14. Sulzmann, M., Thiemann, P.: Derivatives and partial derivatives for regular shuffle expressions. J. Comput. System Sci. **104**, 323–341 (2019)

Sweep Complexity Revisited

Szilárd Zsolt Fazekas[1]([✉])(iD) and Robert Mercaş[2](iD)

[1] Department of Mathematical Science and Electrical-Electronic-Computer
Engineering, Akita University, Akita, Japan
szilard.fazekas@ie.akita-u.ac.jp
[2] Department of Computer Science, Loughborough University, Loughborough,
England
R.G.Mercas@lboro.ac.uk

Abstract. We study the sweep complexity of DFA in one-way jumping mode answering several questions posed earlier. This measure is the number of times in the worst case that such machines have to return to the beginning of their input after having skipped some of the symbols. The class of languages accepted by these machines strictly includes the regular class and constant sweep complexity allows exactly the acceptance of regular languages. However, we show that there exist machines with higher than constant complexity still only accepting regular languages and that in general the sweep complexity of an automaton does not distinguish between accepting regular and non-regular languages. We establish separation results for asymptotic classes defined by this complexity measure and give a surprising exponential/logarithmic relation between factors of certain inputs which can be verified by such machines.

Keywords: automata · deterministic · one-way jumping · sweep complexity

1 Introduction

In roughly the last three decades, several non-classical models of automata have been introduced to study the effect of processing inputs with simple machines in a non-sequential way. Such models include restarting automata [10], jumping automata [12], input revolving automata [4] and automata with translucent letters [13]. However, these models are either strictly more powerful or accept a class incomparable with the regular one.

One-way jumping finite automata (OWJFA) were introduced [5] to study the power of deterministic finite automata (DFA) performing non-sequential processing without completely discarding structural information about the inputs à la jumping automata. The resulting model is, in a sense, a minimal extension of finite automata. Machines are specified in exactly the same way as DFA allowing partial transition functions. The only change is the behaviour of the machine when encountering a letter for which the current state has no outgoing transition

This work was supported by JSPS KAKENHI Grant Number JP23K10976.

defined. In the classical case such inputs are rejected, but in one-way jumping mode the letters are skipped temporarily to be processed later. The relative order of the skipped symbols is maintained, and the automaton moves back to the beginning after each pass (called *sweep* here), seeing only the symbols previously skipped. Therefore one can also view this model as a DFA with an input tape which works as a restricted queue, or one that reads and erases symbols from a circular tape always jumping clockwise to the nearest letter for which it has a defined transition from the current state. When the transition function is complete, no symbols are skipped, so the machine behaves as ordinary DFA, which means that the class of languages accepted by DFA in one-way jumping mode trivially includes all regular languages.

Various properties of the accepted language class [1] and the status of fundamental decidability questions have been settled [2]. More powerful machines with this new processing mode have also been investigated, such as nondeterministic finite automata [3,6], two-way finite automata [7], pushdown automata and linear bounded automata [6]. While the language classes defined by the models have no nontrivial closure properties under usual language operations, the accepting power and decidability issues raised some intriguing problems.

Except for linear bounded automata, the machine models mentioned above become more powerful when they are allowed to jump to the nearest symbol readable in the current state, which is not surprising. However, it has proven challenging to get a clear picture of just how powerful the new processing mode is, even in the simplest case when one starts from DFA. Such automata can accept all regular languages and the language class defined by them is incomparable with the context-free class, but included in the context-sensitive class and in $\mathrm{DTIME}(n^2)$ [1]. The separation results make use of combinations of a handful of regular languages together with a very simple type of non-regular languages which contain words having letter counts in a certain ratio, e.g., the frequently used $L_{ab} = \{w \in \{a, b\}^* \mid w$ contains as many a's as b's$\}$ accepted by the machine $\mathcal{A}$ in Fig. 1 (with states **1**, or **2** final). While this was enough to establish virtually all separations of interest, it left a significant gap in our understanding of the model: can such machines accept any ('interesting') non-regular languages apart from the ones which establish linear relationships among letter counts?

In this work we answer the question above, building on the investigation of sweep complexity of DFA in one-way jumping mode. Sweep count can be viewed as a measure of non-regular resources used by a machine posing the natural question of how much of this resource is needed to be able to accept non-regular languages? It has been shown that constant sweep complexity does not increase the accepting power of the machines [9] and that superconstant sweep complexity requires cycles containing 'complementary deficient' states [8]. In the latter paper it was conjectured that, in fact, any automaton with higher than constant sweep complexity accepts a non-regular language. In Sect. 3 we refute that conjecture by exhibiting a small DFA accepting a regular language while processing some inputs of length n in $\Omega(\log n)$ sweeps. We also show that there is no non-trivial upper bound on the sweep complexity of regular languages, that is, there are machines with linear complexity accepting regular languages.

A natural question regarding the new complexity measure is whether there exists a meaningful hierarchy which does not collapse to the extremes of $\mathcal{O}(1)$ and $\mathcal{O}(n)$. The aforementioned example shows that automata with logarithmic complexity exist, which answers another question posed earlier. Furthermore, following the line of computational complexity theory, we set out to explore whether the language classes defined through asymptotic complexity form a true hierarchy, that is whether there are languages which can be accepted by a machine with $\mathcal{O}(f(n))$ complexity but not by any with $o(f(n))$ complexity, for various functions $f(n)$. In Sect. 4 we demonstrate that such a hierarchy exists by presenting languages with $\Theta(\log n)$ and $\Theta(n)$ sweep complexity, respectively.

Finally we mention that sweep complexity as an idea has been studied in other contexts, too: an interesting and thorough investigation of a similar flavor established infinite hierarchies in terms of sweep count for iterated uniform finite transducers [11], although that model is significantly more powerful than ours, so the techniques used there do not translate here as far as we can tell.

2   Preliminaries

We consider words over a finite alphabet, e.g., $\Sigma = \{a, b\}$. The set of all words over Σ is Σ^*, which includes the empty word ε.

A *DFA* is a quintuple $M = (Q, \Sigma, R, \mathbf{s}, F)$, where Q is the finite set of states, Σ is the finite input alphabet, $\Sigma \cap Q = \emptyset$, $R : Q \times \Sigma \to Q$ is the transition function, $\mathbf{s} \in Q$ is the start state, and $F \subseteq Q$ is the set of final states. Elements of R are referred to as (transition) rules of M and we write $\mathbf{p}y \to \mathbf{q} \in R$ instead of $R(\mathbf{p}, y) = \mathbf{q}$. A configuration of M is a string in $Q \times \Sigma^*$.

A DFA transitions from a configuration $\mathbf{p}w$ to a configuration $\mathbf{q}w'$ if $w = aw'$ and $\mathbf{p}a \to \mathbf{q} \in R$, with $\mathbf{p}, \mathbf{q} \in Q$, $w, w' \in \Sigma^*$ and $a \in \Sigma$. By extending the meaning of $\to$ we denote this by $\mathbf{p}w \to \mathbf{q}w'$ and the reflexive and transitive closure of $\to$ by $\to^*$. A word w is *accepted* by a DFA M if there exists $\mathbf{f} \in F$, such that $\mathbf{s}w \to^* \mathbf{f}$. The language accepted by M is $\{w \in \Sigma^* \mid \exists \mathbf{f} \in F : \mathbf{s}w \to^* \mathbf{f}\}$.

One-way jumping automata

The *one-way jumping relation* (denoted by $\circlearrowright$) between configurations from $Q\Sigma^*$, was originally defined in [5]. Here we follow the slightly different definition of [8] which does not change the accepting power of the model, but is more convenient.

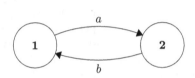

position :	0	1	2	3	4	5	6
input	a	d	c	b	c	b	a
after sweep 1	ε	d	c	b	c	b	ε
after sweep 2	ε	d	c	ε	c	ε	ε
after sweep 3	ε	d	ε	ε	ε	ε	ε
after sweep 4	ε	ε	ε	ε	ε	ε	ε

Fig. 1. The only two-state ROWJFA satisfying Lemma 1

Fig. 2. The computation table for *adcbcba* by the machine in Example 1.

A tuple $M = (Q, \Sigma, R, \mathbf{s}, F)$ representing a *deterministic right one-way jumping automaton* (ROWJFA) is defined the same way as a DFA, where the configurations are also elements of the set $Q \times \Sigma^*$. Let $\Sigma_p = \{b \in \Sigma \mid \exists \mathbf{q} \in Q$ such that $\mathbf{p}b \to \mathbf{q} \in R\}$ be the set of all of the letters from Σ for which we have a transition defined from state $\mathbf{p}$. A jumping transition (or jump, for short), denoted $\circlearrowleft$, is defined between configurations $\mathbf{p}ax$ and $\mathbf{p}xa$ if state $\mathbf{p}$ cannot read the letter a, formally:

$$\mathbf{p}ax \circlearrowleft \mathbf{p}xa, \text{ if } a \in \Sigma \setminus \Sigma_p.$$

A ROWJFA can transition from configuration $\mathbf{p}ax$ to configuration $\mathbf{q}y$, which we denote by $\mathbf{p}ax \vdash \mathbf{q}y$, if

(*i*) $\mathbf{p}ax \to \mathbf{q}y$, where $x = y$ and $\mathbf{p}a \to \mathbf{q} \in R$, as defined earlier, or

(*ii*) $\mathbf{p}ax \circlearrowleft \mathbf{p}xa$, when $a \in \Sigma \setminus \Sigma_p, \mathbf{p} = \mathbf{q}$ and $xa = y$.

A word w is accepted by M if $\mathbf{s}w \vdash^* \mathbf{f}$. The language accepted by M is defined by $L(M) = \{x \in \Sigma^* \mid \exists \mathbf{f} \in F : \mathbf{s}x \vdash^* \mathbf{f}\}$.

While some texts define DFA having complete transition functions, our DFA allow partially defined ones. Indeed, the pairs $(\mathbf{p}, a) \in Q \times \Sigma$ for which no transition is defined enable the ROWJFA to perform a jump as opposed to rejecting the input as a DFA would. Hence, a ROWJFA with a complete transition function is just a DFA.

Sweeps are contiguous sequences of transitions on a given input, consisting of the steps from reading or jumping over the leftmost remaining input letter to reading or jumping over the rightmost one. If a position is jumped over, then the input symbol in that position is processed in a later sweep. The number of sweeps needed to process the whole input is the number of times the automaton reaches the last position of the original input word or, equivalently, one more than the maximum number of times any position is jumped over.

For an intuitive picture of sweeps, consider the computation of a ROWJFA M on input w as a table with rows representing the k sweeps needed to process w and columns representing positions in the input word. Cell i, j in the table contains either a letter or a symbol representing that the letter has been *read*, e.g., ε. Once a letter has been marked *read* and erased it stays that way, so each column is a word of the form $a^\ell \varepsilon^{k-\ell} (= a^\ell)$ for some $a \in \Sigma$ and $1 \leq \ell \leq k$.

Example 1. Consider the automaton M_1 in Fig. 3 and the input $adcbcba$, processed in the order $aabbccd$. The ROWJFA jumps over the letter d three times before processing it, hence the number of sweeps is four. Moreover, its computation table is described in Fig. 2.

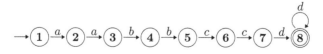

Fig. 3. ROWJFA M_1 accepting all w with $|w|_a = |w|_b = |w|_c = 2$ and $|w|_d \geq 1$.

In order to be able to analyze the boundary between regular and non-regular languages accepted by the one-way jumping model, as well as to quantify the use of resources beyond the capabilities of classical DFA, when it is the case, the following complexity measure was proposed [8], which gives us the number of sweeps performed by a machine in the 'worst case' for an input of length n.

Let M be a ROWJFA and $w \in L(M)$, and let

$$\mathbf{p}_0 w \vdash \mathbf{p}_1 w_1 \vdash \mathbf{p}_2 w_2 \vdash \cdots \vdash \mathbf{p}_m, \text{ where } \mathbf{p}_0 = s \text{ and } \mathbf{p}_m \in F,$$

be the computation of M on the input w. Sweep 1 consists of $\mathbf{p}_0 w \vdash^* \mathbf{p}_{|w|} w_{|w|}$, and we say that sweep 1 ends in configuration $\mathbf{p}_{|w|} w_{|w|}$. Then, for all $i \geq 1$, if sweep i ends in configuration $\mathbf{p}_{s_i} w_{s_i}$, then sweep $i + 1$ is the sequence of configurations $\mathbf{p}_{s_i} w_{s_i} \vdash^* \mathbf{p}_{s_i+|w_{s_i}|} w_{s_i+|w_{s_i}|}$. The last sweep ends in configuration $\mathbf{p}_m$, that is, when all input symbols have been read. We define

$$E(M, w) = \{\text{the number of sweeps performed by } M \text{ on } w\}.$$

When $w \notin L(M)$, then we set $E(M, w) = 0$. The *sweep complexity* of a machine M is a function $sc_M : \mathbb{N} \to \mathbb{N}$, with $sc_M(n)$ being the maximum number of sweeps M makes on processing inputs $w \in L(M)$ of length n, formally:

$$sc_M(n) = \max\{E(M, w) \mid w \in \Sigma^n\}.$$

In a sense the "most non-regular" word (using the largest amount of non-classical resources) of each length is considered. With this in mind, we can define complexity classes in the usual manner: the class $\text{SWEEP}(f(n))$ consists of languages accepted by some one-way jumping machine with sweep complexity $\mathcal{O}(f(n))$.

Observe that the sweep complexity of a machine can be defined to also take into account the sweep count of rejected words. However, this allows to 'artificially' increase the sweep complexity of machines with complexity $o(n)$ without affecting regularity. Let A be a machine accepting a regular language and B a non-regular language with sweep complexities $f(n)$ and $g(n)$, respectively, such that $f(n) \in o(g(n))$. Then we can construct a ROWJFA accepting $aL(A)$ with sweep complexity $g(n)$ by adding a new initial state from which reading a takes us to the initial state of A while reading b takes us to the initial state of B. We set all states of B non-final and this way we get that on inputs starting with b the machine performs B's computations but never accepts anything. Moreover, $aL(A)$ is regular if and only if $L(A)$ was (see Fig. 4).

Each machine considered up to the point when the above measures were introduced [8] had either constant or, the maximal possible, linear sweep complexity, so it seemed that there is a gap between them. Moreover, the examples with linear complexity accepted non-regular languages, while as the theorem below states, the constant complexity languages are exactly the regular languages.

Theorem 1 ([9]). *ROWJFA with $\mathcal{O}(1)$ sweep complexity accept regular languages.*

The sufficient condition above was conjectured to be also necessary for regularity in general, evidenced by the known examples at that point.

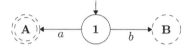

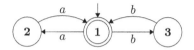

Fig. 4. Artificially increasing the automaton's complexity by adding non-functional states (all final states in **A**).

Fig. 5. ROWJFA $\mathcal{B}$ accepts $\{w \in \{a,b\}^* \mid |w|_a$ and $|w|_b$ are even$\}$ with sweep complexity $\Theta(\log n)$.

Next, we investigate the apparent gap between constant and linear complexities and show that the presumed condition above is not necessary for regularity. Our search for machines with non-constant sweep complexity is directed by the following structural lemma, which says that such machines need to have two 'complementary deficient states' in a cycle.

Lemma 1 ([8]). *If a ROWJFA has sweep complexity $\omega(1)$ then its state diagram has a closed walk with states $\mathbf{p}$ and $\mathbf{q}$, such that $\mathbf{p}au \to^* \mathbf{q}bv \to^* \mathbf{p}$ for $a, b \in \Sigma$, $u, v \in \Sigma^*$ and $\mathbf{p}$ has no transition defined for b, while $\mathbf{q}$ has no transition for a.*

3   Regular Languages with Non-constant Sweep Complexity

In this section we show that there is no sweep complexity separation between regular and non-regular languages by exhibiting automata which accept regular languages while requiring superconstant number of sweeps.

Consider first the automaton $\mathcal{B}$ with states $\{1, 2, 3\}$ where 1 is initial and final, and transitions are $\{1a \to 2, 2a \to 1, 1b \to 3, 3b \to 1\}$, described in Fig. 5.

Proposition 1. $L(\mathcal{B})$ *is regular.*

Proof. We claim that $L(\mathcal{B}) = \{w \in \{a,b\}^* \mid |w|_a$ and $|w|_b$ are even$\}$. This is obviously a regular language (i.e., Fig. 8 where **00** is the final state).

The computation for a word w is rejecting if it finishes in either **2** or **3**. However, the only time that the machine ends up in state **2** is when it reads an odd number of a's, and, similarly, it ends in **3** when it reads an odd number of b's. Since both of these types of words are rejected, we conclude.   □

Theorem 2. *The sweep complexity of $\mathcal{B}$ is $\Theta(\log n)$.*

Proof. Firstly, observe that in any sweep, while in **1** or **2**, the automaton fully reads any block of a's, and, similarly, while in **1** or **3**, the automaton fully reads any block of b's. Thus, the number of sweeps necessary to process a word w consisting of $2n$ unary blocks is never higher than that of processing the word $(ab)^n$. Now, for the inputs $(ab)^n$ (and $(ba)^n$), starting with the first b (respectively, a) every third symbol is jumped over while the rest is read. This means that from an arbitrary word with k unary blocks, after one sweep at

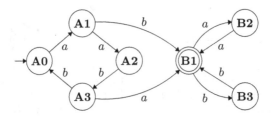

Fig. 6. ROWJFA $\mathcal{C}$ accepts $\{w \in \{a,b\}^* \mid |w|_a$ and $|w|_b$ are odd$\}$ with sweep complexity $\Theta(n)$.

most $\lfloor \frac{k}{3} \rfloor + 1$ blocks remain. This immediately gives us that the machine makes at most logarithmically many sweeps. As for the other side, consider an input $w = (ab)^{6^k}$. Per the previous argument, after $i \leq$ sweeps the remaining input will be $(ab)^{\frac{6^k}{3^i}}$ or $(ba)^{\frac{6^k}{3^i}}$ depending on the parity of i, so the number of sweeps is at least $\log_3 \frac{|w|}{2} = k$. Eventually, the input is accepted according to Proposition 1, so the sweep complexity of $\mathcal{B}$ is also $\Omega(\log n)$.   □

The above results showcase the existence of ROWJFAs that accept regular languages while performing a logarithmic number of sweeps. Next we construct of a ROWJFA that accepts a regular language while requiring a linear number of sweeps in the worst case. Consider the automaton $\mathcal{C}$ in Fig. 6 defined as

$$\mathcal{C} = \{\{\mathbf{A0, A1, A2, A3, B1, B2, B3}\}, \{a,b\}, R, \mathbf{A0}, \{\mathbf{B1}\}\},$$

where the transitions from R are given by the edges in the figure.

Proposition 2. *The sweep complexity of $\mathcal{C}$ is $\Theta(n)$.*

Proof. To see that the complexity is $\Omega(n)$, consider the word $a^{2n+1}b^{2n+1}$, for $n > 1$. In this case, from $\mathbf{A0}$ we go first to $\mathbf{A2}$ where we jump over all the remaining a's, then we move back to $\mathbf{A0}$ where we jump over all the remaining b's, and we are left with $a^{2n-1}b^{2n-1}$ to process. After the nth sweep, we are only left with ab to process, which takes us from $\mathbf{A0}$ to $\mathbf{B1}$, and we accept.

For the $\mathcal{O}(n)$ complexity, observe that the above computation is indeed the longest possible. Once we reach $\mathbf{B1}$ we either accept or reject a word in at most $\mathcal{O}(\log n)$ sweeps, same as in Theorem 2. Of course, this part also directly follows from the fact that all ROWJFA process their inputs in $\mathcal{O}(n)$ sweeps.   □

Proposition 3. *$L(\mathcal{C})$ is regular.*

Proof. We show that $L(\mathcal{C}) = \{w \in \{a,b\}^* \mid |w|_a$ and $|w|_b$ are odd$\}$. This is obviously a regular language (i.e., Fig. 8 where **11** is the final state).

To show that indeed $L(\mathcal{C})$ is the language containing every binary word that has odd number of a's and b's, first note that the right hand side automaton consisting only of the **B**-labelled states, accepts every language that has an even number of a's and b's, as shown by Proposition 1.

To reach **B1** we have to read exactly one a and one b starting from either **A0** or **A2**. Since from the start state **A0** we can reach **A0** or **A2** by processing an even number of a's and b's, possibly with jumps, our conclusion follows.   □

As a consequence of Propositions 2 and 3, we know that the class of regular languages has no upper bound in terms of sweep complexity, since the sweep complexity of any is in $\mathcal{O}(n)$. The left hand cycle in the automata $\mathcal{C}$ described in Fig. 6 also showcases that while the conditions from Lemma 1 are necessary for non-regularity (as it requires superconstant complexity), they are not sufficient.

4   Separation Results for the Language Classes SWEEP($\log n$) and SWEEP(n)

Consider the prolongable morphism $\varphi(a) = abab$, $\varphi(b) = b$ starting from the word ab. We get $\varphi(ab) = ababb$, $\varphi^2(ab) = \varphi(ababb) = ababbababbb$, etc. The infinite word $\phi = \lim_{n \to \infty} \varphi^n(ab) = ababbababbb\ldots$ is a fixed point of ϕ. It is easy to see that in ϕ all a's stand alone, that is, we never have blocks of a's longer than 1, and the lengths of the blocks of b's are $1, 2, 1, 3$, and so on[1]. When applying φ, each a introduces a new block of b's of length 1 and extends a block of b's by one, while the number of a's doubles. Thus every other block of b's gets longer by one on each application of φ, because of the a preceding it. A simple inductive argument shows that the last block of b's in $\varphi^n(ab)$ has length $n + 1$, and is preceded by 2^n occurrences of a's, separated by blocks of b's.

Lemma 2. *Consider the morphism $\varphi : \{a, b\}^* \to \{a, b\}^*$ given by $\varphi(a) = abab$, $\varphi(b) = b$. The following statements hold for any $n \geq 1$:*

(i) $\varphi^n(ab) \in (ababb^+)^+$;
(ii) if $\varphi^n(ab) = ab^{k_1} \cdots ab^{k_m}$, then $\varphi^{n+1}(ab) = abab^{k_1+1}abab^{k_2+1} \cdots abab^{k_m+1}$;
(iii) $\varphi^n(ab) = ab^{k_1} \cdots ab^{k_m}$, where $m = 2^n$, $k_m = n + 1$ and $k_{2i-1} = 1$ for all $i \in \{1, \ldots, 2^{n-1}\}$.

Proof. When $n = 1$, then $\varphi(ab) = ababb$, so for $n = 1$ all three claims hold. Suppose they hold for n. By (ii) and (iii) we have that $\varphi^{n+1}(ab)$ has the form $abab^{k_1+1}abab^{k_2+1} \cdots abab^{k_m+1}$, satisfying (i) for $n + 1$. Then,

$$\varphi^{n+2}(ab) = \varphi(abab^{k_1+1} \cdots abab^{k_m+1}) = \varphi(ab)\varphi(ab^{k_1+1}) \cdots \varphi(ab)\varphi(ab^{k_m+1})$$
$$= (abab^{1+1})(abab^{k_1+2}) \cdots (abab^{1+1})(abab^{k_m+2})$$

From this we can conclude that (ii) also holds for $n + 1 \geq 1$. Further, by the equation above we have $\varphi^{n+1}(ab) = ab^{\ell_1} \cdots ab^{\ell_{m'}}$ with $m' = 2m = 2 \cdot 2^n = 2^{n+1}$. Finally, because of (ii) we also get that $\ell_{m'} = k_m + 1 = n + 2$ and $\ell_{2i-1} = 1$ for all $i \in \{1, \ldots, 2^n\}$.   □

In what follows we analyze the language accepted by the automaton $\mathcal{D} = (\{1, 2, 3\}, \{a, b\}, \{1a \to 2, 2a \to 2, 2b \to 3, 3b \to 1\}, 1, \{3\})$, described in Fig. 7.

---

[1] The sequence $\{c(n)\}_{n=1}^{\infty}$ given by the lengths of b blocks is A001511 in OEIS; its most relevant characterization for us is that $c(n) - 1$ is the number of trailing zeros in the binary expansion of n, since this means that $c(n) - 1$ is $\log n$ for powers of 2.

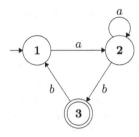

Fig. 7. ROWJFA $\mathcal{D}$ accepts a non-regular language with $\Theta(\log n)$ sweeps.

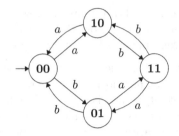

Fig. 8. DFA accepting words with even (for **00** final state) or odd (for **11** final state) number of a's and b's.

Lemma 3. *For any $n \geq 0$, the ROWJFA $\mathcal{D}$ accepts $\varphi^n(ab)$ in $n+1$ sweeps.*

Proof. We show that the machine accepts $\varphi^n(ab)$, for any $n \geq 0$. From state **1** after reading/jumping through a factor of the form $abab b^+$ the automaton gets back to state **1**. In fact, $1abab^k w \vdash^* 1wab^{k-1}$, for any $k \geq 1$, so in one sweep the factor $abab^k$ is reduced to ab^{k-1}. From Lemma 2 we can see that we can write $\varphi^{n+1}(ab) = abab^{k_1+1}abab^{k_2+1} \cdots abab^{k_m+1}$, which means that one sweep of $\mathcal{D}$ acts as the inverse of φ on those words when starting from state **1**, that is,

$$1\varphi^{n+1}(ab) = 1abab^{k_1+1}abab^{k_2+1} \cdots abab^{k_m+1} \vdash^* 1ab^{k_1}ab^{k_2} \cdots ab^{k_m} = 1\varphi^n(ab).$$

This means that in n sweeps the machine reduces $\varphi^n(ab)$ to $\varphi^0(ab)$. Finally, for $n = 0$, we have $\varphi^0(ab) = ab$, which is accepted by $\mathcal{D}$ in a single sweep. □

Lemma 4. *The ROWJFA $\mathcal{D}$ accepts a non-regular language.*

Proof. By Lemma 3 we know that for any n the machine accepts $\varphi^n(ab)$, which means that for arbitrarily long unary factors consisting of b's, there is some word in $L(\mathcal{D})$ having such a factor as a suffix. Our strategy is to first establish a non-linear relation between the length of those unary factors and the length of the preceding factors in all words accepted by $\mathcal{D}$. Then, by a pumping argument we show that a classical finite automaton cannot verify such a non-linear relation, therefore $L(\mathcal{D})$ cannot be regular.

Claim 1. *Words of the form wb^n are only accepted if $|w| \in \Omega(2^{\frac{n}{2}})$.*
Proof of Claim 1: In any sweep, any block of a's which the automaton starts to read is read and erased completely through a sequence of transitions $1a^k bu \rightarrow^* 2bu$. For the automaton to jump over a block of a's, it needs to arrive to its start in state **3**. Then it jumps over it to the next b, after which it starts and reads completely the following block of a's, as described earlier. This means that the machine can never jump over two consecutive blocks of a's. From here we get that if at the beginning of the sweep the number of a blocks was ℓ, then after the sweep it is at most $\lfloor \frac{\ell}{2} \rfloor + 1$.

Furthermore, in each sweep, each block of b's is reduced by at most 2. This means that the automaton needs at least $\frac{n}{2}$ sweeps to read a block b^n, in each of which it reduces the number of a blocks by half (or more). Thus we can conclude that in order to accept a word with a suffix b^n, we have to start out with at least $2^{\frac{n}{2}}$ blocks of a's preceding it. $\qquad\qquad$ ▽

Claim 2. No finite automaton can accept $L(\mathcal{D})$.

Proof of Claim 2: Suppose the opposite, i.e., that there exists some complete DFA $\mathcal{F}$ having N states such that $L(\mathcal{F}) = L(\mathcal{D})$. We know that there are words in the language with arbitrarily long suffixes of b's, so there is a $wb^m \in L(\mathcal{F})$ for some word w and exponent $m > N$. By a usual pumping argument, this means that there exists some ℓ with $0 < \ell < N$ such that $wb^{m+i\cdot\ell} \in L(\mathcal{F})$ for any $i \geq 0$. However, for a large enough i this contradicts Claim 1, as the block of b's can outgrow any upper bound in terms of the length of $|w|$. $\qquad\qquad$ ▽

Our result follows as a result of Claims 1 and 2. $\qquad\qquad$ □

Lemma 5. *The sweep complexity of $\mathcal{D}$ is $\Theta(\log n)$.*

Proof. As $|\varphi^n(ab)| = 2^{n+1} + 2^n - 1$, by Lemma 3 we have that the sweep complexity of $\mathcal{D}$ is $\Omega(\log n)$, so what remains to show is that it is also $\mathcal{O}(\log n)$.

We first note that within a sweep all blocks of a's separated by bb are fully processed (including any prefix of a's), while for any symbols a that were jumped over, the entire block that they were part of it was jumped over. Following the argument in the proof of Claim 1 of Lemma 4, in each sweep the number of blocks of a's is reduced by at least half, which means that after $\mathcal{O}(\log n)$ sweeps there are no more blocks of a on the tape. Then, the machine either accepts in one sweep or it rejects the input. This leads to our conclusion. $\qquad\qquad$ □

The results of Lemmas 4 and 5 mean that we have separation between SWEEP(1) and SWEEP($\log n$).

Theorem 3. *SWEEP(1) $\subsetneq$ SWEEP($\log n$)*

Proof. Lemma 5 says $L(\mathcal{D}) \in$ SWEEP($\log n$). By Theorem 1 we know that SWEEP(1) is included in the class of regular languages. Finally, by Lemma 4 we have that $L(\mathcal{D})$ is not regular which means that $L(\mathcal{D}) \notin$ SWEEP(1). $\qquad$ □

Lemma 6. *Any automaton which accepts $L_{ab} = \{w \in \{a, b\}^* \mid |w|_a = |w|_b\}$ has sweep complexity $\Theta(n)$.*

Proof. We know that every machine has sweep complexity $\mathcal{O}(n)$, so it is enough to show that it is not possible to accept L_{ab} with sublinear sweep complexity. For that we assume that such an automaton, say $\mathcal{F} = (Q, \Sigma, R, s, F)$ exists, and derive a contradiction.

If $\mathcal{F}$ had linear sweep complexity, then it could have computations on infinitely many inputs in which all sweeps process a constant number of symbols. However, with sublinear complexity we get that for any constant C and for all long enough inputs $w \in L_{ab}$, during the processing of w at least one sweep reads more than C symbols. We also know that $a^n b^n \in L_{ab}$ for any $n \geq 0$. Let $C = 2|Q|$ where $|Q|$ is the number of states of $\mathcal{F}$ and consider an input $w = a^m b^m$ with m large enough that the machine reads more than C symbols in some sweep while processing w. The remaining input at the beginning of that sweep is $a^k b^\ell$ for some k, ℓ such that $k + \ell > C$. During the sweep the machine reads $a^{k'} b^{\ell'}$ where $k' + \ell' > C$. This means that either $k' > |Q|$ or $\ell' > |Q|$. Without loss of generality we can assume $k' > |Q|$. This gives us that while reading $a^{k'}$ the automaton must visit some state **p** at least twice while reading only a's, so we get that $\mathbf{p}a^r \to^* \mathbf{p}$ for some $r > 0$. But then, by a usual pumping argument the machine also needs to accept $a^{n+r} b^n \notin L_{ab}$ contradicting our assumption that $L(\mathcal{F}) = L_{ab}$ and concluding the proof. □

Theorem 4. *For any $f : \mathbb{N} \to \mathbb{N}$ with $f(n) \in o(n)$ we have $SWEEP(f(n)) \subsetneq SWEEP(n)$.*

Proof. By Lemma 6 we know that $L_{ab} \notin \mathrm{SWEEP}(f(n))$ for any sublinear function $f(n)$. The two-state automaton $\mathcal{A}$ accepts the language with sweep complexity $\Theta(n)$. This is easy to see when considering the worst-case inputs of the form $a^n b^n$ for $n \geq 0$. □

5 Concluding Remarks

Apart from the complexity considerations listed below we think the proof of Lemma 4 contains a detail worth emphasizing: the automaton can verify a logarithmic/exponential relation between two factors of suitably chosen inputs! We found this very surprising since we still basically deal with DFA which cannot store information and cannot 'choose' which symbols to read or jump over[2].

We presented automata for all pairings of regular and non-regular languages with logarithmic and linear worst case sweep complexity. This way we disproved the conjecture on the constant sweep requirement for regularity [9] and answered several questions regarding sweep complexity posed in [8]:

1. Is the language of each machine with $\omega(1)$ complexity non-regular? NO, by Sect. 3.
2. Is there a machine with sweep complexity between constant and linear, that is, $\omega(1)$ and $o(n)$? YES, by Theorem 2 (and Lemma 5).
3. Is there a *language* with sweep complexity between constant and linear, that is, all machines accepting it have superconstant complexity and at least one has sublinear? YES, by Theorem 3.
4. Is there an upper bound in terms of sweep complexity on machines accepting regular languages? NO, by Propositions 2 and 3.
5. Are machines less complex in the case of binary alphabets, given that the complementary deficient pairs of Lemma 1 are predetermined? NO, illustrated by the fact that all results have been obtained over a binary alphabet.

These coarser forms of Questions 2 and 3 have been answered here, but for a complete picture one would want to know whether there exist machines

---

[2] Iterated uniform finite transducers can also verify such relationships, albeit their computing power is much stronger. [11].

with arbitrary (constructible) sublinear complexity and its equivalent for languages. The most obvious choices for such a study would probably be complexities $\Theta(\log^k n)$ and $\Theta(n^\epsilon)$, for constants $k > 1$ and $\epsilon < 1$. Another angle related to Question 4 is to study the lower bound of non-regularity: logarithmic complexity can produce non-regular languages, but can we do it with less of this 'non-regular' resource? In the case of Question 5, our answer may be refined, as there may by some sublinear $f(n)$ such that the machines of $\Theta(f(n))$ complexity all accept regular or all accept non-regular languages, although we have not seen anything that indicates such perplexing behaviour.

Another interesting direction relates to our original motivation in looking at the complexity of these automata, deciding regularity. The question more generally becomes, is it decidable given a machine or language and a function $f(n)$, whether the machine/language has $\Theta(f(n))$ complexity (or its one-sided variants with $\mathcal{O}$ and Ω)? We suspect that the answer is yes at least in the case of constant and linear functions but have no idea about the logarithmic and more complicated cases.

References

1. Beier, S., Holzer, M.: Properties of right one-way jumping finite automata. Theoret. Comput. Sci. **798**, 78–94 (2019)
2. Beier, S., Holzer, M.: Decidability of right one-way jumping finite automata. Int. J. Found. Comput. Sci. **31**(6), 805–825 (2020)
3. Beier, S., Holzer, M.: Nondeterministic right one-way jumping finite automata. Inf. Comput. **284**, 104687 (2022). selected papers from DCFS 2019
4. Bensch, S., Bordihn, H., Holzer, M., Kutrib, M.: On input-revolving deterministic and nondeterministic finite automata. Inf. Comput. **207**(11), 1140–1155 (2009)
5. Chigahara, H., Fazekas, S.Z., Yamamura, A.: One-way jumping finite automata. Int. J. Found. Comput. Sci. **27**(3), 391–405 (2016)
6. Fazekas, S.Z., Hoshi, K., Yamamura, A.: The effect of jumping modes on various automata models. Nat. Comput. **21**(1), 17–30 (2021). https://doi.org/10.1007/s11047-021-09844-4
7. Fazekas, S.Z., Hoshi, K., Yamamura, A.: Two-way deterministic automata with jumping mode. Theor. Comput. Sci. **864**, 92–102 (2021)
8. Fazekas, S.Z., Mercaş, R., Wu, O.: Complexities for jumps and sweeps. J. Autom. Lang. Comb. **27**(1–3), 131–149 (2022)
9. Fazekas, S.Z., Yamamura, A.: On regular languages accepted by one-way jumping finite automata. In: 8th Workshop on Non-Classical Models of Automata and Applications, Short Papers, pp. 7–14 (2016)
10. Jančar, P., Mráz, F., Plátek, M., Vogel, J.: Restarting automata. In: Reichel, H. (ed.) FCT 1995. LNCS, vol. 965, pp. 283–292. Springer, Heidelberg (1995). https://doi.org/10.1007/3-540-60249-6_60
11. Kutrib, M., Malcher, A., Mereghetti, C., Palano, B.: Descriptional complexity of iterated uniform finite-state transducers. Inf. Comput. **284**, 104691 (2022)
12. Meduna, A., Zemek, P.: Jumping finite automata. Int. J. Found. Comput. Sci. **23**(7), 1555–1578 (2012)
13. Nagy, B., Otto, F.: Finite-state acceptors with translucent letters. In: BILC 2011–1st International Workshop on AI Methods for Interdisciplinary Research in Language and Biology, ICAART 2011, pp. 3–13 (2011)

The Pumping Lemma for Regular Languages is Hard

Hermann Gruber[1], Markus Holzer[2(✉)], and Christian Rauch[2]

[1] Planerio GmbH, Gewürzmühlstr. 11, 80538 München, Germany
h.gruber@planerio.de
[2] Institut für Informatik, Universität Giessen, Arndtstr. 2, 35392 Giessen, Germany
{holzer,christian.rauch}@informatik.uni-giessen.de

Abstract. We investigate the computational complexity of the PUMP-ING-PROBLEM, that is, for a given finite automaton A and a value p, to decide whether the language $L(A)$ satisfies a previously fixed pumping lemma w.r.t. the value p. Here we concentrate on two different pumping lemmata from the literature. It turns out that this problem is intractable, namely, it is already coNP-complete, even for deterministic finite automata (DFAs), and it becomes PSPACE-complete for nondeterministic finite state devices (NFAs), for at least one of the considered pumping lemmata. In addition we show that the minimal pumping constant for the considered particular pumping lemmata cannot be approximated within a factor of $o(n^{1-\delta})$ with $0 \leq \delta \leq 1/2$, for a given n-state NFA, unless the Exponential Time Hypothesis (ETH) fails.

1 Introduction

The syllabus on courses of automata theory and formal languages certainly contains the introduction of pumping lemmata for regular and context-free languages in order to prove non-regularity or non-context freeness, respectively, of certain languages. Depending on the preferences of the course instructor and the used monograph, different variants of pumping lemmata are taught. For instance, consider the following pumping lemma, or iteration lemma, that can be found in [9, page 70, Theorem 11.1], which describes a necessary condition for languages to be regular.

Lemma 1. *Let L be a regular language over Σ. Then, there is a constant p (depending on L) such that the following holds: If $w \in L$ and $|w| \geq p$, then there are words $x \in \Sigma^*$, $y \in \Sigma^+$, and $z \in \Sigma^*$ such that $w = xyz$ and $xy^t z \in L$ for $t \geq 0$—it is then said that y can be* pumped *in w.*

A lesser-known pumping lemma, attributed to Jaffe [8], characterizes the regular languages, by describing a necessary and sufficient condition for languages to be regular. For other pumping lemmata see, e.g., the annotated bibliography on pumping [10]:

© The Author(s), under exclusive license to Springer Nature Switzerland AG 2023
B. Nagy (Ed.): CIAA 2023, LNCS 14151, pp. 128–140, 2023.
https://doi.org/10.1007/978-3-031-40247-0_9

Lemma 2. *A language L is regular if and only if there is a constant p (depending on L) such that the following holds: If $w \in \Sigma^*$ and $|w| = p$, then there are words $x \in \Sigma^*$, $y \in \Sigma^+$, and $z \in \Sigma^*$ such that $w = xyz$ and*[1]

$$wv = xyzv \in L \iff xy^t zv \in L$$

for all $t \geq 0$ and each $v \in \Sigma^*$.

For a regular language L the value of p in Lemma 1 can always be chosen to be the number of states of a finite automaton, regardless whether it is deterministic or nondeterministic, accepting L. Consider the unary language $a^n a^*$, where all values p with $0 \leq p \leq n$ do *not* satisfy the properties of Lemma 1, but $p = n+1$ does. A closer look on some example languages reveals that sometimes a much smaller value suffices. For instance, consider the language

$$L = a^* + a^* bb^* + a^* bb^* aa^* + a^* bb^* aa^* bb^*,$$

which is accepted by a (minimal) deterministic finite automaton with five states, the sink state included—see Fig. 1. Already for $p = 1$ the statement of Lemma 1 is satisfied since regardless whether the considered word starts with a or b, this letter can be readily pumped. Thus, the minimal pumping constant satisfying the statement of Lemma 1 for the language L is 1, because the case $p = 0$ is equivalent to $L = \emptyset$ [3]. This already shows that the minimal pumping constant w.r.t. Lemma 1 is non-recursively related to the deterministic state complexity of a language. For the pumping constant w.r.t. Lemma 2 the situation is even more involved. Here the value of p can only be chosen to be the number of states of a deterministic finite automaton accepting the language in question. In fact, the gap between the nondeterministic state complexity and the minimal pumping constant p satisfying Lemma 2 for a specific language can be arbitrarily large—cf. Theorem 3. Moreover, the relation between the minimal pumping constant p w.r.t. Lemma 2 is related to the deterministic state complexity n of a language L over the alphabet Σ by the inequality $p \leq n \leq \sum_{i=0}^{p-1} |\Sigma|^i$, which was recently shown in [5]. A careful inspection of the language L mentioned above reveals that the minimal pumping constant p when considering Jaffe's pumping lemma is equal to five. Any word of length at least five uses a loop-transition in its computation—see Fig. 1—and hence the letter on this loop-transition can be pumped in the word under consideration. Thus $\mathtt{mpe}(L) \leq 5$. On the other hand, the word $baba$ cannot be pumped such that the Myhill-Nerode equivalence classes are respected, because the word xz belongs to a different equivalence class than xyz, for every decomposition of $baba$ into x, y, and z with $|y| \geq 1$—again, see Fig. 1. Therefore, $\mathtt{mpe}(L) > 4$ and thus, $\mathtt{mpe}(L) = 5$. This example shows that the minimal pumping constant is *not* equivalent to the length of the shortest accepting path in (the minimal) automaton accepting the language. The authors experienced that this is the most common misconception of the minimal pumping constant when discussing this for the first time.

---

[1] Observe that the words $w = xyz$ and $xy^t z$, for all $t \geq 0$, belong to the same Myhill-Nerode equivalence class of the language L. Thus, one can say that the pumping of the word y in w *respects equivalence classes.*

Fig. 1. The minimal deterministic finite automaton A accepting the language L.

This gives rise to the following definition of a *minimal pumping constant*: for a regular language L let $\mathrm{mpc}(L)$ ($\mathrm{mpe}(L)$, respectively) refer to the minimal number p satisfying the conditions of Lemma 1 (Lemma 2, respectively). Recently, in [3,5] minimal pumping constants w.r.t. the above two lemmata were investigated from a descriptional complexity perspective. Here we focus more on the computational complexity of pumping, a problem that to our knowledge was not considered so far. This is even more surprising, since pumping lemmata are omnipresent in theoretical courses in automata theory and formal languages. We will consider the following problem related to the pumping lemma stated above:

LANGUAGE-PUMPING-PROBLEM or for short PUMPING-PROBLEM:
  INPUT: a finite automaton A and a natural number p, i.e., an encoding $\langle A, 1^p \rangle$.
  OUTPUT: Yes, if and only if the statement from Lemma 1 holds for the language $L(A)$ w.r.t. the value p.

A similar definition applies when considering the condition of Lemma 2 instead. We summarize our findings on the computational complexity of the PUMPING-PROBLEM in Table 1.

2   Preliminaries

We assume the reader to be familiar with the basics in computational complexity theory [11]. In particular we recall the inclusion chain: P $\subseteq$ NP $\subseteq$ PSPACE. Here P (NP, respectively) denotes the class of problems solvable by deterministic (nondeterministic, respectively) Turing machines in polytime, and PSPACE refers to the class of languages accepted by deterministic or nondeterminis-

Table 1. Complexity of the PUMPING-PROBLEM for variants of finite state devices.

PUMPING-PROBLEM w.r.t. ...	
Lemma 1	Lemma 2
DFA	coNP-complete
NFA	coNP-hard contained in Π_2^P	PSPACE-compl.
	No det. $2^{o\left(s^\delta\right)}$-time approx. within $o(s^{1-\delta})$, unless ETH fails.	

tic Turing machines in polynomial space [14]. As usual, the prefix co refers to the complement class. For instance, coNP is the class of problems that are complements of NP problems. Moreover, recall the complexity class Π_2^P from the polynomial hierarchy, which can be described by polynomial time bounded oracle Turing machines. Here $\Pi_2^P = \mathrm{coNP}^{\mathrm{NP}}$, where coNP^A is the set of decision problems solvable in polynomial time by a universal Turing machine with an oracle for some complete problem in class A. The class NP^{A} is defined analogously. Completeness and hardness are always meant with respect to deterministic many-one logspace reducibilities ($\leq_{m}^{\log}$) unless stated otherwise.

Next we fix some definitions on finite automata—cf. [4]. A *nondeterministic finite automaton* (NFA) is a quintuple $A = (Q, \Sigma, \cdot, q_0, F)$, where Q is the finite set of *states*, Σ is the finite set of *input symbols*, $q_0 \in Q$ is the *initial state*, $F \subseteq Q$ is the set of *accepting states*, and the *transition function* $\cdot$ maps $Q \times \Sigma$ to 2^Q. Here 2^Q refers to the powerset of Q. The *language accepted* by the NFA A is defined as $L(A) = \{\, w \in \Sigma^* \mid (q_0 \cdot w) \cap F \neq \emptyset \,\}$, where the transition function is recursively extended to a mapping $Q \times \Sigma^* \to 2^Q$ in the usual way. An NFA A is said to be *partial deterministic* if $|q \cdot a| \leq 1$ and *deterministic* (DFA) if $|q \cdot a| = 1$ for all $q \in Q$ and $a \in \Sigma$. In these cases we simply write $q \cdot a = p$ instead of $q \cdot a = \{p\}$. Note that every partial DFA can be made complete by introducing a non-accepting sink state that collects all non-specified transitions. For a DFA, obviously every letter $a \in \Sigma$ induces a mapping from the state set Q to Q by $q \mapsto q \cdot a$, for every $q \in Q$. Finally, a finite automaton is *unary* if the input alphabet Σ is a singleton set, that is, $\Sigma = \{a\}$, for some input symbol a.

The *deterministic state complexity of a finite automaton* A with state set Q is referred to as $\mathsf{sc}(A) := |Q|$ and the *deterministic state complexity of a regular language* L is defined as

$$\mathsf{sc}(L) = \min\{\, \mathsf{sc}(A) \mid A \text{ is a } DFA \text{ accepting } L, \text{i.e., } L = L(A) \,\}.$$

A similar definition applies for the *nondeterministic state complexity of a regular language* by changing DFA to NFA in the definition, which we refer to as $\mathsf{nsc}(L)$. It is well known that

$$\mathsf{nsc}(L) \leq \mathsf{sc}(L) \leq 2^{\mathsf{nsc}(L)},$$

for every regular language L.

A finite automaton is *minimal* if its number of states is minimal with respect to the accepted language. It is well known that each minimal DFA is isomorphic to the DFA induced by the Myhill-Nerode equivalence relation. The *Myhill-Nerode* equivalence relation $\sim_L$ for a language $L \subseteq \Sigma^*$ is defined as follows: for $u, v \in \Sigma^*$ let $u \sim_L v$ if and only if $uw \in L \iff vw \in L$, for all $w \in \Sigma^*$. The equivalence class of u is referred to as $[u]_L$ or simply $[u]$ if the language is clear from the context and it is the set of all words that are equivalent to u w.r.t. the relation $\sim_L$, i.e., $[u]_L = \{\, v \mid u \sim_L v \,\}$. Therefore we refer to the automaton induced by the Myhill-Nerode equivalence relation $\sim_L$ as the minimal DFAfor the language L. On the other hand there may be minimal non-isomorphic NFAs for L.

3   The Complexity of Pumping

Before we start with the investigation of the complexity of pumping problems we list some simple facts about the minimal pumping constants known from the literature: (1) $\mathtt{mpc}(L) = 0$ if and only if $L = \emptyset$, (2) for every non-empty finite language L we have $\mathtt{mpc}(L) = 1 + \max\{\,|w| \mid w \in L\,\}$, (3) $\mathtt{mpc}(L) = 1$ implies for the empty word λ that $\lambda \in L$, and (4) $\mathtt{mpe}(L) = 1$ if and only if $L = \emptyset$ or $L = \Sigma^*$. Also the inequalities $\mathtt{mpc}(L) \leq \mathtt{sc}(L) \leq \mathtt{nsc}(L)$ hold—see, e.g., [3,5]. For a finite languages we have $\mathtt{mpe}(L) = 2 + \max\{|w| \mid w \in L\}$ if each w with $|w| = \max\{|w| \mid w \in L\}$ is contained in L. Otherwise $\mathtt{mpe}(L) = 1 + \max\{|w| \mid w \in L\}$ holds. The relation of $\mathtt{mpe}(L)$ and the state complexities is more subtle, namely for a regular language over the alphabet Σ it was shown in [5] that

$$\mathtt{mpe}(L) \leq \mathtt{sc}(L) \leq \sum_{i=0}^{\mathtt{mpe}(L)-1} |\Sigma|^{i}.$$

On the other hand, $\mathtt{mpe}(L)$ and $\mathtt{nsc}(L)$ are incomparable in general as we see next—the automaton used to prove the second statement of the following Theorem is shown in Fig. 2:

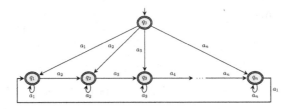

Fig. 2. The nondeterministic finite automaton A_n accepting $L_n = L(A_n)$ satisfying $\mathtt{mpe}(L_n) = 3$ and $\mathtt{nsc}(L_n) \geq n$.

Theorem 3. *The following statements hold: (1) There is a family of unary regular languages $(L_i)_{i \geq 2}$ such that the inequality $\mathtt{nsc}(L_i) < \mathit{mpe}(L_i)$ holds. (2) There is a family of regular languages $(L_i)_{i \geq 4}$ over a growing size alphabet such that $3 = \mathit{mpe}(L_i) < i \leq \mathtt{nsc}(L_i)$.*

Observe, that the automaton depicted in Fig. 2 is a partial DFA, where only the non-accepting sink state is missing. It is worth mentioning that this finite state device shows that even for DFAs the $\mathtt{mpe}$-measure and the *longest path* of the automaton, that is, a simple[2] directed path of maximum length starting in the initial state of the automaton, are different measures in general. Nevertheless, the following observation is immediate by Jaffe's proof, cf. [8]:

Lemma 4. *Let A be a DFA and $L := L(A)$. Then $\mathit{mpe}(L) \leq \ell_A + 1$, where ℓ_A is the length, i.e., number of transitions, of the longest path of the automaton A. If L is a unary language, then $\mathit{mpe}(L) = \mathtt{sc}(L)$.* ☐

---

[2] Here a path is called *simple* if it does not have repeated states/vertices.

3.1   The PUMPING-PROBLEM for NFAs

We start our journey with results on the complexity of the PUMPING-PROBLEM for NFAs. First we consider the problem w.r.t. Lemma 1 and later in the subsection w.r.t. Lemma 2.

Theorem 5. *Given an NFA A and a natural number p, it can be decided in Π_2^P whether for the language $L(A)$ the statement of Lemma 1 holds for the value p.*

In order to prove the theorem, we establish an auxiliary lemma.

Lemma 6. *Given an NFA $A = (Q, \Sigma, \cdot, q_0, F)$ and a word w over Σ, the language inclusion problem for w^* in $L(A)$ is coNP-complete. The variant of the problem where A is deterministic is L-complete.*

Now we are ready to prove the upper bound for the PUMPING-PROBLEM:

Proof (of Theorem 5). We show that the pumping problem for NFAs belongs to Π_2^P. Let $\langle A, p \rangle$ be an input instance of the problem in question, where Q is the state set of A. We construct a coNP Turing machine M with access to a coNP oracle: first the device M deterministically verifies whether $p \geq |Q|$, and if so halts and accepts. Otherwise the computation universally guesses ($\forall$-states) a word w with $p \leq |w| < |Q|$. On that particular branch M checks deterministically if w belongs to $L(A)$. If this is *not* the case the computation halts and accepts. Otherwise, M deterministically cycles through all valid decompositions $w = xyz$ with $|y| \geq 1$. Then it constructs a finite automaton B accepting the language quotient $(x^{-1} \cdot L(A)) \cdot z^{-1}$. Here, if A is deterministic, then so is B. Then M decides whether $y^* \subseteq L(B)$ with the help of the coNP oracle—compare Lemma 6. If $y^* \subseteq L(B)$, then the cycling through the valid decompositions is stopped, and the device M halts and accepts. Notice that the latter is the case iff $xy^*z \subseteq L(A)$. Otherwise, i.e., if $y^* \not\subseteq L(B)$, the Turing machine M continues with the next decomposition in the enumeration cycle. Finally, if the cycle computation finishes, the Turing machine halts and rejects, because no valid decomposition of w was found that allows for pumping. In summary, the Turing machine operates universally, runs in polynomial time, and uses a coNP oracle. Thus, the containment within Π_2^P follows.   □

Next we show that the problem in question is coNP-hard and gives us a nice non-approximability result under the assumption of the so-called *Exponential-Time Hypothesis (ETH)* [2,7]: there is no deterministic algorithm that solves 3SAT in time $2^{o(n+m)}$, where n and m are the number of variables and clauses, respectively.

Note that the unary regular language $T_p = a^{p-1}a^*$ satisfies $\mathtt{mpc}(T_p) = p$. The languages T_p will be a basic building block for our reduction. We build upon the classical coNP-completeness proof of the inequality problem for regular expressions without star given in [6, Thm. 2.3]. We modify the reduction a bit, since we want to deal with only one parameter in the analysis to come, and that is the number of clauses.

Theorem 7. *Let φ be a formula in 3DNF with n variables and m clauses. Then a nondeterministic finite automaton A_φ can be computed in time $O(m^2)$ such that the language $Z = L(A_\varphi)$ is homogeneous[3] and Z equals $\{0,1\}^{3m}$ if and only if φ is a tautology. Furthermore, A_φ has $O(m^2)$ states.*

The last $3m - n$ positions of the words in the homogeneous language used for the above reduction do not convey any information; they simply serve the purpose of avoiding a parameterization by n. In order to finish our reduction, we embed Z into the language $Y = Z \cdot \# \cdot \Sigma^* + \{0,1\}^{3m} \cdot \# \cdot T_p$, for some carefully chosen p, where $\#$ is a new letter not contained in Σ. This reduction runs in polynomial time.

Lemma 8. *Let φ be a formula in 3DNF with n variables and m clauses and let A_φ be the NFA constructed in Theorem 7. Furthermore, let*

$$Y = Z \cdot \# \cdot \Sigma^* + \{0,1\}^{3m} \cdot \# \cdot T_p,$$

for $p \geq 2$. Then the minimal pumping constant of Y w.r.t. Lemma 1 is equal to $3m + 2$, if $Z = \{0,1\}^{3m}$, and is equal to $3m + 1 + p$ otherwise.

As a direct corollary we obtain:

Corollary 9. *Given an NFA A and a natural number p in unary, it is coNP-hard to decide whether for the language $L(A)$ the statement from Lemma 1 holds for the value p.* □

Next we prove the following main result:

Theorem 10. *Let A be an NFA with s states, and let δ be a constant such that $0 < \delta \leq 1/2$. Then no deterministic $2^{o(s^\delta)}$-time algorithm can approximate the minimal pumping constant w.r.t. Lemma 1 within a factor of $o(s^{1-\delta})$, unless ETH fails.*

Proof. We give a reduction from the 3DNF tautology problem as in Lemma 8. That is, given a formula φ in 3DNF with n variables and m clauses, we construct an NFA A that accepts the language $Y = Z \cdot \# \cdot \Sigma^* + \{0,1\}^{3m} \cdot \# \cdot T_p$. Here, the set Y features some carefully chosen parameter p, which will be fixed later on. For now, we only assume $p \geq 2$.

By Lemma 8, the reduction is correct in the sense that if φ is a tautology, then the minimal pumping constant w.r.t. Lemma 1 is strictly smaller than in the case where it admits a falsifying assignment.

Observe that the running time of the reduction is linear in the number of states of the constructed NFA describing Y. It remains to estimate that number of states. Recall from Theorem 7 that the number of states in the NFA A_φ is of order $O(m^2)$. The set $Z \cdot \# \cdot \Sigma^*$ thus admits an NFA with $O(m^2)$ states; and the language $\{0,1\}^{3m} \cdot \# \cdot T_p$ can be accepted by an NFA with $O(m+p)$ states. Altogether, the number of states is in $O(m^2 + p)$.

---

[3] A language $L \subseteq \Sigma^*$ is *homogeneous* if all words in L are of same length.

Now we need to fix the parameter p in our reduction; let us pick $p = m^{\frac{1}{\delta}}$, where it is understood that we round up to the next integer. Given that δ is constant, for large enough m, we can ensure that $p \geq 2$. So this is a valid choice for the parameter p—in the sense that the reduction remains correct.

Towards a contradiction with the ETH, assume that there is an algorithm A_δ approximating the pumping constant within $o(s^{1-\delta})$ running in time $2^{o(s^\delta)}$. Then algorithm A_δ could be used to decide whether $Z = \{0,1\}^{3m}$ as follows: the putative approximation algorithm A_δ returns a cost C that is at most $o(s^{1-\delta})$ times the optimal cost C^*, that is, $C = o(s^{1-\delta}) \cdot C^*$.

We consider two cases: if $Z = \{0,1\}^{3m}$, then the pumping constant is in $O(m)$ by Lemma 8. In this case, the hypothetical approximation algorithm A_δ returns a cost C with

$$C = o(m \cdot s^{1-\delta}) = o\left(m \cdot O\left(m^2 + m^{\frac{1}{\delta}}\right)^{1-\delta}\right)$$

$$= o\left(m \cdot O\left(m^{\frac{1}{\delta}}\right)^{1-\delta}\right) = o(m^{\frac{1}{\delta}}) = o\left(m^{\frac{1}{\delta}} + m\right) = o\left(m + p\right);$$

where in the second step of the above calculation, we used the fact that $s = O(m^2 + p) = O(m^2 + m^{\frac{1}{\delta}})$; in the third step, we applied the inequality $\frac{1}{\delta} \geq 2$ to see that the term $m^{\frac{1}{\delta}}$ asymptotically dominates the term m^2; the fourth step is a simple term transformation; and the last two steps apply these facts in reverse order.

On the other hand, in case Z is not full, then Lemma 8 states that the pumping constant is in $\Omega(m + p)$. Using the constants implied by the O-notation, the size returned by algorithm A_δ could thus be used to decide, for large enough m, whether Z is full, and thus by Theorem 7 whether the 3DNF formula φ is a tautology.

It remains to show that the running time of A_δ in terms of the number of clauses m is in $2^{o(m)}$, which contradicts the ETH. Recall that $s = O(m^2 + p)$ and $p = m^{\frac{1}{\delta}}$ with $\frac{1}{\delta} \geq 2$; we thus can express the running time of the algorithm A_δ in terms of the number of clauses m, namely,

$$2^{o(s^\delta)} = 2^{o((m^2+p)^\delta)} = 2^{o\left((m^{\frac{1}{\delta}})^\delta\right)} = 2^{o(m)},$$

which yields the desired contradiction to the ETH.   □

A careful inspection of Lemma 8 and Theorem 10 reveals that both results remain valid if one considers the minimal pumping constant w.r.t. Lemma 2, since $\mathtt{mpe}(T_p) = p$ as the interested reader may easily verify.

Although, we have to leave open the exact complexity of the PUMPING-PROBLEM for NFAs w.r.t. Lemma 1—coNP-hard and contained in Π_2^P, we can give a precise answer if we consider Jaffe's pumping lemma instead. First we establish an auxiliary theorem.

Theorem 11. *Given a DFA $A = (Q, \Sigma, \cdot, q_0, F)$ and a deterministic or non-deterministic finite automaton B, deciding whether every word $w \in L(B)$ describes the same equivalence class w.r.t. the Myhill-Nerode relation $\sim_{L(A)}$, is NL-complete. If the automaton A is an NFA, the problem becomes PSPACE-complete.*

This allows us to prove the following PSPACE-completeness:

Theorem 12. *Given an NFA A and a natural number p in unary, it is PSPACE-complete to decide whether for the language $L(A)$ the statement from Lemma 2 holds for the value p.*

3.2   The PUMPING-PROBLEM for DFAs

Here we find for both pumping lemmata under consideration that the corresponding PUMPING-PROBLEM for DFAs is coNP-complete. First let us prove the upper bound:

Theorem 13. *Given a DFA A and a natural number p in unary. To decide whether for the language $L(A)$ the statement from Lemma 1 holds for the value p can be solved in coNP. The same upper bound applies if Lemma 2 is considered.*

In fact, both problems are coNP-hard. To this end, we utilize the construction of [13] of a directed (planar) graph from a 3SAT instance φ that has a Hamiltonian cycle if and only if φ is satisfiable. Assume that $\varphi = \bigwedge_{i=1}^{m} C_i$ is a 3SAT formula with n variables $x_1, x_2, \ldots, x_n$ and m clauses. Without loss of generality we may assume that every variable occurs at most four times in φ and no variable appears in only one polarity (pure literal). Let us briefly recall the construction of [13], slightly adapted to our needs,[4] which is illustrated in Fig. 3a for the formula

$$\varphi = (x_1 \vee \bar{x}_2) \wedge (\bar{x}_1 \vee x_2 \vee x_3) \wedge (\bar{x}_2 \vee \bar{x}_3).$$

The conversion of the given skeleton graph into a directed graph is achieved through the utilization of couplings as demonstrated in Fig. 3b and clause gadgets depicted in Fig. 3c—clause gadgets for a single literal or two literals are constructed straight-forwardly. We require that the value of x_i (or $\bar{x}_i$) is *true* if a given Hamiltonian s-t-path of the directed graph indicated in Fig. 3a contains the left edge from every pair of edge assigned to x_i (or $\bar{x}_i$, respectively). Otherwise the value is assumed to be *false*.

The directed graph G_φ that has been constructed will possess a Hamiltonian s-t-path if and only if the Boolean formula φ is satisfiable. Additionally, the Hamiltonian s-t-path must satisfy the following conditions:

---

[4] Instead of a Hamiltonian cycle we ask for a Hamiltonian s-t-path.

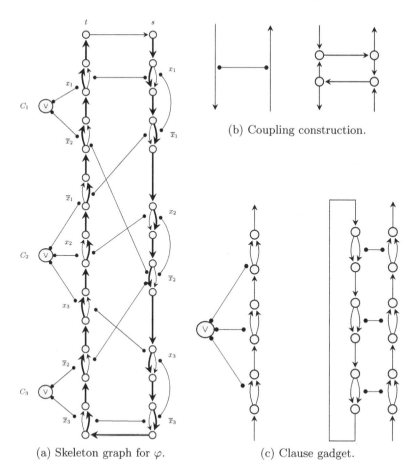

(a) Skeleton graph for φ.   (c) Clause gadget.

(b) Coupling construction.

Fig. 3. Schematic drawing of the skeleton graph constructed for the 3SAT instance φ. The obtained direct graph has a Hamiltonian s-t-path if and only if φ is satisfiable. Here, formula φ is satisfiable, because $\varphi|_{x_1=1,x_2=0,x_3=1} = 1$. A Hamiltonian s-t-path (without traversing the coupling connections) is indicated with a boldface line.

1. The s-t-path must pass through all left edges assigned to x_i (or $\overline{x}_i$), or all right edges assigned to x_i ($\overline{x}_i$, respectively).
2. It is not permitted for the s-t-path to pass through both a left (right) edge of x_i and a left (right) edge of $\overline{x}_i$ at the same time.
3. There must be at least one left edge present in every clause C_j.

Now we are ready to state our next theorem:

Theorem 14. *Given a DFA A and a natural number p, it is* coNP-*hard to decide whether for the language $L(A)$ the statement of Lemma 1 (Lemma 2, respectively) holds for the value p.* □

Proof. Let φ be a 3SAT formula with n variables and m clauses. Then by the above construction we obtain a skeleton graph and in turn a directed graph G_φ that has a Hamiltonian s-t-path if and only if φ is satisfiable. It remains to construct a *partial* DFA A_φ out of the skeleton/directed graph by giving an appropriate labeling of the edges such that the Hamiltonian s-t-path can only be pumped trivially. The vertices of the directed graph become the states of the automaton A_φ and the edges become transitions with appropriate labels described below. Moreover, the initial state of A_φ is the state s and the sole accepting state is set to t.

In the skeleton graph, an edge labeled a (or b, respectively) is coupled by the construction illustrated in Fig. 4 with an edge labeled b (or a, respectively).

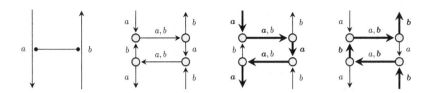

Fig. 4. Coupling of two edges labeled a and b and two possible traversals.

Observe, that in the resulting graph both connecting edges (these are the back and forth edges) carry the labels a and b simultaneously. Then a left (right, respectively) bended edge of a variable x_i is labeled a (b, respectively). For $\overline{x}_i$ this is exactly the other way around. The determinism of the automaton together with the coupling thus induces the a-b-labeling of all other bended edges. Finally, every (vertical) straight edge, except for the edge connecting t and s, is labeled by the letter $\#$. The t-s-edge is labeled with all letters, i.e., a, b, and $\#$. Thus, the input alphabet is equal to $\Sigma = \{a, b\} \cup \{\#\}$. This completes the description of the partial DFA A_φ—see Fig. 5 for a partial drawing of the automaton.

First we show that A_φ is minimal, if completed.

Claim 1. The partial DFA A_φ is bideterministic[5] and if completed, i.e., by introducing a non-accepting sink state that collects all non-specified transitions, it is the minimal (ordinary) DFA.

Next we consider the minimal pumping constants induced by the language accepted by A_φ.

Claim 2. Let L be the language accepted by the partial DFA A_φ. Then we have $\mathtt{mpe}(L) = \mathtt{sc}(A_\varphi)$ if and only if the Boolean formula φ is satisfiable. The same holds true for the measure $\mathtt{mpc}(L)$.

---

[5] A finite automaton A is *bideterministic* [1,12] if it is both partially deterministic and partially co-deterministic and has a sole accepting state. Here A is *partially co-deterministic* if the reversed automaton obtained by reversing the transitions of A is partially deterministic.

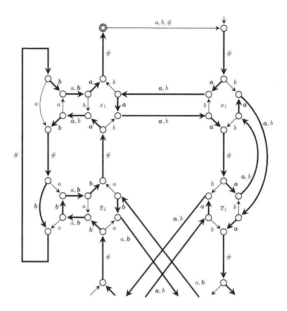

Fig. 5. Part of the partial DFA constructed from the skeleton graph with the help of coupling and clause gadgets. Here $n = 3$ and $m = 3$. A Hamiltonian s-t-path p is indicated with boldface transitions and any word w that fits to p is depicted with boldface letters on the transitions.

The complete automaton for A_φ and $\mathsf{sc}(A_\varphi)$ is a no instance of the PUMPING-PROBLEM w.r.t. Lemma 1 (Lemma 2, respectively) if and only if the Boolean formula φ is unsatisfiable. Thus the PUMPING-PROBLEM for DFAs is coNP-hard.
□

Putting the results together we get:

Corollary 15. *Given a DFA A and a natural number p, it is* coNP-*complete to decide whether for the language $L(A)$ the statement of Lemma 1 (Lemma 2, respectively) holds for the value p.*
□

References

1. Angluin, D.: Inference of reversible languages. J. ACM **29**(3), 741–765 (1982)
2. Cygan, M., et al.: Lower bounds based on the exponential-time hypothesis. In: Parameterized Algorithms, pp. 467–521. Springer, Cham (2015). https://doi.org/10.1007/978-3-319-21275-3_14
3. Dassow, J., Jecker, I.: Operational complexity and pumping lemmas. Acta Inform. **59**, 337–355 (2022)
4. Harrison, M.A.: Introduction to Formal Language Theory. Addison-Wesley (1978)
5. Holzer, M., Rauch, C.: On Jaffe's pumping lemma, revisited. In: DCFS. LNCS, Springer, Potsdam, Germany (2023), accepted for publication

6. Hunt, III, H.B.: On the time and tape complexity of languages I. In: Proceedings of the 5th Annual ACM Symposium on Theory of Computing. pp. 10–19. ACM, Austin, Texas, USA (1973)
7. Impagliazzo, R., Paturi, R., Zane, F.: Which problems have strongly exponential complexity? J. Comput. System Sci. **63**(4), 512–530 (2001)
8. Jaffe, J.: A necessary and sufficient pumping lemma for regular languages. SIGACT News. **10**(2), 48–49 (1978)
9. Kozen, D.C.: Automata and Computability. Undergraduate Texts in Computer Science, Springer, New York (1997). https://doi.org/10.1007/978-1-4612-1844-9
10. Nijholt, A.: YABBER–yet another bibliography: Pumping lemma's. An annotated bibliography of pumping. Bull. EATCS. **17**, 34–53 (1982)
11. Papadimitriou, C.H.: Computational Complexity. Addison-Wesley (1994)
12. Pin, J.E.: On reversible automata. In: Simon, I. (eds.) LATIN 1992. LATIN 1992. LNCS, vol. 583, pp. 401–416. Springer, Berlin, Heidelberg (1992). https://doi.org/10.1007/BFb0023844LATIN
13. Plesník, J.: The NP-completeness of the Hamiltonian cycle problem in planar digraphs with degree bound two. Inform. Process. Lett. **8**(4), 199–201 (1978)
14. Savitch, W.J.: Relationships between nondeterministic and deterministic tape complexities. J. Comput. System Sci. **4**(2), 177–192 (1970)

M-equivalence of Parikh Matrix
over a Ternary Alphabet

Joonghyuk Hahn, Hyunjoon Cheon, and Yo-Sub Han[✉]

Department of Computer Science, Yonsei University, Seoul 03722, Republic of Korea
{greghahn,hyunjooncheon,emmous}@yonsei.ac.kr

Abstract. The Parikh matrix, an extension of the Parikh vector for words, is a fundamental concept in combinatorics on words. We investigate M-unambiguity that identifies words with unique Parikh matrices. While the problem of identifying M-unambiguous words for a binary alphabet is solved using a palindromicly amicable relation, it is open for larger alphabets. We propose substitution rules that establish M-equivalence and solve the problem of M-unambiguity for a ternary alphabet. Our rules build on the principles of the palindromicly amicable relation and enable tracking of the differences of length-3 ordered scattered-factors. We characterize the set of M-unambiguous words and obtain a regular expression for the set.

Keywords: Parikh Matrix · Parikh Vector · M-unambiguity · M-equivalence

1 Introduction

Parikh [10] introduced a concept of mapping words to vectors. The resulting vector is called a Parikh vector, by counting the occurrences of each letter [6,7]. Mateescu et al. [9] extended the Parikh vector to the Parikh matrix that captures more complex numeric properties, by considering occurrences of scattered-factors.

Given an ordered alphabet $\Sigma = \{a_1, a_2, \ldots, a_k\}$, a Parikh matrix M over Σ is a $(k + 1) \times (k + 1)$ upper triangular matrix, where its main diagonal fills with all 1's, the second diagonal counts the occurrences of length-1 scattered-factors, the third diagonal counts length-2 ordered scattered-factors, and so on. For instance, given a word $w = 00121$ over $\Sigma = \{0, 1, 2\}$, the Parikh vector of w is $(|w|_0, |w|_1, |w|_2) = (2, 2, 1)$ and the Parikh matrix of w is

$$\begin{pmatrix} 1 & |w|_0 & |w|_{01} & |w|_{012} \\ 0 & 1 & |w|_1 & |w|_{12} \\ 0 & 0 & 1 & |w|_2 \\ 0 & 0 & 0 & 1 \end{pmatrix} = \begin{pmatrix} 1 & 2 & 4 & 2 \\ 0 & 1 & 2 & 1 \\ 0 & 0 & 1 & 1 \\ 0 & 0 & 0 & 1 \end{pmatrix}.$$

Note that the second diagonal (in red) is the Parikh vector of w. Parikh matrices provide a simple and intuitive approach that computes the occurrences of scattered-factors.

© The Author(s), under exclusive license to Springer Nature Switzerland AG 2023
B. Nagy (Ed.): CIAA 2023, LNCS 14151, pp. 141–152, 2023.
https://doi.org/10.1007/978-3-031-40247-0_10

We say that two words u and v are M-equivalent if u and v have the same Parikh matrix [5,9]. Atanasiu et al. [3] identified a family of binary words having the same Parikh matrix and characterized M-equivalence over binary words by the concept of palindromicly amicable property. This characterization on M-equivalence allows the identification of words with unique Parikh matrices—the injectivity problem [2,12]. Specifically, given a word u, if there is a distinct M-equivalent word v, then u is M-ambiguous; otherwise, u is M-unambiguous [8]. The injectivity problem is to find M-unambiguous words over a given ordered alphabet. Mateescu and Salomaa [8] constructed a regular expression for M-unambiguous words and solved the injectivity problem over a binary alphabet. However, it has been a challenging problem to extend this result to a larger alphabet.

Researchers partially characterized M-equivalence and M-unambiguity over a ternary alphabet [1,4,13]. Şerbănuţă and Şerbănuţă [4] enumerated all M-unambiguous words and proposed patterns that identify M-unambiguous words over a ternary alphabet. However, the pattern regular expression is incorrect; it misses some M-unambiguous words such as $bcbabcbabc$. Nevertheless, their work has laid a foundation for further research on M-unambiguity and M-equivalence [1,13]. However, a complete and simple characterization towards M-equivalence and M-unambiguity over a general alphabet remains elusive [11,14]. Even for a ternary alphabet, a comprehensive characterization of M-equivalence such as palindromicly amicable property has been open for decades.

We propose substitution rules that maintain the occurrences of length-1 or -2 ordered scattered-factors and keep track of the occurrences of length-3 ordered scattered-factors. Our three substitution rules can represent all words over a given ternary alphabet. We introduce $\cong$-relation that establishes M-equivalence based on the substitutions and compute the language of M-unambiguous words over a ternary alphabet.

We explain some terms and notations in Sect. 2. We introduce substitution rules and an equivalence relation that characterizes M-equivalent words in Sect. 3. Based on the proposed rules and relations, we compute a regular expression for M-ambiguous words and characterize M-unambiguity in Sect. 4. We conclude our paper with a brief summary and a few questions in Sect. 5.

2   Preliminaries

Let $\mathbb{N}$ denote the set of all nonnegative integers and $\mathbb{Z}$ denote the set of all integers. We use $\binom{u}{k}$ to denote the *binomial coefficient* for $u \geq k \in \mathbb{N}$. An *alphabet* Σ_k is a finite set of k letters and $|\Sigma_k| = k$ is the number of letters in Σ_k. We use Σ generally when the alphabet size k is not important. Without loss of generality, we use nonnegative integers as alphabet letters (e.g., $\Sigma_3 = \{0,1,2\}$). A *word* u over Σ is a finite sequence of letters in Σ. Let $|u|$ be the length of u. The symbol λ denotes the *empty word* whose length is 0. Given a word w, we use w^R to denote its reversal; $w = a_1 a_2 \cdots a_n$ and $w^R = a_n a_{n-1} \cdots a_1$. The *Kleene star* Σ^* of an alphabet Σ is the set of all words over Σ. An ordered alphabet $(\Sigma_k, <)$ consists of an alphabet $\Sigma_k = \{a_1, a_2, \ldots, a_k\}$ and a strict total order $<$ on Σ_k. We often denote the ordered alphabet by $\Sigma_k = \{a_1 < a_2 < \cdots < a_k\}$. If a total order $<$ is clear in the context, we simply use Σ_k to denote an ordered alphabet.

Given two words u and $v \in \Sigma^*$, we say that v is a *factor* of u if $u = \alpha v \beta$ for some $\alpha, \beta \in \Sigma^*$. Similarly, we say that v is a *scattered-factor* of u if there exist $u_0, u_1, \ldots, u_n$ and $v_1, v_2, \ldots, v_n \in \Sigma^*$ such that $v = v_1 v_2 \cdots v_n$ and $u = u_0 v_1 u_1 \cdots u_{n-1} v_n u_n$. We denote by $|u|_v$ the number of distinct occurrences of a nonempty word v as a scattered-factor in u. For instance, if $u = 0110$ and $v = 01$, then v is both a factor and a scattered-factor of u and $|u|_v = 2$.

We now present definitions that are directly related to our problem on the Parikh matrix.

Definition 1. *Let $\Sigma_k = \{a_1 < a_2 < \cdots < a_k\}$ be an ordered alphabet. The Parikh mapping is a monoid morphism $\Psi : \Sigma_k^* \to \mathbb{N}^k$ defined as $\Psi(w) = (|w|_{a_1}, |w|_{a_2}, \ldots, |w|_{a_k})$. Then, $\Psi(w)$ for $w \in \Sigma_k^*$ is the Parikh vector of w.*

The extension of the Parikh mapping to the Parikh matrix mapping considers a (upper) *unitriangular matrices* of nonnegative integers. A unitriangular matrix is a square matrix $m = (m_{i,j})_{1 \le i,j \le k}$ such that (1) $m_{i,j} \in \mathbb{N}$, (2) $m_{i,j} = 0$ for all $1 \le j < i \le k$, and (3) $m_{i,i} = 1$ for all $1 \le i \le k$. The set of all unitriangular matrices of dimension $k \ge 1$ is denoted by $\mathcal{M}_k$.

Definition 2. *Let $\Sigma_k = \{a_1 < a_2 < \cdots < a_k\}$ be an ordered alphabet. The Parikh matrix mapping is a monoid morphism $\Psi_{\Sigma_k} : \Sigma_k^* \to \mathcal{M}_{k+1}$ defined as follows. For $a_t \in \Sigma_k$, if $\Psi_{\Sigma_k}(a_t) = (m_{i,j})_{1 \le i,j \le k+1}$, then $m_{i,i} = 1$ for $1 \le i \le k+1$, $m_{t,t+1} = 1$, and all the other entries are zero. Then, $\Psi_{\Sigma_k}(w)$ for $w \in \Sigma_k^*$ is the Parikh matrix of w.*

Proposition 1. *[9, Theorem 3.1] Let $\Sigma_k = \{a_1 < a_2 < \cdots < a_k\}$ be an ordered alphabet. We denote by $a_{i,j}$ the word $a_i a_{i+1} \cdots a_j$ for $1 \le i \le j \le k$. For $w \in \Sigma_k^*$, its Parikh matrix $\Psi_{\Sigma_k}(w)$ has the following properties:*

1. *$m_{i,j} = 0$, for all $1 \le j < i \le k + 1$,*
2. *$m_{i,i} = 1$, for all $1 \le i \le k + 1$,*
3. *$m_{i,j+1} = |w|_u$ where $u = a_{i,j}$ for all $1 \le i \le j \le k$.*

Note that the Parikh matrix $\Psi_\Sigma(w)$ for $w \in \Sigma^*$ satisfies the associativity of matrix multiplication and $\Psi_\Sigma(w)$ can be constructed from the Parikh matrices of factors of w. For instance when $w = uv$, we have $\Psi_\Sigma(w) = \Psi_\Sigma(u) \cdot \Psi_\Sigma(v)$.

Example 1. Consider $w = 0110$ over a binary alphabet $\Sigma_2 = \{0 < 1\}$. As an example for Proposition 1,

$$\Psi_{\Sigma_2}(0110) = \Psi_{\Sigma_2}(0)\Psi_{\Sigma_2}(1)\Psi_{\Sigma_2}(1)\Psi_{\Sigma_2}(0) = \begin{pmatrix} 1 & 2 & 2 \\ 0 & 1 & 2 \\ 0 & 0 & 1 \end{pmatrix} = \begin{pmatrix} 1 & |0110|_0 & |0110|_{01} \\ 0 & 1 & |0110|_1 \\ 0 & 0 & 1 \end{pmatrix}.$$

3 M-equivalence

We discuss words with the equivalent Parikh matrices. For instance, the following words u and $v \in \Sigma^*$ have the same Parikh matrix.

$$\Psi_{\Sigma_3}(u) = \begin{pmatrix} 1 & |u|_0 & |u|_{01} & |u|_{012} \\ 0 & 1 & |u|_1 & |u|_{12} \\ 0 & 0 & 1 & |u|_2 \\ 0 & 0 & 0 & 1 \end{pmatrix} = \begin{pmatrix} 1 & |v|_0 & |v|_{01} & |v|_{012} \\ 0 & 1 & |v|_1 & |v|_{12} \\ 0 & 0 & 1 & |v|_2 \\ 0 & 0 & 0 & 1 \end{pmatrix} = \Psi_{\Sigma_3}(v).$$

This equivalence relation is called M-equivalence [5,9].

Definition 3. *Given two words w and w' over an ordered alphabet Σ, we define w and w' to be M-equivalent if $\Psi_\Sigma(w) = \Psi_\Sigma(w')$, and denote it by $w \equiv_M w'$.*

Researchers have studied how the changes in a word affect its Parikh matrix and when the Parikh matrix does not change. Proposition 2 illustrates substitutions of factors that do not change the Parikh matrix over arbitrary alphabets.

Proposition 2. [3, **Proposition 3.1**] *Let $\Sigma_k = \{a_1 < a_2 < \cdots < a_k\}$ be an ordered alphabet and $1 \le i, j \le k$. Then, the following equations hold:*

1. *If $|i - j| \ge 2$, then $\Psi_{\Sigma_k}(a_i a_j) = \Psi_{\Sigma_k}(a_j a_i)$.*
2. *If $|i - j| = 1$, then $\Psi_{\Sigma_k}(a_i a_j a_j a_i) = \Psi_{\Sigma_k}(a_j a_i a_i a_j)$.*

Proposition 2 is a necessary condition to establish M-equivalence but is not sufficient because they are not applicable to every word such as 10101, which is M-equivalent to 01110. Atanasiu et al. [3] proposed *palindromicly amicable* property that identifies M-equivalent words over a binary alphabet.

Definition 4. [3] *Let $\Sigma_2 = \{0 < 1\}$. Two words $\alpha, \beta \in \Sigma_2^*$ are palindromicly amicable if the following two statements hold:*

1. *α and β are palindromes,*
2. *$\Psi(\alpha) = \Psi(\beta)$.*

For $x, y \in \Sigma_2^*$ over $\Sigma_2 = \{0 < 1\}$, $x \equiv_{pa} y$ if a nonempty factor $\alpha \in \Sigma_2^*$ of x and a nonempty factor $\beta \in \Sigma_2^*$ of y are palindromicly amicable. We denote by $\equiv_{pa}^*$, the reflexive and transitive closure of $\equiv_{pa}$.

Proposition 3. [3, **Proposition 3.4**] *For $x, y \in \Sigma_2^*$ over $\Sigma_2 = \{0 < 1\}$,*

1. *$\equiv_{pa}^*$ is an equivalence relation.*
2. *If $x \equiv_{pa}^* y$, then for all $u \in \Sigma_2^*$, $ux \equiv_{pa}^* uy$ and $xu \equiv_{pa}^* yu$.*

Theorem 1. [3, **Theorem 3.1**] *For $\Sigma_2 = \{0 < 1\}$ and $x, y \in \Sigma_2^*$, $x \equiv_M y$ if and only if $x \equiv_{pa}^* y$.*

Theorem 1 is based on the palindromicly amicable relation between x and y. If we can compute y by substituting factors from x based on Proposition 2, then x

and y are M-equivalent. This is because the relation keeps the same value of $|x|_{01}$ and the Parikh vector also does not change. Theorem 1, however, does not hold for an alphabet with three or more letters. For instance, let $\Sigma_3 = \{0 < 1 < 2\}$ and $x = 10122101$, $y = 01122110$. It is easy to see that $x \equiv^*_{pa} y$ but $x \neq_M y$.[1]

We first consider the following conditions that are satisfied for words x and y over a ternary alphabet to be M-equivalent:

1. $\Psi(x) = \Psi(y)$,
2. $|x|_{01} = |y|_{01}$, $|x|_{12} = |y|_{12}$, and $|x|_{012} = |y|_{012}$.

Certain substitutions preserve the value of 01-,12-, and 012-occurrences, implying that the substitutions also do not change the Parikh matrix. We investigate what these substitutions are.

Proposition 4. [2, **Theorem 13**] *For $\Sigma_3 = \{0 < 1 < 2\}$, the following statements hold for $\alpha, \beta, u \in \Sigma_3^*$.*

1. *If $w = \alpha 02\beta$ and $w' = \alpha 20\beta$, then $w \equiv_M w'$.*
2. *If $w = \alpha 01u10\beta$ and $w' = \alpha 10u01\beta$ where $|u|_2 = 0$, then $w \equiv_M w'$.*
3. *If $w = \alpha 12u21\beta$ and $w' = \alpha 21u12\beta$ where $|u|_0 = 0$, then $w \equiv_M w'$.*

While Proposition 4 suggests useful substitution rules that preserve the Parikh matrix, the substitution rules are not applicable to all the words. We cannot apply the second rule to $w = \alpha 01u10\beta$ such that $|u|_2 > 0$. Likewise, the third rule is not applicable to $w = \alpha 12u21\beta$ such that $|u|_0 > 0$. For instance, we cannot apply any substitutions in Proposition 4 to an M-ambiguous word $u = 0101210121$. On the other hand, for w that we can apply substitutions in Proposition 4, we cannot enumerate all w' that are M-equivalent to w. For $u = 1002101112$, we cannot compute $u' = 0101210121$, which is M-equivalent to u by Proposition 4. If we design equivalence relations that maintain the same Parikh matrix for a given word $u \in \Sigma_3^*$, then any relations should preserver the value of $|u|_{01}$, $|u|_{12}$, and $|u|_{012}$. This leads us to design an equivalence relation that considers the following:

1. For all M-ambiguous words, the relation should be applicable.
2. Given an M-ambiguous word u, all M-equivalent words to u should be computed.

We relax the constraint that a single substitution rule should preserve the Parikh matrix value and allow the value of 012-occurrences to change. We suggest Proposition 4.

Proposition 5. *For $\Sigma_3 = \{0 < 1 < 2\}$, and $u, \alpha, \beta \in \Sigma_3^*$, the followings are substitution rules that satisfy $\Psi(w) = \Psi(w')$, $|w|_{01} = |w|_{01}$, and $|w|_{12} = |w'|_{12}$:*

1. *If $w = \alpha 02\beta$ and $w' = \alpha 20\beta$, then $|w|_{012} = |w'|_{012}$.*
2. *If $w = \alpha 01u10\beta$ and $w' = \alpha 10u01\beta$, then $|w|_{012} = |w'|_{012} + |u|_2$.*
3. *If $w = \alpha 12u21\beta$ and $w' = \alpha 21u12\beta$, then $|w|_{012} = |w'|_{012} - |u|_0$.*

---

[1] $x \equiv_{pa} 10211201 \equiv_{pa} 11200211 \equiv_{pa} 02111120 \equiv_{pa} y$.

Proof. Three substitution rules satisfy $\Psi(w) = \Psi(w')$ because each rule does not change the numbers of 0's, 1's, and 2's. The first substitution rule does not change the numbers of 01's and 12's. The second and the third substitution rules change 01 to 10 (respectively, 12 to 21) and also change 10 to 01 (respectively, 21 to 12), which keeps the same numbers of 01's and 12's at the end. For the occurrences of 012, Proposition 5 can be deduced by computing $|w|_{012}$ and $|w'|_{012}$.

In the first substitution rule,

$$|w|_{012} = |\alpha 02\beta|_{012} = |\alpha|_{012} + |\beta|_{012} + |\alpha|_{01} \times (|02|_2 + |\beta|_2) + (|\alpha|_0 + |02|_0) \times |\beta|_{12},$$

$$|w'|_{012} = |\alpha 20\beta|_{012} = |\alpha|_{012} + |\beta|_{012} + |\alpha|_{01} \times (|20|_2 + |\beta|_2) + (|\alpha|_0 + |20|_0) \times |\beta|_{12}.$$

It is easy to verify that $|02|_0$ and $|02|_2$ are the same to $|20|_0$ and $|20|_2$, respectively. Therefore, we know that $|w|_{012} = |w'|_{012}$.

For the second substitution rule, the substitution only occurs in the factor $01u10$ in w. We only have to keep track of 012 occurrences in $01u10$ of w and $10u01$ of w'. While $|01u10|_{012} = |u|_{012} + |01|_0 \times |u|_{12} + |01|_{01} \times |u|_2$, after the substitution, $|10u01|_{012} = |u|_{012} + |10|_0 \times |u|_{12} + |10|_{01} \times |u|_2 = |u|_{012} + |u|_{12}$. Therefore, the second substitution rule reduces the occurrences of 012 by $|u|_2$. Similarly, we can show that the third substitution rule increases the occurrences of 012 by $|u|_0$. □

We employ the second and third substitution rules of Proposition 5 to keep track of the occurrences of 012 and furthermore, analyze when $|w|_{012} = |w'|_{012}$. For instance, given $w, w' \in \Sigma_3^*$, Fig. 1 demonstrates that $|w|'_{012} = |w|_{012} - |\alpha|_2 + |\beta|_0$ when applied with the substitution rules of Proposition 5. Thus, when $|\alpha|_2 = |\beta|_0$, the Parikh matrices of w and w' are the same.

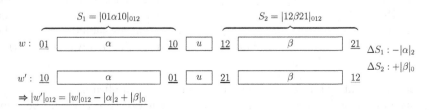

Fig. 1. An illustration of substitutions maintaining M-equivalence for a word $w = 01\alpha10u12\beta21$.

Figure 1 demonstrates one of the four scenarios where a single substitution step involving two replacements maintains the identical $|w|_{012}$ values, thereby preserving the Parikh matrix. Additionally, there are cases where the swapping pairs overlap. Figure 2 further illustrates cases of alternating sequences, with 01 followed by 12 and 10 followed by 21.

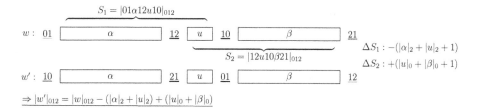

Fig. 2. An illustration of substitutions for a word w such that $w = 01\alpha 12u10\beta 21$, where 12 occurs before 10.

While Figs. 1 and 2 illustrate words with the same Parikh matrix by Proposition 5, there are other M-ambiguous words $w \in \Sigma_3^*$ that are not identified by Proposition 5, for instance, 012102021. Figure 3 depicts patterns of such words.

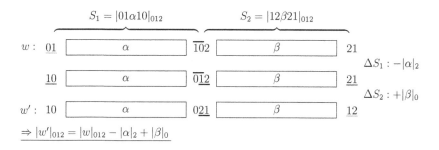

Fig. 3. An illustration of substitutions for a word w such that $w = 01\alpha 102\beta 21$ where w cannot maintain M-equivalence with a single substitution.

For M-equivalent words with such patterns, we develop substitution rules from Proposition 5 and introduce an equivalence relation of $(w, \Delta|w|_{012})$, the pair of a word w and a relative occurrence of 012.

Definition 5. *Given an ordered ternary alphabet $\Sigma_3 = \{0 < 1 < 2\}$, let $\cong$ be the minimal symmetric relation on $\Sigma_3^* \times \mathbb{Z}$ satisfying:*

1. *$(\alpha 02\beta, k) \cong (\alpha 20\beta, k)$,*
2. *$(\alpha 01u10\beta, k) \cong (\alpha 10u01\beta, k - |u|_2)$ for all u such that $|u|_2 \leq 1$,*
3. *$(\alpha 12u21\beta, k) \cong (\alpha 21u12\beta, k + |u|_0)$ for all u such that $|u|_0 \leq 1$,*

where $\alpha, \beta, u \in \Sigma_3^*$ and $k \in \mathbb{Z}$. Then, a $\cong$-sequence $(u_1, k_1), (u_2, k_2), \ldots, (u_n, k_n)$ is a sequence of pairs satisfying $(u_i, k_i) \cong (u_{i+1}, k_{i+1})$. Note that $\cong^*$ is the reflexive and transitive closure of $\cong$.

It is easy to verify that the minimal symmetric relation $\cong$ keeps the same values of $|u|_{01}$ and $|u|_{12}$ based on Proposition 4 and Definition 5. Note that, for binary words u and v over $\Sigma_2 = \{0 < 1\}$, we have $u \equiv_{pa}^* v$ if and only if $(u, k) \cong^* (v, k)$.

Proposition 6. *The followings always hold.*

1. $(\alpha u \beta, k) \cong^* (\alpha v \beta, l)$ *if and only if* $(u, k) \cong^* (v, l)$ *and*
2. $(u, k) \cong^* (v, l)$ *implies* $(u, k + c) \cong^* (v, l + c)$ *for any integer* c.

We can show the first property of Proposition 6 by induction on the number of $\cong$ applications, and the second property is immediate.

Lemma 1. *For two M-equivalent words u and v over a ternary alphabet $\Sigma_3 = \{0, 1, 2\}$, and two integers k and l such that $k \leq l$, let $S = [(u, k) \cong \cdots \cong (v, l)]$ be a $\cong$-sequence from (u, k) to (v, l). Then, for any integer t between k and l ($k \leq t \leq l$), there exists a pair $(w, t) \in S$ for some $w \in \Sigma_3^*$.*

Proof. For the sake of contradiction, assume that there exists t such that $(u', t) \notin S$. Then, there must be two pairs $(u_i, t - 1) \cong (u_{i+1}, t + 1)$. However, by Definition 5, $|k_{i+1} - k_i| \leq 1$ for $(u_i, k_i) \cong (u_{i+1}, k_{i+1})$. This leads to a contradiction that such pairs of $(u_i, t-1)$ and $(u_{i+1}, t+1)$ do not exist. Therefore, the statement holds.   □

Theorem 2. *Let Σ be an ordered ternary alphabet Σ_3. For two words $u, v \in \Sigma_3^*$ and two integers k, l, we have a $\cong$-sequence $S = [(u, k) \cong^* (v, l)]$ such that, for all pairs $(\alpha_1, \beta_1), (\alpha_2, \beta_2) \in S$, $|\beta_1 - \beta_2| \leq 1$ if and only if*

1. $|u|_{012} - |v|_{012} = k - l$,
2. $|k - l| \leq 1$, *and*
3. $|u|_x = |v|_x$ *for $x \in \{0, 1, 2, 01, 12\}$.*

Proof. [only-if direction] We prove the statement by induction on the length of a $\cong$-sequence. For two words $u, v \in \Sigma_3^*$ and two integers k, l satisfying $(u, k) \cong^0 (v, l)$ or $(u, k) \cong^1 (v, l)$, it is trivial to see that the two conditions hold. Note that, in the case of $\cong^0$, it is immediate that $u = v$ and $k = l$. Suppose that if $(u_i, k_i) \cong^i (v_i, l_i)$ then the two conditions hold for $2 \leq i < N$. For $(u, k) \cong^N (v, l)$, there exist two positive integers i, j such that $i + j = N$ and a pair (w, m) such that $(u, k) \cong^i (w, m) \cong^j (v, l)$. Thus, the statement holds for $\cong^*$.

From the statement, we know that if $(u_1, k_1) \cong^* (u_n, k_n)$, there is a $\cong$-sequence whose length is bounded above by $\binom{|u|}{|u|_0} \cdot \binom{|u| - |u|_0}{|u|_1}$. In other words, there always exists a finite $\cong$-sequence.

[if direction] Since it is trivial when $u = v$, we assume that $u \neq v$. We prove by induction on the length of u and v. When $|u| = |v| \leq 3$, we prove the claim by checking every pair of words.

Our induction hypothesis (IH) is that, for $N \geq 4$, if we have two words u and v, which satisfy the two preconditions and $|u| = |v| < N$, then, we establish $(u, k) \cong^* (v, l)$. Consider two words u and v of length N. Due to the aforementioned property 1 of the relation $\cong^*$ from Proposition 6, we assume that u and v do not have common nonempty prefixes and suffixes: $u = u_p u' u_s$ and $v = v_p v' v_s$ such that $u_p, u_s, v_p, v_s \in \Sigma_3$, $u_p \neq v_p$ and $u_s \neq v_s$.

If and only if u and v are in forms of the followings, $(u, k) \cong (v, l)$ holds:

1. $u = 01x10$, $v = 10x01$ ($|x|_2 \le 1$), or
2. $u = 21y12$, $v = 12y21$ ($|y|_0 \le 1$).

$u = 02$ and $v = 20$ never occur due to the restriction on the length.

We now show that there exists a string x such that $S = [(u, k) \cong^* (x, c) \cong^* (v, l)]$ where $x = ax'b$ and $a, b \in \Sigma_3$. Then, only one of the followings holds:

1. All x's between (u, k) and (v, l) satisfies $u_p \ne a \ne v_p$ and $u_s \ne b \ne v_s$, or
2. at least one x between (u, k) and (v, l) satisfies $a \in \{u_p, v_p\}$ or $b \in \{u_s, v_s\}$.

In the first case, we subdivide $S = [(u, k) \cong (x_1, c_1) \cong^* (x_2, c_2) \cong (v, l)]$ into three subsequences $S_1 = [(u, k) \cong (x_1, c_1)]$, $S_2 = [(x_1, c_1) \cong^* (x_2, c_2)]$ and $S_3 = [(x_2, c_2) \cong (v, l)]$. It must be the case that $x_1 = ax_1'b$ and $x_2 = ax_2'b$ where $u_p \ne a \ne v_p$ and $u_s \ne b \ne v_s$. Then, the base cases cover S_1 and S_3 and we can apply the first claim of Proposition 6 IH on S_2. Note that all strings in S_2 have common prefixes and suffixes.

Otherwise, we can subdivide S into two subsequences $S_1' = [(u, k) \cong^* (x, c)]$ and $S_2' = [(x, c) \cong^* (v, l)]$. Without loss of generality, let $a = u_p$. Then, u and x have a common prefix. We can detach the common prefix and IH applies on $u'u_s$ and $x'b$ thus the sequence S_1' is covered. Note that Theorem 2 also applies on S_2' because $|x|_{012} - |v|_{012} = (|u|_{012} - k + c) - |v|_{012} = (|u|_{012} - |v|_{012}) - k + c = (k - l) - k + c = c - l$ and the occurrences of length-1 and length-2 ordered scattered-factors are the same. $|c - l| \le 1$ is trivial when $k = l$. When $k \ne l$, $c = k$ or $c = l$ by Lemma 1 and thus $|c - l| \le 1$. Therefore, $(u, k) \cong^* (v, l)$.   □

Theorem 2 provides a characterization for M-equivalence over a ternary alphabet. The following result is immediate from Theorem 2.

Corollary 1. *For a ternary alphabet Σ_3 and two words $u, v \in \Sigma_3^*$, $u \equiv_M v$ if and only if $(u, 0) \cong^* (v, 0)$.*

4 M-unambiguity

We investigate another property of the Parikh matrix, M-*unambiguity*. Recall that a word $w \in \Sigma^*$ is M-*unambiguous* if there is no word $w' \ne w$ such that $w \equiv_M w'$. Otherwise, w is M-*ambiguous*. Atanasiu et al. [3] established the family of M-ambiguous words over a binary alphabet. Then, Mateescu and Salomaa [8] first presented a regular expression of an M-unambiguous language over a binary alphabet.

Theorem 3. [8, **Theorem 3**] *For a binary alphabet $\Sigma_2 = \{0 < 1\}$, a word $w \in \Sigma_2^*$ is M-unambiguous if and only if*

$$w \in L(0^*1^* + 1^*0^* + 0^*10^* + 1^*01^* + 0^*101^* + 1^*010^*).$$

The regular expression in Theorem 3 is sufficient to identify M-unambiguous words over a binary alphabet. However, we cannot apply the same result to M-unambiguous words over a ternary alphabet. Şerbănuţă and Şerbănuţă [4] presented a collection of regular expressions of M-unambiguous words by enumerating all words for a ternary alphabet[2]. Based on Corollary 1, we establish an intuitive approach that computes a regular expression for M-ambiguous words and identifies M-unambiguous words.

Theorem 4. *Given a ternary alphabet $\Sigma_3 = \{0 < 1 < 2\}$, let $L \subseteq \Sigma_3^*$ be a regular language defined by the union of the following regular expressions.*

$$E_1 = \Sigma_3^* \cdot (02 + 01(0 + 1)^*10 + 10(0 + 1)^*01 + 12(1 + 2)^*21 + 21(1 + 2)^*12) \cdot \Sigma_3^*$$
$$E_2 = \Sigma_3^* \cdot (01\Sigma_3^*2\Sigma_3^*10\Sigma_3^*10\Sigma_3^*2\Sigma_3^*01) \cdot \Sigma_3^*$$
$$E_3 = \Sigma_3^* \cdot (01\Sigma_3^*2\Sigma_3^*10\Sigma_3^*12\Sigma_3^*0\Sigma_3^*21) \cdot \Sigma_3^*$$
$$E_4 = \Sigma_3^* \cdot (21\Sigma_3^*0\Sigma_3^*12\Sigma_3^*10\Sigma_3^*2\Sigma_3^*01) \cdot \Sigma_3^*$$
$$E_5 = \Sigma_3^* \cdot (21\Sigma_3^*0\Sigma_3^*12\Sigma_3^*12\Sigma_3^*0\Sigma_3^*21) \cdot \Sigma_3^*$$
$$E_6 = \Sigma_3^* \cdot (01\Sigma_3^*12\Sigma_3^*10\Sigma_3^*21) \cdot \Sigma_3^*$$
$$E_7 = \Sigma_3^* \cdot (10\Sigma_3^*21\Sigma_3^*01\Sigma_3^*12) \cdot \Sigma_3^*$$

and L^R be its reversal language $\{w^R \mid w \in L\}$.
    Then, $L_{amb} = L \cup L^R$ is the set of all M-ambiguous words over Σ_3.

Proof. Let X be the set of all M-ambiguous words over Σ_3 and we show that $X = L_{amb}$. We prove that X is equivalent to L_{amb}.

$[X \subseteq L_{amb}]$. Suppose that there exists $u \in X \setminus L_{amb}$ and let $v \neq u$ be M-equivalent to u. Since $u \equiv_M v$, $(u, 0) \cong^* (v, 0)$ and thus, $(v, 0)$ is derived from $(u, 0)$ by a sequence of $\cong$ applications from Definition 5. For all the string patterns in Definition 5, we can easily find them in L_{amb}. For instance, E_2 contains $\alpha 01u10\beta$ as a prefix where $\alpha, \beta, u \in \Sigma_3$. Likewise, we can find the other string patterns of Definition 5. This contradicts that there exists u with distinct patterns that are not in L_{amb}. Therefore, $X \subseteq L_{amb}$.

$[L_{amb} \subseteq X]$. Suppose that there exists $u \in L_{amb} \setminus X$. This implies that u is M-unambiguous. Since $u \in L_{amb}$, we can derive v with s of Definition 5. We investigate when u is included in one of E_i of L_{amb}. When $u \in E_1$, we first examine when u contains 02 as a factor. By the first $\cong$-relation in Definition 5, u is M-ambiguous. There is also $u \in E_1$ that contains factors that are palindromicly amicable of a binary alphabet u is M-ambiguous by Theorem 1. Thus, $u \in E_1$ is M-ambiguous. Similarly, we can prove in the same way for the reversal of E_1.

For $u \in E_i$ for $2 \leq i \leq 7$, we inspect the change of 012 occurrence values by the second and the third $\cong$-relations of Definition 5. We show one of the cases when $u \in E_6$. When $u \in E_6$, the following holds for some $v \in \Sigma_3^*$:

$$\overbrace{(u, k) \cong^* \underbrace{(u', k - |\alpha 12\beta|_2)}_{01\alpha12\beta10 \rightarrow 10\alpha12\beta01} \cong^* (v, k - |\alpha 12\beta|_2 + |\beta 01\gamma|_0)}^{12\beta01\gamma21 \rightarrow 21\beta01\gamma12}.$$

---

[2] The regular expression is incorrect since it misses some M-unambiguous words illustrated in Fig. 4 in Sect. 4.

Without loss of generality, let $0 < |\alpha 12\beta|_2 \leq |\beta 01\gamma|_0$. Then, $k - |\alpha 12\beta|_2 < k \leq k - |\alpha 12\beta|_2 + |\beta 01\gamma|_0$ and by Lemma 1, there exists (v', k) such that $(u, k) \cong^* (v', k) \cong^* (v, l)$ and $u \neq v'$. This leads to a contradiction that u is M-unambiguous because $(u, k) \cong^* (v', k)$ implies that v' is M-equivalent to u. We can prove similarly for E_2, E_3, E_4, E_5, E_7. Thus, $L_{amb} \subseteq X$.   □

Theorem 4 establishes an identification for M-ambiguous words over a ternary alphabet. Then, the following result is immediate.

Corollary 2. *For $\Sigma_3 = \{0 < 1 < 2\}$ and $u \in \Sigma_3^*$, we have that u is M-unambiguous if and only if $u \notin L_{amb}$.*

Using the regular expression in Theorem 4, we find all M-unambiguous words that are missing in Şerbănuţă and Şerbănuţă [4]. Figure 4 is the minimal DFA for such missing M-unambiguous words.

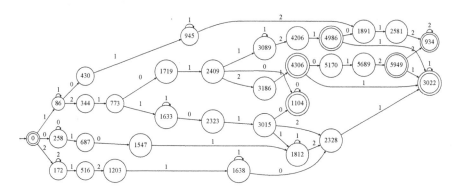

Fig. 4. An FA for M-unambiguous words missing in Şerbănuţă and Şerbănuţă [4].

5   Conclusions

We have presented a polished and complete characterization of M-equivalence and M-unambiguity over a ternary alphabet using $\cong^*$-relation. While the problem was solved for a binary alphabet, the larger-alphabet case has been open. We have presented key characteristics of M-equivalence and M-unambiguity over a ternary alphabet based on our substitution rules and $\cong^*$-relation. This result facilitates exploring further combinatorial properties of M-equivalent words.

Our equivalence relation is well-defined for a ternary alphabet but it can also be developed with further substitution rules for larger alphabets. We plan to extend $\cong$-relation to arbitrary alphabets and continue working towards establishing equivalent relations to M-equivalence and M-unambiguity. We also aim to address open problems related to other properties of Parikh matrices, such as ME-equivalence, strong M-equivalence, and weak M-relation [11,14].

Acknowledgments. We thank all the reviewers for their valuable comments. This research was supported by the NRF grant funded by MIST (NRF-RS-2023-00208094). The first two authors contributed equally.

References

1. Atanasiu, A.: Parikh matrix mapping and amiability over a ternary alphabet. In: Discrete Mathematics and Computer Science, pp. 1–12 (2014)
2. Atanasiu, A., Atanasiu, R., Petre, I.: Parikh matrices and amiable words. Theoret. Comput. Sci. **390**(1), 102–109 (2008)
3. Atanasiu, A., Martín-Vide, C., Mateescu, A.: On the injectivity of the Parikh matrix mapping. Fund. Inform. **49**(4), 289–299 (2002)
4. Şerbănuţă, V.N., Şerbănuţă, T.F.: Injectivity of the Parikh matrix mappings revisited. Fund. Inform. **73**(1–2), 265–283 (2006)
5. Fossé, S., Richomme, G.: Some characterizations of Parikh matrix equivalent binary words. Inf. Process. Lett. **92**(2), 77–82 (2004)
6. Ibarra, O.H., Ravikumar, B.: On the Parikh membership problem for FAs, PDAs, and CMs. In: Proceedings of the 8th Language and Automata Theory and Applications, pp. 14–31 (2014)
7. Karhumäki, J.: Generalized Parikh mappings and homomorphisms. Inf. Control **47**, 155–165 (1980)
8. Mateescu, A., Salomaa, A.: Matrix indicators for subword occurrences and ambiguity. Int. J. Found. Comput. Sci. **15**(2), 277–292 (2004)
9. Mateescu, A., Salomaa, A., Salomaa, K., Yu, S.: A sharpening of the Parikh mapping. RAIRO Inform. Theorique et Appl. **35**(6), 551–564 (2001)
10. Parikh, R.: On context-free languages. J. ACM **13**(4), 570–581 (1966)
11. Poovanandran, G., Teh, W.C.: On M-equivalence and strong M-equivalence for Parikh matrices. Int. J. Found. Comput. Sci. **29**(1), 123–138 (2018)
12. Salomaa, A.: On the injectivity of Parikh matrix mappings. Fund. Inform. **64**(1–4), 391–404 (2005)
13. Teh, W.C.: On core words and the Parikh matrix mapping. Int. J. Found. Comput. Sci. **26**(1), 123–142 (2015)
14. Teh, W.C., Subramanian, K.G., Bera, S.: Order of weak M-relation and Parikh matrices. Theoret. Comput. Sci. **743**, 83–92 (2018)

Operational Complexity in Subregular Classes

Michal Hospodár[ID] and Galina Jirásková[(✉)][ID]

Mathematical Institute, Slovak Academy of Sciences, Grešákova 6, 040 01 Košice, Slovakia
{hospodar,jiraskov}@saske.sk

Abstract. The state complexity of a regular operation is a function that assigns the maximal state complexity of the language resulting from the operation to the sizes of deterministic finite automata recognizing the operands of the operation. We study the state complexity of intersection, union, concatenation, star, and reversal on the classes of combinational, singleton, finitely generated left ideal, symmetric definite, star, comet, two-sided comet, ordered, star-free, and power-separating languages. We get the exact state complexities in all cases. The complexity of all operations on combinational languages is given by a constant function. The state complexity of the considered operations on singleton languages is $\min\{m, n\}$, $m + n - 3$, $m + n - 2$, $n - 1$, and n, respectively, and on finitely generated left ideals, it is $mn - 2$, $mn - 2$, $m + n - 1$, $n + 1$, and $2^{n-2} + 2$. The state complexity of concatenation, star, and reversal is $m2^n - 2^{n-1} - m + 1$, n, 2^n and $m2^{n-1} - 2^{n-2} + 1$, $n + 1$, $2^{n-1} + 1$ for star and symmetric definite languages, respectively. We also show that the complexity of reversal on ordered and power-separating languages is $2^n - 1$, which proves that the lower bound for star-free languages given by [Brzozowski, Liu, Int. J. Found. Comput. Sci. 23, 1261–1276, 2012] is tight. In all the remaining cases, the complexity is the same as for regular languages. Except for reversal on finitely generated left ideals and ordered languages, all our witnesses are described over a fixed alphabet.

1 Introduction

The state complexity of a regular operation is the number of states that is sufficient and necessary in the worst case for a deterministic finite automaton (DFA) to accept the language resulting from the operation, considered as a function of the sizes of DFAs for operands. The tight upper bounds on the complexity of union, intersection, concatenation, star, and reversal were presented by Maslov [12], Yu, Zhuang and Salomaa [17], and Leiss [11], and to describe witness languages, a binary alphabet was used.

Supported by the Slovak Grant Agency for Science (VEGA), contract 2/0096/23 "Automata and Formal Languages: Descriptional and Computational Complexity".
M. Hospodár—This publication was supported by the Operational Programme Integrated Infrastructure (OPII) for the project 313011BWH2: "InoCHF - Research and development in the field of innovative technologies in the management of patients with CHF", co-financed by the European Regional Development Fund.

B. Nagy (Ed.): CIAA 2023, LNCS 14151, pp. 153–165, 2023.
https://doi.org/10.1007/978-3-031-40247-0_11

If operands have some special properties, then the complexity of an operation may be significantly smaller. For example, the state complexity of star on prefix-free languages is n, while it is $\frac{3}{4}2^n$ in the regular case. On the other hand, the complexity of intersection, union, concatenation, and star on star-free languages is the same as in the regular case [5]. This led to the investigation of operational complexity in several subclasses of regular languages. Operations on unary languages were studied by Yu et al. [17] and Pighizzini and Shallit [15], and on finite languages by Câmpeanu et al. [6]. The classes of prefix- and suffix-free languages were examined by Han et al. [7,8], and the classes of bifix-, factor-, and subword-free, star-free, ideal, and closed languages by Brzozowski et al. [2–5], Except for reversal on star-free languages, the state complexity of basic regular operations in each of the above mentioned classes was determined. Bordihn, Holzer, and Kutrib [1] considered also some other subclasses, including combinational, finitely generated left ideal, symmetric definite, star, comet, two-sided comet, ordered, and power-separating languages, and investigated the complexity of the conversion of nondeterministic finite automata (NFAs) to deterministic ones. The closure properties of these classes, as well as nondeterministic operational complexity in them, were studied by Olejár et al. [9,13].

Here we continue this research and study the state complexity of basic regular operations in several classes from [1]. For each considered operation and each considered class, we get a tight upper bound on its complexity. To get the upper bounds, we examine minimal DFAs for languages in considered classes. We show that every minimal DFA for a finitely generated left ideal must have two states that can be distinguished only by the empty string. This gives upper bounds for union, intersection, and reversal on finitely generated left ideals. We also show that the set of all non-final states in a minimal DFA for a power-separating language cannot be reachable in the reversed automaton. This gives an upper bound $2^n - 1$ for reversal on power-separating languages, and shows that the same lower bound for star-free languages [5, Theorem 7] is tight since every star-free language is power-separating. To get upper bounds for concatenation, we carefully inspect the reachable and distinguishable states in the subset automaton of an NFA for the concatenation of given languages. Finally, the star of a star language is the same language, and if a given language is symmetric definite, then its star differs from it only in the empty string. The corresponding upper bounds n and $n + 1$ for the star operation follow. To get lower bounds, we sometimes use witnesses known from the literature, and just prove that they belong to a considered class. However, most often, we describe appropriate witnesses so that the corresponding product automata for union and intersection, or subset automata of NFAs for concatenation, star, or reversal have a desired number of reachable and pairwise distinguishable states.

2   Preliminaries

We assume that the reader is familiar with the basic notions in formal languages and automata theory, and for details, we refer to [16].

Let Σ be a finite non-empty alphabet of symbols. Then Σ^* denotes the set of all strings over Σ including the empty string ε. For a finite set S, the notation $|S|$ denotes the size of S and 2^S denotes the power set of S. For two non-negative integers i, j, the set $\{i, i+1, \ldots, j\}$ is denoted by $[i, j]$.

A *nondeterministic finite automaton with multiple initial states* (MNFA) is a quintuple $A = (Q, \Sigma, \cdot, I, F)$ where Q is a finite non-empty set of states, Σ is a finite non-empty input alphabet, $\cdot : Q \times \Sigma \to 2^Q$ is the transition function, $I \subseteq Q$ is the set of initial states, and $F \subseteq Q$ is the set of final states. The transition function can be extended to the domain $2^Q \times \Sigma^*$ in the natural way. For states p, q and a symbol a, we sometimes write (p, a, q) whenever $q \in p \cdot a$. The language accepted by A is the set of strings $L(A) = \{w \in \Sigma^* \mid I \cdot w \cap F \neq \emptyset\}$.

The *reverse* of an MNFA $A = (Q, \Sigma, \cdot, I, F)$ is the MNFA $A^R = (Q, \Sigma, \cdot^R, F, I)$ where $q \cdot^R a = \{p \mid q \in p \cdot a\}$. We say that a subset S of Q is *reachable* in A if there exists a string w such that $S = I \cdot w$. A subset S is *co-reachable* in A if it is reachable in A^R. We usually omit $\cdot$ and write just juxtaposition qw instead of $q \cdot w$. If $S \subseteq Q$, then $Sw = \{qw \mid q \in S\}$ and $wS = \{q \mid qw \in S\}$.

An MNFA is a *nondeterministic finite automaton* (NFA) if $|I| = 1$. An NFA is a *deterministic finite automaton* (DFA) if $|qa| = 1$ for each state q and each symbol a. We usually write $pa = q$ instead of $pa = \{q\}$. A non-final state q of a DFA is called *dead* if $qa = q$ for each symbol a. A DFA is *minimal* if all its states are reachable and pairwise distinguishable. The *state complexity of a regular language* L, $\mathrm{sc}(L)$, is the number of states in a minimal DFA recognizing L. The *state complexity of a k-ary regular operation* f is the function from $\mathbb{N}^k$ to $\mathbb{N}$ defined as $(n_1, n_2, \ldots, n_k) \mapsto \max\{\mathrm{sc}(f(L_1, L_2, \ldots, L_k)) \mid \mathrm{sc}(L_i) \leq n_i \text{ for each } i.\}$

Every MNFA $A = (Q, \Sigma, \cdot, I, F)$ can be converted to an equivalent deterministic automaton $\mathcal{D}(A) = (2^Q, \Sigma, \cdot, I, \{S \in 2^Q \mid S \cap F \neq \emptyset\})$. The DFA $\mathcal{D}(A)$ is called the *subset automaton* of A.

Finally, we recall the definitions of language classes considered in this paper. A language $L \subseteq \Sigma^*$ is:

- *combinational* (class abbreviation CB): if $L = \Sigma^* H$ for $H \subseteq \Sigma$;
- *singleton* (SGL): if it consists of one string;
- *finitely generated left ideal* (FGLID): if $L = \Sigma^* H$ for some finite language H;
- *left ideal* (LID): if $L = \Sigma^* L$;
- *symmetric definite* (SYDEF): if $L = G \Sigma^* H$ for some regular languages G, H;
- *star* (STAR): if $L = L^*$;
- *comet* (COM): if $L = G^* H$ for some regular languages G, H with $G \notin \{\emptyset, \{\varepsilon\}\}$;
- *two-sided comet* (2COM): if $L = E G^* H$ for some regular languages E, G, H with $G \notin \{\emptyset, \{\varepsilon\}\}$;
- *ordered* (ORD): if it is accepted by a (possibly non-minimal) DFA with ordered states such that $p \preceq q$ implies $p \cdot \sigma \preceq q \cdot \sigma$ for each symbol σ;
- *star-free* (STFR): if L is constructible from finite languages, concatenation, union, and complementation (equivalently, if L has an aperiodic DFA);
- *power-separating* (PSEP): if for every w in Σ^* there exists an integer k such that $\bigcup_{i \geq k} \{w^i\} \subseteq L$ or $\bigcup_{i \geq k} \{w^i\} \subseteq L^c$.

We have CB $\subsetneq$ FGLID $\subsetneq$ LID $\subsetneq$ SYDEF, STAR $\setminus \{\{\varepsilon\}\} \subsetneq$ COM $\subsetneq$ 2COM, and SGL $\subsetneq$ ORD $\subsetneq$ STFR $\subsetneq$ PSEP [1,13].

3   Combinational and Singleton Languages

We start with two simple subregular classes: the class of combinational languages and the class of singleton languages. The following two theorems present our results on the state complexity of basic operations in these two classes.

Theorem 1. *Let K and L be combinational languages over Σ. Then we have* $\mathrm{sc}(K), \mathrm{sc}(L) \leq 2$ *and* $\mathrm{sc}(K \cap L), \mathrm{sc}(K \cup L) \leq 2$, $\mathrm{sc}(KL) \leq 3$, $\mathrm{sc}(L^*) \leq 2$, *and* $\mathrm{sc}(L^R) \leq 3$. *All these upper bounds are tight if $|\Sigma| \geq 2$ (star, reversal) or if $|\Sigma| \geq 1$ (intersection, union, concatenation).*

Proof. Let $K = \Sigma^* G$ and $L = \Sigma^* H$ with $G, H \subseteq \Sigma$. If H is empty, then L is empty. Otherwise, L is accepted by a two-state DFA $(\{1, 2\}, \Sigma, \cdot, 1, \{2\})$ with $1a = 2a = 1$ if $a \notin H$ and $1a = 2a = 2$ if $a \in H$. We only prove the result for star. Each string in Σ^* either ends with a symbol in H, or ends with a symbol in $\Sigma \backslash H$, or is empty. Since $L = L^+$ we have $L^* = \{\varepsilon\} \cup L = (\Sigma^* (\Sigma \backslash H))^c$, so L^* is the complement of a combinational language. The upper bound 2 is met by $(a + b)^* a$. $\qquad \square$

Theorem 2. *Let K and L be singleton languages accepted by DFAs with m and n states, respectively. Then* $\mathrm{sc}(K \cap L) \leq \min\{m, n\}$, $\mathrm{sc}(K \cup L) \leq m + n - 3$, $\mathrm{sc}(KL) \leq m + n - 2$, $\mathrm{sc}(L^*) \leq n - 1$, *and* $\mathrm{sc}(L^R) \leq n$. *All these bounds are tight, with witnesses described over an alphabet of size at least 1 (intersection, concatenation, reversal) or 2 (union, star).*

Proof. We only prove the result for star. The minimal DFA for a singleton language $\{v\}$ has $|v| + 2$ states, including the dead state. Next, the language $\{v\}^*$ is accepted by a DFA with $|v| + 1$ states, possibly including the dead state. This gives the upper bound $n - 1$, which is met by the language $\{a^{n-2}\}$ over the binary alphabet $\{a, b\}$. $\qquad \square$

4   Intersection and Union

In this section, we examine the intersection and union operations in subregular classes. Recall that the state complexity of these two operations is mn with binary witnesses [12,17]. Our aim is to show that the complexity of these operations is $mn - 2$ in the class of finitely generated left ideals and mn in the remaining classes. Recall that if a DFA for a language L has a unique final state which is the initial one, then $L = L^*$, so L is a star language. To get the upper bound in class FGLID, we use the following property.

Lemma 3. *In the minimal DFA for a finitely generated left ideal different from $\emptyset$ and Σ^*, there exist two states that are distinguishable only by the empty string.*

Proof. Let A be a minimal DFA for a finitely generated left ideal L with the initial state 1. Assume, to get a contradiction, that any two distinct states of A can be distinguished by a non-empty string. Since $L \notin \{\emptyset, \Sigma^*\}$, there is a symbol a

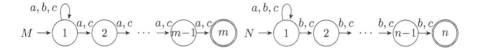

Fig. 1. Finitely generated left ideal witnesses for union $(mn - 2)$.

such that $1a \neq 1$. Now, the states 1 and $1a$ are distinguishable by a non-empty string w_1. We must have $1w_1 \notin F$ and $1aw_1 \in F$, because $1w_1 \in F$ and $1aw_1 \notin F$ would mean that $w_1 \in L$ and $aw_1 \notin L$, so L would not be a left ideal. The resulting states are again distinguishable by a non-empty string w_2, and again, we must have $1w_1w_2 \notin F$ and $1aw_1w_2 \in F$. Inductively, the states $1w_1w_2 \cdots w_{i-1}$ and $1aw_1w_2 \cdots w_{i-1}$ are distinguishable by a non-empty string w_i, and we must have $1w_1w_2 \cdots w_{i-1}w_i \notin F$ and $1aw_1w_2 \cdots w_{i-1}w_i \in F$. Since the number of states is finite, there exist j, k with $j < k$ such that $1aw_1w_2 \cdots w_k = 1aw_1w_2 \cdots w_j$. This means that $aw_1w_2 \cdots w_j(w_{j+1}w_{j+2} \cdots w_k)^* \subseteq L$, and on the other hand we have $w_1w_2 \cdots w_j(w_{j+1}w_{j+2} \cdots w_k)^* \subseteq L^c$. It follows that the minimal generator $L \setminus \Sigma L$ of the left ideal L is infinite, a contradiction. □

Theorem 4. *Let K and L be finitely generated left ideals over Σ accepted by an m-state and n-state DFA, respectively. Then $\mathrm{sc}(K \cap L), \mathrm{sc}(K \cup L) \leq mn - 2$. This upper bound is tight for union if $|\Sigma| \geq 3$, and for intersection if $|\Sigma| \geq 10$.*

Proof. Let A and B be minimal DFAs for K and L, respectively. By Lemma 3, there exist states p, p' of A and q, q' of B that are distinguished only by the empty string. It follows that $pa = p'a$ and $qa = q'a$ for each input symbol a. Then, in the product automaton $A \times B$, the pairs $(p, q), (p, q'), (p', q), (p', q')$ are sent to the same state by each input symbol, so, they may be distinguished only by ε. This gives the upper bound $mn - 2$ for union and intersection.

We can prove that this upper bound for union and intersection is met by languages accepted by NFAs from Figs. 1 and 3, respectively; notice that equivalent DFAs shown in Figs. 2 and 4 have m and n states as well. □

Theorem 5. *The state complexity of intersection and union on the classes of symmetric definite, ordered, power-separating, and star languages is mn.*

Proof. The upper bound mn is met by the intersection of binary left ideal languages $K = \{w \in \{a, b\}^* \mid |w|_a \geq m - 1\}$ and $L = \{w \in \{a, b\}^* \mid |w|_b \geq n - 1\}$

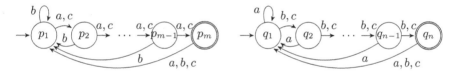

Fig. 2. The reachable parts of the subset automata $\mathcal{D}(M)$ and $\mathcal{D}(N)$ of NFAs M and N from Fig. 1; we have $p_i = [1, i]$ and $q_j = [1, j]$.

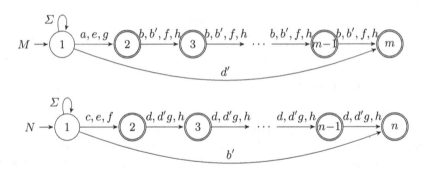

Fig. 3. Finitely generated left ideal witnesses for intersection $(mn - 2)$.

[3, Theorem 7], and by the union of quaternary left ideals from [3, Theorem 8]. Every left ideal is symmetric definite. Next, the languages K and L are ordered (so also power-separating), and these classes are closed under complementation. This gives the complexity mn with binary witnesses on ORD and PSEP. Finally, to get star witnesses, we consider the minimal DFAs, the first of which counts the number of a's modulo m, and the second the number of b's modulo n.   □

Remark 6. Since STAR $\subseteq$ COM $\subseteq$ 2COM, the previous theorem gives complexity mn for intersection and union on comet and two-sided comet languages.   □

5   Concatenation

Now we examine the concatenation operation, the state complexity of which is $m2^n - 2^{n-1}$ with binary witnesses [12]. We first consider the concatenation operation on finitely generated left ideal, symmetric definite, and star languages, and show that the resulting complexities are always smaller than the regular

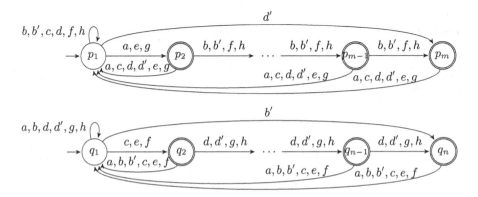

Fig. 4. The reachable parts of the subset automata $\mathcal{D}(M)$ and $\mathcal{D}(N)$ of NFAs M and N from Fig. 3; we have $p_i = \{1, i\}$ and $q_j = \{1, j\}$.

one. On the other hand, in the remaining classes, the complexity is the same as in the regular case. To get an upper bound for SYDEF, we use the following observation, cf. [14, Theorem 9.2].

Lemma 7. *Every minimal DFA for a symmetric definite language contains a state $\bar{q}$ such that each accepting computation goes through $\bar{q}$ and a loop on every input symbol can be added in $\bar{q}$ without changing the language.* ☐

Remark 8. As shown in [3, Theorem 9], the upper bound on the complexity of concatenation on left ideals is $m + n - 1$. This upper bound is shown to be met by unary languages a^*a^{m-1} and a^*a^{n-1} in the proof of [3, Theorem 9]. These unary languages are finitely generated left ideals. This gives the same complexity of concatenation on FGLID. ☐

Theorem 9. *Let $K, L \subseteq \Sigma^*$ be accepted by an m-state and n-state DFA, respectively. If K and L are symmetric definite, then $\mathrm{sc}(KL) \leq m2^{n-1} - 2^{n-2} + 1$. If K and L are star languages, then $\mathrm{sc}(KL) \leq m2^n - 2^{n-1} - m + 1$. Both upper bounds are tight if $|\Sigma| \geq 3$.*

Proof. Let $A = (Q_A, \Sigma, \cdot_A, s_A, F_A)$ and $B = (Q_B, \Sigma, \cdot_B, s_B, F_B)$ be the minimal DFAs for K and L, respectively.

(a) By Lemma 7, there is a state $\bar{q}$ of B such that every accepting computation of B goes through the state $\bar{q}$ and, moreover, we can add a loop on every input symbol in $\bar{q}$ without changing the language; denote the resulting NFA by B'. Construct the MNFA N for KL from automata A and B' by adding the transition (q, a, s_B) whenever $qa \in F_A$, the initial set is $\{s_A\}$ if $s_A \notin F_A$ and $\{s_A, s_B\}$ otherwise, and the set of final states is F_B. Each reachable subset of the subset automaton of N is of the form $\{p\} \cup S$ with $p \in Q_A$ and $S \subseteq Q_B$; let us represent it by the pair (p, S). Recall that $p \in F_A$ implies $s_B \in S$ for each reachable state (p, S).

If $\bar{q}$ is the initial state of B, then L is a left ideal, and the complexity of KL is at most $m + n - 1$; cf. [3, Theorem 9]. If $\bar{q}$ is final, then in the corresponding subset automaton, each state $(p, S \cup \{\bar{q}\})$ is equivalent to the state $(p, \{\bar{q}\})$. This gives the upper bound $m2^{n-1} - 2^{n-2} + 1$. Finally, let $\bar{q}$ be neither initial nor final. Let Q_1 be the set of states in Q_B reachable without going through $\bar{q}$ and let $Q_2 = Q_B \setminus (Q_1 \cup \{\bar{q}\})$. Let $S_1 \subseteq Q_1$ and $S_2 \subseteq Q_2$. Every accepting computation of B goes through $\bar{q}$, and $\bar{q}$ has a loop on every input symbol in B'. Let us show that each state $(p, S_1 \cup \{\bar{q}\} \cup S_2)$ is equivalent to $(p, \{\bar{q}\} \cup S_2)$. It is enough to shown that each string accepted in B' from a state in S_1 is also accepted from a state in $\{\bar{q}\} \cup S_2$. Let a string w be accepted from S_1. Then the computation on w must go through $\bar{q}$, so $w = uv$ where u leads a state in S_1 to the state $\bar{q}$ and v is accepted from $\bar{q}$. Since $\bar{q}$ has a loop on each symbol in B', the string uv is also accepted from $\bar{q}$. Hence w is accepted from $(p, \{\bar{q}\} \cup S_2)$. This gives an upper bound $(m - 1)(2^{|Q_1|} + 2^{|Q_2|}) + 2^{|Q_1|-1} + 2^{|Q_2|} + 1 \leq (m - 1)(2^{n-1}) + 2^{n-2} + 1$.

For tightness, consider the DFAs A and B from Fig. 5. The automaton A recognizes a left ideal since, after adding a loop on each input symbol in its initial state and performing determinization and minimization, we get the isomorphic

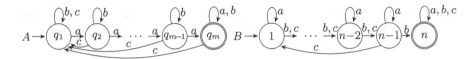

Fig. 5. Symmetric definite witnesses for concatenation $(m2^{n-1} - 2^{n-2} + 1)$.

automaton. The automaton B recognizes a right ideal since it has a loop on each input symbol in its unique final state. Thus both automata recognize symmetric definite languages, and we can show that they meet the desired upper bound.

(b) If K and L are star languages, then $\varepsilon \in K$, so the initial state of the DFA for K is final. It follows that the MNFA for KL has two initial states, and so the initial state of the corresponding subset automaton is $(s_A, \{s_B\})$. This means that the states $(p, \emptyset)$ are unreachable. This gives the desired upper bound.

For tightness, consider the DFAs A and B from Fig. 6. Since the initial state is the unique final state, both automata recognize star languages.

Construct the MNFA M for $L(A)L(B)$ from A and B by adding the transitions $(q_1, b, 1)$ and $(q_m, a, 1)$ and making the state q_1 non-final; the set of initial states is $\{q_1, 1\}$. Notice that in the reversed automaton M^R, each singleton $\{j\}$ is reached from $\{1\}$ by a string in cb^*, and each singleton $\{q_i\}$ is reached by a string in ba^*. It follows that for each state p of M, there is a string w_p such that $\{p\}$ is reached from $\{1\}$ by w_p in M^R. It follows that the string w_p^R is accepted by M from and only from the state p. This means that all states of the subset automaton $\mathcal{D}(M)$ are pairwise distinguishable. Let us denote the (reachable) state $\{q_i\} \cup S$ with $S \subseteq [1, n]$ by (i, S), and let us show that the set of reachable states is $\{(1, S) \mid S \subseteq [1, n] \text{ with } 1 \in S\} \cup \{(i, S) \mid i = 2, 3, \ldots, m \text{ and } S \subseteq [1, n] \text{ with } |S| \geq 1\}$. The initial state is $(1, \{1\})$ and it is sent to $(i, \{j\})$ by $a^{i-1}b^{j-1}$. Next, each (i, S) with $1 \in S$ is reached from $(m, S \setminus \{1\})$ by a^i and each (i, S) with $i \neq 1$, $S \neq \emptyset$, and $\min S = j \geq 2$ is reached from a state $(i, b^{j-1}S)$ with $1 \in b^{j-1}S$ by b^{j-1}. □

Remark 10. If A and B are the binary DFAs shown in Fig. 7, cf. [12], then we have $\mathrm{sc}(L(A)L(B)) = m2^n - 2^{n-1}$ since in the NFA for concatenation, each singleton set is co-reachable, and the reachability of the desired number of states is shown similarly as in Theorem 9(b). Since $L(A) = b^*L(A)$ and $L(B) = a^*L(B)$, automata A and B recognize comet languages. Moreover we have $\mathrm{COM} \subseteq 2\mathrm{COM}$, so the state complexity of concatenation on comets and two-sided comets is the same as in the class of regular languages. □

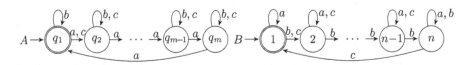

Fig. 6. Star witnesses for concatenation $(m2^n - 2^{n-1} - m + 1)$.

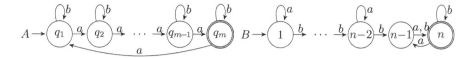

Fig. 7. Comet witnesses for concatenation ($m2^n - 2^{n-1}$).

Remark 11. The star-free witness languages that meet the regular upper bound for concatenation are described in [5, Theorem 2], and they are recognized by the DFAs shown in Fig. 8. Notice that both these DFAs are ordered. Since we have ORD ⊆ STFR ⊆ PSEP, the state complexity of concatenation on ordered and power-separating languages is $m2^n - 2^{n-1}$.   □

6   Star and Reversal

In this section, we consider the star and reversal operations. The state complexity of star is $\frac{3}{4}2^n$ with binary witnesses [17]. Our aim is to show that the complexity of star in our classes is always the same as in the regular case, except for classes of finitely generated left ideal, symmetric definite, and star languages, where it is $n + 1$, $n + 1$, and n, respectively. Recall that FGLID ⊆ SYDEF.

Theorem 12. *Let $n \geq 3$. Let L be a symmetric definite language accepted by an n-state DFA. Then L^* is accepted by a DFA with $n + 1$ states. This upper bound is met by the finitely generated left ideal $L = (a + b)^*a^{n-1}$.*

Proof. If $L \in$ SYDEF then $L = G\Sigma^*H$. Hence $L^+ = LL^* = G\Sigma^*H(G\Sigma^*H)^* = G\Sigma^*(HG\Sigma^*)^*H = G\Sigma^*H = L$. It follows that $L^* = \{\varepsilon\} \cup L$. This gives the upper bound $n + 1$. The reader may verify that $L = (a + b)^*a^{n-1}$ is accepted by an n-state DFA and $\mathrm{sc}(\{\varepsilon\} \cup L) = n + 1$.   □

Remark 13. If L is a star language, then $L = L^*$, so $\mathrm{sc}(L^*) = \mathrm{sc}(L)$.   □

Remark 14. The regular witness for star [17, Theorem 3.3] is recognized by the DFA A from Fig. 9. Since $L(A) = b^*L(A)$, the language $L(A)$ is a comet, so also a two-sided comet. Thus the complexity of star on COM and 2COM is $\frac{3}{4}2^n$.   □

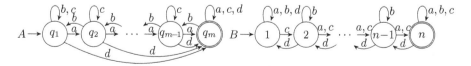

Fig. 8. Ordered witnesses for concatenation ($m2^n - 2^{n-1}$).

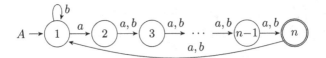

Fig. 9. A comet witness for star $\left(\frac{3}{4}2^n\right)$.

Remark 15. The regular upper bound $\frac{3}{4}2^n$ is met by the quaternary star-free language [5, Theorem 6] which is accepted by the DFA shown in Fig. 10. Since this DFA is ordered, and every ordered language is power-separating, the state complexity of star on ORD and PSEP is the same as in the regular case.   □

We conclude our investigation with the reversal operation, the state complexity of which is 2^n with binary witnesses [11]. First, using Lemmas 3 and 7, we get tight upper bounds for finitely generated left ideals and symmetric definite languages. Then, by proving that the set of all non-final states of a DFA accepting a power-separating language cannot be co-reachable, we get the tight upper bound $2^n - 1$ for reversal on ordered, star-free, and power-separating languages, which shows that a lower bound for star-free languages [5, Theorem 7] is tight. Finally, we show that the complexity of reversal on star, comet, and two-sided comet languages is the same as in the regular case.

Theorem 16. *Let L be a finitely generated left ideal over Σ accepted by an n-state DFA. Then $\mathrm{sc}(L^R) \leq 2^{n-2} + 2$, and this bound is tight if $|\Sigma| \geq 2^{n-2} + 2$.*

Proof. Let A be the minimal DFA for L. By Lemma 3, A has two states q, q' with $qa = q'a$ for each input a, and such that exactly one of them, say q, is final. Since L is a left ideal, adding a loop on each symbol in the initial state of A does not change the language; denote the resulting NFA by N. Thus all final sets of the subset automaton of N^R are equivalent. Moreover, except for possible initial set $\{q\}$, each reachable set contains either both q, q', or none of them.

For tightness, let $\Sigma = \{a, b\} \cup \{c_S \mid S \subseteq [2, n-1]\}$. Consider an n-state NFA $N = ([1, n], \Sigma, \cdot, 1, \{n\})$ where $1 \cdot a = \{1, 2\}$, $1 \cdot b = \{1\}$ and $i \cdot b = \{i+1\}$ if $2 \leq i \leq n-1$, $1 \cdot c_S = \{1\}$ and $i \cdot c_S = \{n\}$ if $i \in S$, and all the remaining transitions go to the empty set. The NFA N recognizes a finitely generated left ideal since state 1 has a loop on each symbol and for any other transition (i, σ, j) we have $i < j$. Since the subset automaton of N has n reachable states, the language $L(N)$ is accepted by an n-state DFA. In the reverse N^R, the initial set

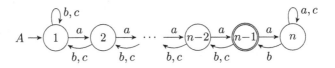

Fig. 10. An ordered witness for star; transitions (i, d, n) for each i are not shown $\left(\frac{3}{4}2^n\right)$.

is $\{n\}$, and it is sent to a subset S of $[2, n-1]$ by c_S. The set $\{1\}$ is reached from $\{2\}$ by a. For distinguishability, notice that $\{1\}$ is the unique final state. If $S, T \subseteq [2, n]$ and $s \in S \setminus T$, then $b^{s-2}a$ sends S to the final set $\{1\}$ and it sends T to the empty set. $\square$

Theorem 17. *Let $L \subseteq \Sigma^*$ be a symmetric definite language accepted by an n-state DFA. Then* $\mathrm{sc}(L^R) \leq 2^{n-1} + 1$, *and this upper bound is tight if* $|\Sigma| \geq 3$.

Proof. We get an upper bound in a similar way as in the proof of Theorem 9(a). The ternary left ideal from [3, Theorem 11(2), Fig. 17] meets this bound. $\square$

Theorem 18. *Let L be a power-separating language accepted by an n-state DFA. Then* $\mathrm{sc}(L^R) \leq 2^n - 1$, *and this bound is met by an ordered language over an alphabet of size $n - 1$.*

Proof. Let $A = (Q, \Sigma, \cdot, s, F)$ be a minimal DFA for a power-separating language L. Our aim is to show that the set $Q \setminus F$ of all non-final states of A cannot be co-reachable in A. Assume that $Q \setminus F$ is co-reachable by a string w. Since w sends F to $Q \setminus F$ in A^R, the string w must be rejected by A from each final state, and accepted from each non-final state, so there is no $k \geq 0$ such that $\bigcup_{i \geq k} \{w^i\} \subseteq L$ or $\bigcup_{i \geq k} \{w^i\} \subseteq L^c$. This means that L is not power-separating, a contradiction.

To get tightness, let $\Sigma = \{a, b\} \cup \{c_i \mid i = 3, 4, \ldots, n-1\}$. Consider the n-state DFA $A = ([1, n], \Sigma, \cdot, 1, F)$ where $F = \{i \in [1, n] \mid i \bmod 2 = 0\}$, and for each i in $[1, n]$ and each $j = 3, 4, \ldots, n-1$, let $i \cdot a = i + 1$ if $i \leq n - 1$ and $n \cdot a = n$, $i \cdot b = i - 1$ if $i \geq 2$ and $1 \cdot b = 1$, $i \cdot c_j = i$ if $i \neq j$ and $j \cdot c_j = j - 1$. Then A is ordered. It is shown in [5, Theorem 7] that every DFA for the reverse of $L(A)$ has at least $2^n - 1$ states. $\square$

Remark 19. The DFA A from Fig. 11 differs from Šebej's witness for reversal meeting the upper bound 2^n [10, Theorem 5] only in the set of final states, and Šebej's proof works for this DFA as well. Since the initial state of A is a unique final state, A recognizes a star, so also comet and two-sided comet, language. $\square$

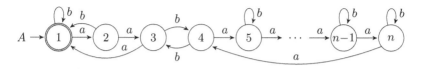

Fig. 11. A star witness for reversal (2^n).

7   Conclusion

We investigated the state complexity of basic regular operations in the classes of combinational, singleton, finitely generated left ideal, symmetric definite, star, comet, two-sided comet, ordered, and power-separating languages. Our results are summarized in Table 1 where the sizes of alphabets used to describe witnesses are given by numbers in parentheses. Except for the ordered and finitely generated left ideal witnesses for reversal, all the remaining witnesses are described over a fixed alphabet, and whenever a binary alphabet is used, it is always optimal. By showing that the upper bound for reversal on power-separating languages is $2^n - 1$, we proved that the same lower bound for reversal on star-free languages from [5, Theorem 7] is tight.

In [1], the classes of definite, generalized definite, locally testable, and strictly locally testable languages were also considered. We leave the operational complexity in these classes for future research. The optimality of alphabet sizes greater than two remains open as well.

Table 1. Operational complexity in subregular classes; we have $\bullet = m2^{n-1} - 2^{n-2} + 1$ and $\circ = m2^n - 2^{n-1} - m + 1$.

class	$K \cap L$	$K \cup L$	KL	L^*	L^R
CB Th. 1	2 (1)	2 (1)	3 (1)	2 (2)	3 (2)
SGL Th. 2	$\min\{m,n\}$ (1)	$m + n - 3$ (2)	$m + n - 2$ (1)	$n - 1$ (2)	n (1)
FGLID	$mn - 2$ (10)	$mn - 2$ (3)	$m + n - 1$ (1)	$n + 1$ (2)	$2^{n-2} + 2$
source		Th. 4	Re. 8	Th. 12	Th. 16
LID	mn (2)	mn (4)	$m + n - 1$ (1)	$n + 1$ (2)	$2^{n-1} + 1$ (2)
source		[3, Th. 8]	[3, Th. 9]	[3, Th. 10]	[3, Th. 11]
SYDEF	mn (2)	mn (4)	$\bullet$ (3)	$n + 1$ (2)	$2^{n-1} + 1$ (3)
source		Th. 5	Th. 9	Th. 12	Th. 17
STAR	mn (2)	mn (2)	$\circ$ (3)	n (1)	2^n (2)
source		Th. 5	Th. 9	Re. 13	Re. 19
COM, 2COM	mn (2)	mn (2)	$m2^n - 2^{n-1}$ (2)	$\frac{3}{4}2^n$ (2)	2^n (2)
source		Re. 6	Re. 10	Re. 14	Re. 19
ORD, PSEP	mn (2)	mn (2)	$m2^n - 2^{n-1}$ (4)	$\frac{3}{4}2^n$ (4)	$2^n - 1$ $(n-1)$
source		Th. 5	Re. 11	Re. 15	Th. 18
STFR	mn (2)	mn (2)	$m2^n - 2^{n-1}$ (4)	$\frac{3}{4}2^n$ (4)	$2^n - 1$ $(n-1)$
source		[5, Th. 1]	[5, Th. 2]	[5, Th. 6]	Th. 18
regular	mn (2)	mn (2)	$m2^n - 2^{n-1}$ (2)	$\frac{3}{4}2^n$ (2)	2^n (2)
source		[17, Th. 4.3]	[12]	[17, Th. 3.3]	[11, Prop. 2]

References

1. Bordihn, H., Holzer, M., Kutrib, M.: Determination of finite automata accepting subregular languages. Theor. Comput. Sci. **410**(35), 3209–3222 (2009). https://doi.org/10.1016/j.tcs.2009.05.019
2. Brzozowski, J., Jirásková, G., Li, B., Smith, J.: Quotient complexity of bifix-, factor-, and subword-free regular languages. Acta Cybern. **21**(4), 507–527 (2014). https://doi.org/10.14232/actacyb.21.4.2014.1
3. Brzozowski, J.A., Jirásková, G., Li, B.: Quotient complexity of ideal languages. Theor. Comput. Sci. **470**, 36–52 (2013). https://doi.org/10.1016/j.tcs.2012.10.055
4. Brzozowski, J., Jirásková, G., Zou, C.: Quotient complexity of closed languages. Theor. Comput. Sci. **54**(2), 277–292 (2013). https://doi.org/10.1007/s00224-013-9515-7
5. Brzozowski, J.A., Liu, B.: Quotient complexity of star-free languages. Int. J. Found. Comput. Sci. **23**(6), 1261–1276 (2012). https://doi.org/10.1142/S0129054112400515
6. Câmpeanu, C., Culik, K., Salomaa, K., Yu, S.: State complexity of basic operations on finite languages. In: Boldt, O., Jürgensen, H. (eds.) WIA 1999. LNCS, vol. 2214, pp. 60–70. Springer, Heidelberg (2001). https://doi.org/10.1007/3-540-45526-4_6
7. Han, Y.S., Salomaa, K.: State complexity of basic operations on suffix-free regular languages. Theor. Comput. Sci. **410**(27–29), 2537–2548 (2009). https://doi.org/10.1016/j.tcs.2008.12.054
8. Han, Y.S., Salomaa, K., Wood, D.: Operational state complexity of prefix-free regular languages. In: Ésik, Z., Fülöp, Z. (eds.) Automata, Formal Languages, and Related Topics, pp. 99–115. University of Szeged, Hungary (2009)
9. Hospodár, M., Mlynárčik, P., Olejár, V.: Operations on subregular languages and nondeterministic state complexity. In: Han, Y., Vaszil, G. (eds.) DCFS 2022. LNCS, vol. 13439, pp. 112–126. Springer, Cham (2022). https://doi.org/10.1007/978-3-031-13257-5_9
10. Jirásková, G., Šebej, J.: Reversal of binary regular languages. Theor. Comput. Sci. **449**, 85–92 (2012). https://doi.org/10.1016/j.tcs.2012.05.008
11. Leiss, E.L.: Succinct representation of regular languages by Boolean automata. Theor. Comput. Sci. **13**, 323–330 (1981). https://doi.org/10.1016/S0304-3975(81)80005-9
12. Maslov, A.N.: Estimates of the number of states of finite automata. Soviet Math. Doklady **11**, 1373–1375 (1970)
13. Olejár, V., Szabari, A.: Closure properties of subregular languages under operations. Int. J. Found. Comput. Sci., online ready. https://doi.org/10.1142/S0129054123450016. Extended abstract in MCU 2022. LNCS, vol. 13419, pp. 126–142. Springer, Cham (2022). https://doi.org/10.1007/978-3-031-13502-6_9
14. Paz, A., Peleg, B.: Ultimate-definite and symmetric-definite events and automata. J. ACM **12**(3), 399–410 (1965). https://doi.org/10.1145/321281.321292
15. Pighizzini, G., Shallit, J.: Unary language operations, state complexity and Jacobsthal's function. Int. J. Found. Comput. Sci. **13**(1), 145–159 (2002). https://doi.org/10.1142/S012905410200100X
16. Sipser, M.: Introduction to the theory of computation. Cengage Learning (2012)
17. Yu, S., Zhuang, Q., Salomaa, K.: The state complexities of some basic operations on regular languages. Theor. Comput. Sci. **125**(2), 315–328 (1994). https://doi.org/10.1016/0304-3975(92)00011-F

When Is Context-Freeness
Distinguishable from Regularity?
an Extension of Parikh's Theorem

Yusuke Inoue[✉], Kenji Hashimoto, and Hiroyuki Seki

Graduate School of Informatics, Nagoya University, Nagoya, Japan
{y-inoue,seki}@sqlab.jp, k-hasimt@i.nagoya-u.ac.jp

Abstract. Parikh's theorem states that for every CFL X, there exists
a regular language X' such that X is Parikh equivalent to X'. As a
corollary of Parikh's theorem, it is known that every unary CFL is regular. This is because if words w and w' are Parikh equivalent, then
their lengths are also equal. In general, if an equivalence relation $\equiv$ is
coarser than Parikh equivalence, it has Parikh property; for every CFL
X, there exists a regular language X' which is equivalent to X in the
sense of $\equiv$. Otherwise, $\equiv$ does not always have Parikh property. In this
paper, focusing on the fact that Parikh equivalence is compatible with
swapping adjacent letters, we propose an equivalence relation compatible with swapping adjacent subwords in a regular language given as a
parameter. We also identify conditions on the parameter language that
guarantee Parikh property for the resulting equivalence relation.

Keywords: Parikh's theorem · finite monoid

1 Introduction

For an alphabet $\Sigma = \{a_1, \ldots, a_n\}$, Parikh vector of a word w is $(v_1, \ldots, v_n) \in \mathbb{N}^\Sigma$
where v_i is the number of occurrences of a_i in w for each $1 \leq i \leq n$. Parikh
equivalence $\equiv_{\mathrm{Pk}}$ is the equivalence relation over the words defined as $w \equiv_{\mathrm{Pk}} w'$
iff Parikh vectors of w, w' are equal. For example, $abb \equiv_{\mathrm{Pk}} bba$ because both have
the same Parikh vector $(1, 2)$ for $\Sigma = \{a, b\}$. We also say that two languages
are Parikh equivalent if their quotients by Parikh equivalence are equal. It is
known as Parikh's theorem that for every context-free language (CFL) X, there
exists a regular language X' which is Parikh equivalent to X [9]. For example,
CFL $\{a^k b a^k \mid k \geq 0\}$ is Parikh equivalent to regular language $b(aa)^*$. Parikh's
theorem is an important and interesting result in automata theory, and many
related studies have been conducted. In [10], Parikh's theorem is proved by using
simultaneous equations on a commutative alphabet. In [5], the classes of regular
languages and CFL are extended to superAFL, and Parikh's theorem is proved as
a corollary of a property of superAFL. There are many other proofs of Parikh's
theorem [3,4]. An extension to weighted models has also been studied [2]. In

B. Nagy (Ed.): CIAA 2023, LNCS 14151, pp. 166–178, 2023.
https://doi.org/10.1007/978-3-031-40247-0_12

[1,6,8], Parikh vectors are extended to Parikh matrices and its properties are studied.

It is well known that every unary CFL is regular, and this fact follows also from Parikh's theorem. This is because if w and w' are Parikh equivalent, then their lengths are also equal. In general, if Parikh equivalence is a refinement of an equivalence relation $\equiv$, then $\equiv$ has *Parikh property*, that is, for every CFL X, there exists a regular language X' which is equivalent to X in the sense of $\equiv$. These cases correspond to "trivial cases" in Fig. 1 ($\equiv_{\mathrm{len}}$ is the equivalence relation defined by the length of words). On the other hand, $\equiv$ does not always have Parikh property if $\equiv \subseteq \equiv_{\mathrm{Pk}}$. For example, the equivalence relation defined based on Parikh matrices does not always have Parikh property. As illustrated by examples of $\equiv_{\mu_1}$ and $\equiv_{\mu_2}$ defined later, the characterization for an equivalence relation to have Parikh property is not yet clear (see Fig. 1).

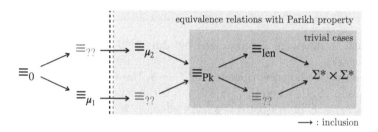

Fig. 1. Equivalence relations and Parikh property.

In this paper, we extend Parikh equivalence using algebraic structures, and study conditions for having Parikh property. In Sect. 2, we define basic concepts such as languages, monoids and transition graphs. In Sect. 3, we define Parikh property, which is the key concept in this paper, and describe our goals. In Sects. 4 and 5, we discuss necessary conditions and sufficient conditions for equivalence relations to have Parikh property, respectively. In Sect. 6, we focus on equivalence relations defined by using finite groups, and mention future work.

2   Preliminaries

Let $\mathbb{N}$ be the set of all non-negative integers. The cardinality of a set S is denoted by $|S|$. Let Σ be a finite alphabet. For a word $w \in \Sigma^*$, the length of w is denoted by $|w|$. For w and a letter $a \in \Sigma$, the number of occurrences of a in w is denoted by $|w|_a$. The empty word is denoted by ε, i.e., $|\varepsilon| = 0$. A language $X \subseteq \Sigma^*$ is *regular* if X is recognized by some finite automaton. We say that X is a *context-free language* (*CFL*) if X is generated by some context-free grammar.

We say that $(S, \odot)$ is a *semigroup* where S is a set and $\odot$ is an associative binary operator on S. A *monoid* is a semigroup having the identity element $e \in S$. An element $s \in S$ of a monoid $(S, \odot)$ is *invertible* if there is an element $s^{-1} \in S$ such that $s \odot s^{-1} = e$. This element s^{-1} is called an *inverse element* of s. An element $\iota \in S$ such that $\iota \odot \iota = \iota$ is called an *idempotent element* of $(S, \odot)$. A *group* is a monoid such that all elements are invertible. Let $|(S, \odot)|$ be the order of a semigroup $(S, \odot)$, i.e., $|(S, \odot)| = |S|$. A monoid $(S, \odot)$ is *cyclic* if $(S, \odot)$ is generated by a single element $s \in S$. With each element s in a finite semigroup, we can associate the finite table $\begin{pmatrix} s_1 \ s_2 \ \cdots \ s_{|S|} \\ t_1 \ t_2 \ \cdots \ t_{|S|} \end{pmatrix}$ where $s_i \odot s = t_i$ for each $1 \le i \le |S|$. We sometimes write the table $\begin{pmatrix} s_1 \ s_2 \ \cdots \ s_{|S|} \\ t_1 \ t_2 \ \cdots \ t_{|S|} \end{pmatrix}$ to represent an element $s \in S$.

For a monoid $(S, \odot)$, a subset $T \subseteq S$ and a surjective homomorphism $\phi : \Sigma^* \to S$, we say that $L = \phi^{-1}(T)$ is recognized by (ϕ, T) with the monoid.

Proposition 1 ([7]). *A language $L \subseteq \Sigma^*$ is regular iff L is recognized by some (ϕ, T) with a finite monoid $(S, \odot)$, a surjective homomorphism $\phi : \Sigma^* \to S$ and $T \subseteq S$.* ∎

In particular, a language L recognized by some (ϕ, T) with a group is called a *group language*.

For a surjective homomorphism $\phi : \Sigma^* \to S$, the *transition graph* of ϕ is the directed graph (S, E) where $E \subseteq S \times \Sigma \times S$ is the set of edges labeled by Σ such that $(s_1, a, s_2) \in E$ iff $s_1 \odot \phi(a) = s_2$. An edge $(s_1, a, s_2) \in E$ is also denoted as $s_1 \xrightarrow{a} s_2$. A *cycle* in a transition graph (S, E) is a path $s_1 \xrightarrow{a_1} s_2 \xrightarrow{a_2} \cdots \xrightarrow{a_{n-1}} s_n$ such that $n \ge 2$ and $s_1 = s_n$. A *simple path* is a path $s_1 \xrightarrow{a_1} \cdots \xrightarrow{a_{n-1}} s_n$ such that $s_1, \ldots, s_n$ are pairwise different, and a *simple cycle* is a cycle $s_1 \xrightarrow{a_1} \cdots \xrightarrow{a_{n-1}} s_n$ such that $s_1 \xrightarrow{a_1} \cdots \xrightarrow{a_{n-2}} s_{n-1}$ is a simple path. A *cycle graph* is a subgraph of (S, E) induced by some subset $V = \{s_1, \ldots, s_n\} \subseteq S$ such that there is a simple cycle $s_1 \xrightarrow{a_1} \cdots \xrightarrow{a_n} s_1$.

The followings are examples of transition graphs (see Fig. 2). For instance, the second from the left is the transition graph of the homomorphism $\phi_2 : \{a, b\}^* \to \{0 = \begin{pmatrix} 0 \ 1 \\ 0 \ 1 \end{pmatrix}, 1 = \begin{pmatrix} 0 \ 1 \\ 1 \ 0 \end{pmatrix}\}$ such that $\phi_2(a) = 0, \phi_2(b) = 1$. Note that we denote the elements of the monoid as $\begin{pmatrix} 0 \ 1 \\ 0 \ 1 \end{pmatrix}$ and $\begin{pmatrix} 0 \ 1 \\ 1 \ 0 \end{pmatrix}$ by using the notation defined above. Each example will be discussed in detail later.

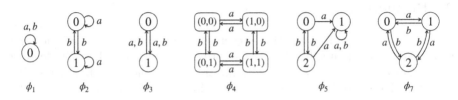

Fig. 2. Examples of transition graphs.

We briefly explain the relationship between transition graphs and finite automata. Let $(S, \odot)$ be a finite monoid, $h : \Sigma^* \to S$ be a monoid homomorphism and T be a subset of S. Let $\mathcal{A}$ be the finite automaton defined by the transition graph (S, E) of ϕ where the initial state and the set of accept states are the identity element e and T, respectively. It is clear that the language L recognized by (ϕ, T) coincides with the language recognized by $\mathcal{A}$. The monoid $(S, \odot)$ is called the *transition monoid* of $\mathcal{A}$. If $\mathcal{A}$ is the minimal automaton recognizing L, $(S, \odot)$ is called the *syntactic monoid* of L.

3 Parikh Property

Let $\equiv \subseteq \Sigma^* \times \Sigma^*$ be an equivalence relation. The equivalence class of $\equiv$ that contains w is denoted by $[w]_\equiv$. For languages $X, X' \subseteq \Sigma^*$, we write $X \equiv X'$ to mean that $\{[w]_\equiv \mid w \in X\} = \{[w]_\equiv \mid w \in X'\}$.

Parikh equivalence $\equiv_{\mathrm{Pk}} \subseteq \Sigma^* \times \Sigma^*$ is the equivalence relation such that $w \equiv_{\mathrm{Pk}} w'$ iff $|w|_a = |w'|_a$ for each $a \in \Sigma$. The following fact is well known.

Theorem 1 (Parikh's Theorem [9]). *For every CFL $X \subseteq \Sigma^*$, there exists a regular language $X' \subseteq \Sigma^*$ such that $X \equiv_{\mathrm{Pk}} X'$.* ∎

In this paper, we generalize and study Parikh equivalence and Parikh's Theorem. For this purpose, we generalize the property of $\equiv_{\mathrm{Pk}}$ stated in Parikh's Theorem to an arbitrary equivalence relation.

Definition 1 (Parikh property). *An equivalence relation $\equiv$ has Parikh property if for every CFL $X \subseteq \Sigma^*$, there exists a regular language $X' \subseteq \Sigma^*$ such that $X \equiv X'$.* ∎

By Parikh's theorem, $\equiv_{\mathrm{Pk}}$ has Parikh property. As a corollary of Parikh's theorem, an equivalence relation $\equiv$ has Parikh property if $\equiv_{\mathrm{Pk}} \subseteq \equiv$. For example, it is well known that the relation $\equiv_{\mathrm{len}} = \{(w, w') \in \Sigma^* \times \Sigma^* \mid |w| = |w'|\}$ also has Parikh property[1] because $\equiv_{\mathrm{Pk}} \subseteq \equiv_{\mathrm{len}}$. On the other hand, an equivalence relation $\equiv$ does not always have Parikh property if $\equiv \subseteq \equiv_{\mathrm{Pk}}$.

Example 1. Let $\Sigma = \{a, b\}$. The smallest equivalence relation $\equiv_0 = \{(w, w) \mid w \in \Sigma^*\}$ does not have Parikh property. In fact, there is no regular language X' such that $X \equiv_0 X'$ for CFL $X = \{a^k b a^k \mid k \geq 0\}$. ∎

Example 2. Let $\Sigma = \{a, b\}$, and let $\mu_1(a) = \begin{bmatrix} 2 & 0 \\ 0 & 1 \end{bmatrix}, \mu_1(b) = \begin{bmatrix} 0 & 3 \\ 1 & 0 \end{bmatrix}$ be 2×2 matrices. We define $\overline{\mu_1} : \Sigma^* \to \mathbb{N}^{2 \times 2}$ by $\overline{\mu_1}(a_1 a_2 \cdots a_n) = \mu_1(a_1)\mu_1(a_2)\cdots\mu_1(a_n)$ for $a_1, \ldots, a_n \in \Sigma$. Let $\equiv_{\mu_1}$ be the equivalence relation such that $w \equiv_{\mu_1} w'$ iff $\overline{\mu_1}(w) = \overline{\mu_1}(w')$. Then,

- $\equiv_{\mu_1} \subseteq \equiv_{\mathrm{Pk}}$ by the following reasons: let $\det(A)$ be the determinant of a matrix A. Note that $\det(\mu_1(a)) = 2$ and $\det(\mu_1(b)) = -3$ are coprime. As $\det(AB) = \det(A)\det(B)$ for any square matrices A and B, we obtain that $u \equiv_{\mu_1} v$ implies $u \equiv_{\mathrm{Pk}} v$. Furthermore,

---

[1] This corollary is sometimes stated as: every CFL $X \subseteq \Sigma^*$ is regular if $|\Sigma| = 1$.

- $\equiv_{\mu_1} \neq \equiv_{Pk}$ because $\begin{bmatrix} 0 & 6 \\ 1 & 0 \end{bmatrix} = \overline{\mu_1}(ab) \neq \overline{\mu_1}(ba) = \begin{bmatrix} 0 & 3 \\ 2 & 0 \end{bmatrix}$.
- Besides, $\equiv_{\mu_1} \not\supseteq \equiv_0$ because $\overline{\mu_1}(abb) = \overline{\mu_1}(bba) = \begin{bmatrix} 6 & 0 \\ 0 & 3 \end{bmatrix}$.

To be precise, $w \equiv_{\mu_1} w'$ iff w can be obtained by repeatedly swapping adjacent subwords of w' that contain even number of b. This is because the multiplication of diagonal matrices is commutative and the product of two 2×2 anti-diagonal matrices is a diagonal matrix. For example, $baaba \equiv_{\mu_1} abaab \neq_{\mu_1} babaa \equiv_{\mu_1} ababa$. In fact, it is not obvious whether $\equiv_{\mu_1}$ has Parikh property or not. We will show that $\equiv_{\mu_1}$ does not have Parikh property later (Example 5). ∎

Example 3. Let $\Sigma = \{a, b\}$, and we define $\equiv_{\mu_2}$ by $\mu_2(a) = \begin{bmatrix} 0 & 2 \\ 1 & 0 \end{bmatrix}, \mu_2(b) = \begin{bmatrix} 0 & 3 \\ 1 & 0 \end{bmatrix}$ in the same way as in Example 2. Then, $w \equiv_{\mu_2} w'$ iff w can be obtained by swapping even-length subwords of w'. For example, $baaa \equiv_{\mu_2} aaba \neq_{\mu_2} abaa$. In contrast to $\equiv_{\mu_1}$, we can show that $\equiv_{\mu_2}$ has Parikh property (Example 10). ∎

In this paper, we study conditions for an equivalence relation $\equiv (\subseteq \equiv_{Pk})$ to have Parikh property. For this purpose, we extend Parikh equivalence. Note that Parikh equivalence is characterized by the condition $u_1 u_2 \equiv_{Pk} u_2 u_1$ for each $u_1, u_2 \in \Sigma^*$. Thus, by using a regular language $L \subseteq \Sigma^*$, we can get a refinement of Parikh equivalence by changing the condition $u_1, u_2 \in \Sigma^*$ to $u_1, u_2 \in L$.

Definition 2. *The equivalence relation defined by a regular language L, denoted by $\equiv_L$, is defined as the smallest equivalence relation satisfying the following conditions: for all $u_1, u_2, v_1, v_2 \in \Sigma^*$,*

- *if $u_1, u_2 \in L$, then $u_1 u_2 \equiv_L u_2 u_1$, and*
- *if $u_1 \equiv_L v_1$ and $u_2 \equiv_L v_2$, then $u_1 u_2 \equiv_L v_1 v_2$.* ∎

As illustrated in the following example, $\equiv_{\Sigma^*} = \equiv_{Pk}$. Therefore, $\equiv_L \subseteq \equiv_{Pk}$ for every language L, and Definition 2 is a natural generalization of Parikh equivalence.

Example 4. Let $\Sigma = \{a, b\}$. See Fig. 2 for ϕ_1, ϕ_2 and ϕ_3.

- Let $\mathcal{C}_1 = \{0\}$ be the trivial group, and $\phi_1(a) = \phi_1(b) = 0$. Then, the language recognized by $(\phi_1, \{0\})$ is $L_1 = \Sigma^*$. The equivalence $\equiv_{L_1}$ is equal to Parikh equivalence $\equiv_{Pk}$.
- Let $\mathcal{C}_2 = \{0 = \begin{pmatrix} 0 & 1 \\ 0 & 1 \end{pmatrix}, 1 = \begin{pmatrix} 0 & 1 \\ 1 & 0 \end{pmatrix}\}$ be the cyclic group with set $\{0, 1\}$, and $\phi_2(a) = 0, \phi_2(b) = 1$. Then, the language recognized by $(\phi_2, \{0\})$ is $L_2 = \{w \mid |w|_b$ is an even number$\}$. The equivalence $\equiv_{L_2}$ is equal to $\equiv_{\mu_1}$ in Example 2.
- Let $\mathcal{C}_2$ be the cyclic group with set $\{0, 1\}$, and $\phi_3(a) = \phi_3(b) = 1$. Then, the language recognized by $(\phi_3, \{0\})$ is $L_3 = \{w \mid w$ is even-length$\}$. The equivalence $\equiv_{L_3}$ is equal to $\equiv_{\mu_2}$ in Example 3. ∎

As in Example 4, we often use a finite monoid and a homomorphism to represent a regular language L (see Proposition 1). In particular, we focus on the inverse images of the identity or idempotent elements because such elements are closely related to the expressive power of regular languages. For example, the usual pumping lemma for regular languages is based on the fact that there is a subword that mapped to an idempotent element by the homomorphism.

The following lemma can be easily shown.

Lemma 1. *Let* Σ, Δ *be finite alphabets,* $L \subseteq \Delta^*$ *be a language and* $h : \Delta^* \to \Sigma^*$ *be a homomorphism. For each* $w, w' \in \Delta^*$, $h(w) \equiv_{h(L)} h(w')$ *if* $w \equiv_L w'$.   ■

4 Necessary Conditions for Having Parikh Property

In this section, we study necessary conditions for an equivalence relation $\equiv_L$ to have Parikh property. As mentioned after Example 4, we focus on the cases where L is the inverse image of a singleton of the identity element or a set of idempotents. In particular, we give a necessary condition for $\equiv_L$ to have Parikh property where L is the inverse image of $\{e\}$ (Corollary 1). For this, we will conduct the case analysis based on whether the monoid is a group or not.

4.1 The Case of Groups

When a monoid is a group, we can derive a useful necessary condition for Parikh property relatively easily because the cancellation property holds: $(s \odot t_1 = s \odot t_2) \Rightarrow t_1 = t_2$ and $(t_1 \odot s = t_2 \odot s) \Rightarrow t_1 = t_2$. (A finite monoid is a group iff the cancellation property holds.)

Theorem 2. *Let* L *be a regular language recognized by* $(\phi, \{e\})$ *with a finite group* $(S, \odot)$. *Let* (S, E) *be the transition graph of* ϕ. *If there are cycle graphs* (V_1, E_1') *and* (V_2, E_2') *of* (S, E) *such that* $V_1 \cap V_2 = \emptyset$, *then* $\equiv_L$ *does not have Parikh property.*

Proof. Let $(V_1, E_1'), (V_2, E_2')$ be the cycle graphs with $V_1 \cap V_2 = \emptyset$ generated by simple cycles $s_1 \xrightarrow{a_1} \cdots \xrightarrow{a_n} s_1$ and $t_1 \xrightarrow{b_1} \cdots \xrightarrow{b_m} t_1$, respectively. As $(S, \odot)$ is a group, any two elements in S are reachable from each other. Thus, there is a simple path $r_1 \xrightarrow{c_1} \cdots \xrightarrow{c_\ell} r_{\ell+1}$ such that $\ell \geq 1, r_1 \in V_1, r_{\ell+1} \in V_2$ and $V_1, V_2, V_3 = \{r_2, \ldots, r_\ell\}$ are pairwise disjoint. Without loss of generality, we can assume that $s_1 = r_1$ and $t_1 = r_{\ell+1}$. Note that $s_1 \odot \phi(c_1 \cdots c_\ell) = t_1$ (see Fig. 3).

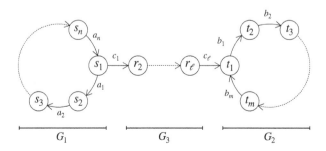

Fig. 3. The transition graph with disjoint cycles.

Let $x = a_1 \cdots a_n$, $y = c_1 \cdots c_\ell$, $z = b_1 \cdots b_m \in \Sigma^*$ and let $w^{(k)} = x^k y z^k \in \Sigma^*$ with $k \in \mathbb{N}$. It is obvious that $x \neq y$, because otherwise $s_1 = s_1 \odot \phi(x) = s_1 \odot \phi(y) = t_1$. We prove the theorem by showing that for CFL $X = \{w^{(k)} \mid k \in \mathbb{N}\}$, there is no regular language X' such that $X \equiv_L X'$.

First, we show that $X = X'$ if $X \equiv_L X'$ by showing that there is no word $w'(\neq w^{(k)})$ such that $w' \equiv_L w^{(k)}$ for each $k \in \mathbb{N}$. By Definition 2, it suffices to show that for $w^{(k)}$, swapping any adjacent subwords $u_1, u_2 \in L$ does not change the whole word $w^{(k)}$. Note that $\phi(u) = e$ iff $u \in L$. Because $s \odot t = s$ iff $t = e$ for each element s in a group, a path labeled with a word $u \in \Sigma^*$ is a cycle iff $\phi(u) = e$. Remember that V_1, V_2, V_3 are pairwise disjoint and $r_1 \xrightarrow{c_1} \cdots \xrightarrow{c_\ell} r_{\ell+1}$ is simple. Therefore, u is a subword of either x^k or z^k if $u \in L$. Hence, we show that every adjacent subwords $u_1, u_2 \in L$ of x^k or z^k are represented as $u_1 = u^{p_1}, u_2 = u^{p_2}$ for some $u \in \Sigma^*, p_1, p_2 \in \mathbb{N}$. We suppose that u_1, u_2 are subwords of x^k. Let $q \le k$ be the smallest number such that u_1 is a subword of x^q, and we can get prefixes x_1, x_2 of x such that $x_1 u_1 = x^{q-1} x_2$. Because $\phi(x_1) = \phi(x_1) \odot e = \phi(x_1 u_1) = \phi(x^{q-1} x_2) = e \odot \phi(x_2) = \phi(x_2)$, we obtain that $x_1 = x_2$ by the following reasons. Both of x_1 and x_2 are prefixes of x, and hence either x_1 is a prefix of x_2 or x_2 is a prefix of x_1. Assume that x_1 is a proper prefix of x_2. Then, for $\overline{x_2}$ such that $x_2 \overline{x_2} = x$, it follows from $\phi(x_2) \circ \phi(\overline{x_2}) = \phi(x) = e$ and $\phi(x_1) = \phi(x_2)$ that $\phi(\overline{x_2} x_1) = \phi(\overline{x_2}) \odot \phi(x_1) = \phi(x_2)^{-1} \odot \phi(x_2) = e$. Hence, there are $i_1 < i_2$ such that $s_{i_2} \xrightarrow{a_{i_2}} \cdots \xrightarrow{a_n} s_1 \xrightarrow{a_1} \cdots \xrightarrow{a_{i_1}} s_{i_1+1}$ is a cycle with $\overline{x_2} x_1 = a_{i_2} \cdots a_n a_1 \cdots a_{i_1}$, and it contradicts the fact that $s_1 \xrightarrow{a_1} \cdots \xrightarrow{a_{n-1}} s_1$ is a simple cycle. The same holds in the case that x_2 is a proper prefix of x_1. Therefore, we can conclude that $x_1 = x_2$ and $u_1 = (\overline{x_2} x_2)^{p_1}$ for some p_1. Because $u_1 u_2$ is a subword of x^k and u_1 ends with the prefix x_2 of x, it follows that $x_2 u_2 = x^{q'} x_3$ for some $q' \in \mathbb{N}$ and some prefix x_3 of x. For the same reasons as u_1, it holds that $u_2 = (\overline{x_2} x_2)^{p_2}$ for some p_2. Similar reasoning works in the case that u_1, u_2 are subwords of z^k, and we obtain that $u_1 = u^{p_1}, u_2 = u^{p_2}$ for some $u \in \Sigma^*, p_1, p_2 \in \mathbb{N}$. Therefore, no word $w'(\neq w^{(k)})$ satisfies $w' \equiv_L w^{(k)}$ for each $k \in \mathbb{N}$, and $X = X'$ if $X \equiv_L X'$. By using a pumping lemma we can show that the language $X = \{x^k y z^k \mid k \in \mathbb{N}\}$ is not regular. $\square$

Example 5. Let $\mathcal{C}_2$ be the cyclic group with set $\{0, 1\}$, and $\phi_2(a) = 0, \phi_2(b) = 1$ (see Example 4). Then, $\equiv_{L_2}$ where L_2 is the language recognized by $(\phi_2, \{0\})$ does not have Parikh property. This is because there are cycles $0 \xrightarrow{a} 0$ and $1 \xrightarrow{a} 1$ in the transition graph, and $\{0\} \cap \{1\} = \emptyset$ (see Fig. 2). In fact, there is no regular language X' such that $X \equiv_{L_2} X'$ for CFL $X = \{a^k b a^k \mid k \in \mathbb{N}\}$.   ∎

Lemma 2. *Let $(S, \odot)$ be a finite group, $\phi : \Sigma^* \to S$ be a surjective homomorphism, and (S, E) be the transition graph of ϕ. If there exists $a \in \Sigma$ such that $\phi(a)$ generates proper subgroup $S' \subsetneq S$, then there are cycle graphs $(V_1, E_1'), (V_2, E_2')$ of (S, E) such that $V_1 \cap V_2 = \emptyset$.*

Proof. Let $s_1 \in S'$ and $s_2 \in S \setminus S'$. Because $(S, \odot)$ is a group, there is an element $t \in S$ such that $t \odot s_1 = s_2$. There are two cycle graphs generated by cosets S' and $tS' = \{t \odot s \mid s \in S'\}$, and S', tS' are disjoint. $\square$

Theorem 3. *Let L be the language recognized by $(\phi, \{e\})$ with a group $(S, \odot)$. If $\equiv_L$ has Parikh property, then $(S, \odot)$ is isomorphic to a cyclic group.*

Proof. By Theorem 2 and Lemma 2, if $\equiv_L$ has Parikh property, then $\phi(a)$ generates $(S, \odot)$ for each $a \in \Sigma$. Because $(S, \odot)$ is generated by a single element, $(S, \odot)$ is cyclic. □

Example 6. Let $\phi_4 : \Sigma^* \to \mathcal{C}_2 \times \mathcal{C}_2$ be the homomorphism defined as $\phi_4(a) = (1, 0)$ and $\phi_4(b) = (0, 1)$. Because the direct product $\mathcal{C}_2 \times \mathcal{C}_2$ is not isomorphic to any cyclic group, $\equiv_{L_4}$ does not have Parikh property where L_4 is the language recognized by $(\phi_4, \{(0, 0)\})$. In fact, there are disjoint cycles $(0, 0) \xrightarrow{a} (1, 0) \xrightarrow{a} (0, 0)$ and $(0, 1) \xrightarrow{a} (1, 1) \xrightarrow{a} (0, 1)$ (see Fig. 2). ■

4.2   The Case of General Monoids

When a monoid is not a group, it is more difficult for $\equiv_L$ to have Parikh property than the case of groups where L is an inverse image of $\{e\}$. This is because any two different words whose prefixes are mapped to non-invertible elements cannot be exchanged when the monoid is not a group.

Theorem 4. *Let L be the regular language recognized by $(\phi, \{e\})$ with a finite monoid $(S, \odot)$ which is not a group. If $|\Sigma| \geq 2$, then $\equiv_L$ does not have Parikh property.*

Proof. If $\phi(a)$ is invertible for each $a \in \Sigma$, then each element $\phi(w) = \phi(a_1 \cdots a_n) \in S$ has the inverse element $\phi(a_n)^{-1} \odot \cdots \odot \phi(a_1)^{-1}$, and $\phi(w)$ is also invertible. Therefore, there is an element $a \in \Sigma$ such that $\phi(a)$ is not invertible because $(S, \odot)$ is not a group. Hence, it holds that $\phi(u) \neq e$ for each word $u \in a\Sigma^*$. Let $b \in \Sigma$ be a letter such that $b \neq a$. Then, $X = X'$ if $X \equiv_L X'$ holds for CFL $X = \{a^k b a^k \mid k \in \mathbb{N}\}$. This is because for each k, there is at most one subword $v \in b\Sigma^*$ of $a^k b a^k$ that can satisfy $\phi(v) = e$. Hence, there is no regular language X' such that $X \equiv_L X'$. □

By Theorems 2 and 4, we obtain a necessary condition for having Parikh property for general monoids.

Corollary 1. *Let L be the regular language recognized by $(\phi, \{e\})$ with a finite monoid $(S, \odot)$. Let (S, E) be the transition graph of ϕ. If there are cycle graphs (V_1, E_1') and (V_2, E_2') of (S, E) such that $V_1 \cap V_2 = \emptyset$, then $\equiv_L$ does not have Parikh property.*

Proof. As a monoid generated by a single element is cyclic, there are at least two letters if there are disjoint cycle graphs (V_1, E_1') and (V_2, E_2'). Therefore, we can conclude the proof by Theorem 2 for the case that $(S, \odot)$ is a group, and by Theorem 4 for the case that $(S, \odot)$ is not a group. □

4.3 Languages as Inverse Images of Sets of Idempotent Elements

As mentioned in the proof of Theorem 2, path $s_1 \xrightarrow{a_1} \cdots \xrightarrow{a_{n-1}} s_n$ is a cycle iff $\phi(a_1 \cdots a_{n-1}) = e$ in the transition graph of a homomorphism ϕ into a group. This does not always hold for general finite monoids because $s \odot t = s$ does not necessarily imply that t is the identity element. Furthermore, even if we take any long word w, we cannot guarantee that there exists a subword u of w such that $\phi(u) = e$. Nevertheless, it is known that for each enough long word w, there exists a subword u of w such that $\phi(u)$ is idempotent (instead of identity).

Proposition 2 ([11]). *Let $(S, \odot)$ be a finite monoid and $\phi : \Sigma^* \to S$ be a surjective homomorphism. There exists $N \in \mathbb{N}$ such that for every word $w = a_1 \cdots a_n \in \Sigma^*$ with $n > N$, there exists $1 \leq i_1 < i_2 \leq N$ such that $\phi(a_{i_1} \cdots a_{i_2})$ is an idempotent element.* ∎

Therefore, when discussing the iteration property of monoids, it is sometimes better to use idempotent elements rather than the identity element. However, the straightforward extension of Corollary 1 does not hold.

Example 7. Let $M = \{s_0 = \begin{pmatrix} 0 & 1 & 2 \\ 0 & 1 & 2 \end{pmatrix}, s_1 = \begin{pmatrix} 0 & 1 & 2 \\ 2 & 2 & 2 \end{pmatrix}, s_2 = \begin{pmatrix} 0 & 1 & 2 \\ 1 & 0 & 2 \end{pmatrix}\}$ be the monoid, and let $\phi_5(a) = s_1, \phi_5(b) = s_2$. Note that s_0 is the identity element and s_0, s_1 are idempotent elements. There are disjoint cycles $s_0 \xrightarrow{b} s_2 \xrightarrow{b} s_0, s_1 \xrightarrow{a} s_1$ and the language L_5 recognized by $(\phi_5, \{s_0, s_1\})$ is the inverse image of the set of idempotent elements (see Fig. 2). However, $\equiv_{L_5}$ has Parikh property for the reasons that will be discussed later (see Example 9). ∎

5 Sufficient Conditions for Having Parikh Property

Theorem 5. *Let $L = D^* \subseteq \Sigma^*$ be a language where D is a finite language. If there exist finite languages $F_1, F_2 \subseteq \Sigma^*$ such that $F_1 L F_2 = \Sigma^*$, then $\equiv_L$ has Parikh property.*

Proof. We first outline the idea of the proof. Let $\Delta = \{\delta_d \mid d \in D\}$ be the alphabet and $h : \Delta^* \to \Sigma^*$ be the homomorphism defined as $h(\delta_d) = d$. Note that $\equiv_{\Delta^*} = \equiv_{Pk} (\subseteq \Delta^* \times \Delta^*)$. It follows from Lemma 1 that $Z \equiv_{Pk} Z' \Rightarrow h(Z) \equiv_{D^*} h(Z')$ for any two languages $Z, Z' \subseteq \Delta^*$. We encode each word $d \in D$ into a letter δ_d by using the inverse of h, and prove the theorem by reducing to Parikh's theorem for languages over Δ.

   Assume that there are finite languages $F_1, F_2 \subseteq \Sigma^*$ such that $F_1 L F_2 = \Sigma^*$. For a given CFL X, we construct a regular language X' such that $X' \equiv_L X$ as follows. First, let $X_{(u,v)} = X \cap u D^* v$ for each $(u, v) \in F_1 \times F_2$. Note that $X_{(u,v)}$ is also a CFL because the class of CFL is closed under intersection with a regular language. Furthermore, $X = \bigcup_{(u,v) \in F_1 \times F_2} X_{(u,v)}$ holds because $F_1 L F_2 = \Sigma^*$. Let $Y_{(u,v)} = \{y \mid uyv \in X_{(u,v)}\}$ and $Z_{(u,v)} = h^{-1}(Y_{(u,v)})$. Because the class of CFL is closed under quotient by a word and inverse homomorphism, $Z_{(u,v)} \subseteq \Delta^*$ is also a CFL. Now, let $Z'_{(u,v)}$ be a regular language such that $Z'_{(u,v)} \equiv_{Pk} Z_{(u,v)}$. Then, let $X'_{(u,v)} = u Y'_{(u,v)} v = u h(Z'_{(u,v)}) v$. Because the class of regular languages is closed under homomorphism and concatenation, $X'_{(u,v)}$ is also a regular language. With Lemma 1, we have $Y_{(u,v)} \equiv_L Y'_{(u,v)}$, and thus $X_{(u,v)} \equiv_L X'_{(u,v)}$. Finally, let $X' = \bigcup_{(u,v) \in F_1 \times F_2} X'_{(u,v)}$. Then, we can conclude that $X' \equiv_L X$ (see Fig. 4). $\qquad\square$

Example 8. Let $\Sigma = \{a, b\}$ and $L_6 = \{a, ba, bb\}^*$. Because $F_1 L_6 F_2 = \Sigma^*$ with $F_1 = \{\varepsilon\}$ and $F_2 = \{\varepsilon, b\}$, the equivalence $\equiv_{L_6}$ has Parikh property. $\qquad\blacksquare$

Example 9. Let $\Sigma = \{a, b\}$ and $L_5 = \Sigma^* \setminus b(bb)^*$. Because $F_1 L_5 F_2 = \Sigma^*$ with $F_1 = \{\varepsilon\}$ and $F_2 = \{\varepsilon, b\}$, the equivalence $\equiv_{L_5}$ has Parikh property, which is also clear from the fact that $\equiv_{L_6} \subseteq \equiv_{L_5}$. $\qquad\blacksquare$

We obtained a sufficient condition for Parikh property by Theorem 5. Next, we focus on the properties of monoids and homomorphisms and construct D, F_1 and F_2 that satisfy the assumption of Theorem 5.

Theorem 6. *Let $(S, \odot)$ be a finite monoid, $\phi : \Sigma^* \to S$ be a surjective homomorphism, and (S, E) be the transition graph of ϕ. If there exists a common element $s \in S$ such that $s \in V$ for each cycle graph (V, E'), then $\equiv_L$ has Parikh property where L is the language recognized by $(\phi, \{t \in S \mid s \odot t = s\})$.*

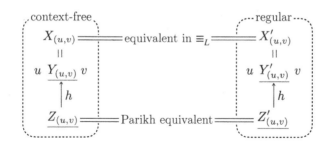

Fig. 4. Constructing regular language X'.

Proof. For each path $s_1 \xrightarrow{a_1} \cdots \xrightarrow{a_{|S|}} s_{|S|+1}$ in (S, E), there are $s_{i_1}, s_{i_2} \in S$ such that $s_{i_1} = s_{i_2}$ with $1 \leq i_1 < i_2 \leq |S| + 1$. Because $s_{i_1} \xrightarrow{a_{i_1}} \cdots \xrightarrow{a_{i_2-1}} s_{i_2}$ is a cycle, $s_j = s$ for some $i_1 \leq j < i_2$. Therefore, each $w \in \Sigma^*$ can be represented as $w = u_1 u_2 \cdots u_k$ where $k \in \mathbb{N}$, $u_\ell \in \Sigma^*$, $|u_\ell| \leq |S|$ with $1 \leq \ell \leq k$, and satisfying the followings: $\phi(u_1) = s$ and $s \odot \phi(u_2) = s \odot \phi(u_3) = \cdots = s \odot \phi(u_{k-1}) = s$. That is, $\Sigma^* = F D_s^* F$ where $F = \bigcup_{k=0}^{|S|-1} \Sigma^k$ and $D_s = \{d \in \Sigma^* \mid |d| \leq |S|, s \odot \phi(d) = s\}$. By Theorem 5, $\equiv_{D_s^*}$ has Parikh property. Note that $D_s^* \subseteq L$ because $s \odot \phi(\varepsilon) = s$ and $s \odot (\phi(d_1) \odot \phi(d_2)) = (s \odot \phi(d_1)) \odot \phi(d_2) = s \odot \phi(d_2) = s$ for each $d_1, d_2 \in D_s^*$. Therefore, $\equiv_{D_s^*} \subseteq \equiv_L$ and $\equiv_L$ also has Parikh property. $\square$

Note that $\{t \in S \mid s \odot t = s\} = \{e\}$ for each $s \in S$ if $(S, \odot)$ is a group. Therefore, the following holds.

Corollary 2. *Let $(S, \odot)$ be a finite group, $\phi : \Sigma^* \to S$ be a surjective homomorphism, and (S, E) be the transition graph of ϕ. If there exists a common element $s \in S$ such that $s \in V$ for each cycle graph (V, E'), then $\equiv_L$ has Parikh property where L is the language recognized by $(\phi, \{e\})$.* ∎

Example 10. Let $L_3 = \{w \mid w \text{ is even-length}\}$ be the language recognized by $(\phi_3, \{e\})$ with $\mathcal{C}_2$ (see Example 4). The transition graph has four simple cycles $\{0 \xrightarrow{a_1} 1 \xrightarrow{a_2} 0 \mid a_1, a_2 \in \{a, b\}\}$ sharing the element 0 (see Fig. 2). Therefore, $\equiv_{L_3}$ has Parikh property. In fact, L_3 can be represented as D^* with $D = \{aa, ab, ba, bb\}$, and $F_1 L_3 F_2 = \Sigma^*$ with $F_1 = \{\varepsilon\}$, $F_2 = \{\varepsilon, a, b\}$. ∎

The condition assumed in Corollary 2 is rather a strong constraint for groups. We can characterize this condition in another way. We say that a surjective homomorphism $\phi : \Sigma^* \to S$ with a monoid $(S, \odot)$ is *letter-symmetric* if $\phi(a) = \phi(b)$ for each $a, b \in \Sigma$.

Lemma 3. *Let (S, E) be the transition graph of ϕ with a surjective homomorphism ϕ into a finite group $(S, \odot)$. If there exists a common element $s \in S$ such that $s \in V$ for each cycle graph (V, E'), then ϕ is letter-symmetric.*

Proof. We give a proof by contradiction. We assume that all cycles share a common element s_1, and there exists a, b such that $\phi(a) \neq \phi(b)$. Let $s_1 \xrightarrow{a_1} \cdots \xrightarrow{a_{n-1}} s_n = s_1$ be a simple cycle in the transition graph and $V_1 = \{s_1, \ldots, s_{n-1}\}$. Because $(S, \odot)$ is a group, $s'_n = s_{n-1} \odot \phi(b) \neq s_n$ for any $b \in \Sigma$ such that $\phi(b) \neq \phi(a_{n-1})$. If there exists s_i with $1 < i < n$ such that $s'_n = s_i$, then there is a cycle $s_i \xrightarrow{a_i} \cdots \xrightarrow{a_{n-2}} s_{n-1} \xrightarrow{b} s_i$ and it contradicts that all cycles share the element $s_1 = s_n$. Therefore, there exists a simple cycle $s_1 \xrightarrow{a_1} \cdots \xrightarrow{a_{n-2}} s_{n-1} \xrightarrow{a_{n-1}} s'_n \xrightarrow{b_1} \cdots \xrightarrow{b_{m-1}} s_1$ such that the set V_2 of vertices of which is larger than V_1. By repeating this procedure, there is an infinite sequence $V_1, V_2, V_3, \cdots \subseteq S$ such that $|V_1| < |V_2| < |V_3| < \cdots$. However, it contradicts that $(S, \odot)$ is a finite group. □

Remark 1. By Lemma 3, the followings are equivalent for a regular group language recognized by (ϕ, T) with a finite group $(S, \odot)$ and the transition graph (S, E) of ϕ:

- There exists a common element $s \in S$ such that $s \in V$ for each cycle graph (V, E') of (S, E).
- For each cycle graphs (V_1, E'_1) and (V_2, E'_2), $V_1 = V_2 = S$.
- ϕ is letter-symmetric.
- The group $(S, \odot)$ is generated by a single element $\phi(a)$ for each a.

6   Subclasses of Regular Group Languages

We obtained some conditions for having Parikh property by the discussion in Sects. 4 and 5. In particular, the detailed analysis was conducted for the case of groups. In this section, we summarize the results and present open problems for group languages. We define predicates $P_0(L), \ldots, P_4(L)$ as follows:

$P_0(L)$: L is a language recognized by $(\phi, \{e\})$ with a surjective homomorphism ϕ into a finite group $(S, \odot)$.

$P_1(L)$: $P_0(L)$ and $(S, \odot)$ is cyclic.

$P_2(L)$: $P_0(L)$ and there are no cycle graphs $(V_1, E_1'), (V_2, E_2')$ such that $V_1 \cap V_2 = \emptyset$ in the transition graph of ϕ.

$P_3(L)$: $P_0(L)$ and $\equiv_L$ has Parikh property.

$P_4(L)$: $P_0(L)$ and ϕ is letter-symmetric.

For each $0 \leq i \leq 4$, we define the class of regular group languages $\mathbf{Gr}_i$ as the class of all languages L that satisfy $P_i(L)$.

Proposition 3. $\mathbf{Gr}_0 \supsetneq \mathbf{Gr}_1 \supsetneq \mathbf{Gr}_2 \supseteq \mathbf{Gr}_3 \supseteq \mathbf{Gr}_4$.

Proof. First, $\mathbf{Gr}_0 \supseteq \mathbf{Gr}_1$ is trivial, and $L_4 \in \mathbf{Gr}_0 \setminus \mathbf{Gr}_1$ (see Example 6). Next, $\mathbf{Gr}_1 \supseteq \mathbf{Gr}_2 \supseteq \mathbf{Gr}_3$ by Theorems 2, 3 and Lemma 2, and $L_2 \in \mathbf{Gr}_1 \setminus \mathbf{Gr}_2$ (see Example 5). Finally, $\mathbf{Gr}_3 \supseteq \mathbf{Gr}_4$ by Corollary 2 and Remark 1. $\square$

We are interested in necessary and sufficient conditions for having Parikh property. If $\mathbf{Gr}_2 \setminus \mathbf{Gr}_3 = \emptyset$ (resp. $\mathbf{Gr}_3 \setminus \mathbf{Gr}_4 = \emptyset$), then $P_2(L)$ (resp. $P_4(L)$) is the necessary and sufficient condition for the case of group language. However, it is possible that both $\mathbf{Gr}_2 \setminus \mathbf{Gr}_3$ and $\mathbf{Gr}_3 \setminus \mathbf{Gr}_4$ are not empty. The following example shows that at least one of $\mathbf{Gr}_2 \setminus \mathbf{Gr}_3$ and $\mathbf{Gr}_3 \setminus \mathbf{Gr}_4$ is not empty.

Example 11. Let $\mathcal{C}_3 = \{0 = \begin{pmatrix} 0 & 1 & 2 \\ 0 & 1 & 2 \end{pmatrix}, 1 = \begin{pmatrix} 0 & 1 & 2 \\ 1 & 2 & 0 \end{pmatrix}, 2 = \begin{pmatrix} 0 & 1 & 2 \\ 2 & 0 & 1 \end{pmatrix}\}$ be the cyclic group with set $\{0, 1, 2\}$, and $\phi_7(a) = 1, \phi_7(b) = 2$. Then, there are no disjoint cycles in the transition graph $(\mathcal{C}_3, E)$. However, there is no element $s \in \mathcal{C}_3$ such that $s \in V$ for each cycle graph (V, E') in $(\mathcal{C}_3, E)$ (see Fig. 2). Therefore, $L_7 \in \mathbf{Gr}_2 \setminus \mathbf{Gr}_4$ where L_7 is the language recognized by $(\phi_7, \{e\})$. It is yet open whether $L_7 \in \mathbf{Gr}_2 \setminus \mathbf{Gr}_3$ or $L_7 \in \mathbf{Gr}_3 \setminus \mathbf{Gr}_4$, i.e., whether $\equiv_{L_7}$ has Parikh property. ∎

We believe that detailed analysis of $\equiv_{L_7}$ should give us a hint for finding a good necessary and sufficient condition for having Parikh property.

7 Conclusion

We extended Parikh equivalence and Parikh's Theorem, and we showed some conditions for equivalence relations to have Parikh property by focusing on monoids and homomorphisms.

Studying the conditions for $\equiv_L$ such that L is the inverse image of a set of idempotent elements is left as future work (see Sect. 4.3). In addition, the problem mentioned in Example 11 will be studied in the future.

References

1. Atanasiu, A., Martín-Vide, C., Mateescu, A.I.: On the injectivity of the Parikh matrix mapping, Fundamenta Informaticae **49**(4), 289–299 (2002)
2. Bhattiprolu, V., Gordon, S., Viswanathan, M.: Extending Parikh's theorem to weighted and probabilistic context-free grammars. In: 14th International Conference on Quantitative Evaluation of Systems, pp. 5–7 (2017)

3. Esparza, J., Ganty, P., Kiefer, S., Luttenberger, M.: Parikh's theorem: a simple and direct automaton construction. Inf. Process. Lett. **111**(12), 614–619 (2011)
4. Goldstine, J.: A simplified proof of Parikh's theorem. Discret. Math. **19**(3), 235–239 (1977)
5. Greibach, S.A.: A generalization of Parikh's semilinear theorem. Discret. Math. **2**(4), 347–355 (1972)
6. Hutchinson, L.K., Mercaş, R., Reidenbach, D.: A toolkit for Parikh matrices. In: 26th International Conference on Implementation and Application of Automata, pp. 116–127 (2022)
7. Lawson, M.V.: Finite Automata. Chapman and Hall/CRC, New York (2003)
8. Mateescu, A., Salomaa, A., Salomaa, K., Yu, S.: A sharpening of the Parikh mapping. Theoret. Inf. Appl. **35**, 551–564 (2001)
9. Parikh, R.J.: On context-free languages. J. ACM **13**(4), 570–581 (1966)
10. Pilling, D.L.: Commutative regular equations and Parikh's theorem. J. London Math. Soc. **2**(6), 663–666 (1973)
11. Pin, J.É.: Mathematical foundations of automata theory. Université Paris, Lecture notes LIAFA (2010)

Enhanced Ternary Fibonacci Codes

Shmuel T. Klein[1] and Dana Shapira[2(⊠)]

[1] Computer Science Department, Bar Ilan University, Ramat Gan, Israel
tomi@cs.biu.ac.il
[2] Computer Science Department, Ariel University, Ariel, Israel
shapird@g.ariel.ac.il

Abstract. Extending previous work on non-binary Fibonacci codes, a new ternary variant is proposed sharing the main features like robustness against errors and ease of encoding and decoding, while improving the compression efficiency relative to other ternary codes. The improvement is based on an increased density of the codes and also shown empirically on large textual examples. A motivation for d-ary codes, with $d > 2$, may be the emergence of future technologies that enable the representation of more than just two values in an atomic storage unit.

Keywords: Data compression · Fibonacci codes · Huffman Coding

1 Introduction

Textual data contains usually much redundancy, and the aim of *lossless data compression* is to eliminate as much as possible of it without losing a single bit, that is, applying a compression algorithm, which is reversible. Many static coders define first an *alphabet* Σ, into the elements of which a given input file T can be parsed, and then collect statistics about the distribution of the elements of Σ, which enable the derivation of an appropriate *code*. The best known such codes are due to Huffman [7] and arithmetic codes [14], both of which are optimal, the first under the assumption that every codeword consists of an integral number of bits, the second without that constraint.

It should also be noted that we use the term *alphabet* in a rather broad sense, without restricting it to just individual letters as $\{a, b, c, \ldots, y, z\}$ or even to ASCII: Σ may include substrings of any type, letter pairs, entire words and fragments, or, ultimately, any set of strings, as long as an exact procedure is used to parse the text T into a well-defined sequence of alphabet elements. For example, for textual files, using words yields much better compression than just single characters [11], so the alphabets used by modern compressors may be large, with hundreds of thousands elements.

Both Huffman and arithmetic coding, as well as many other compression schemes like the popular ones based on the works of Ziv and Lempel, construct a specific code for each given probability distribution, and this can not always be justified. This lead to the development of *universal* codes [2], which are fixed

© The Author(s), under exclusive license to Springer Nature Switzerland AG 2023
B. Nagy (Ed.): CIAA 2023, LNCS 14151, pp. 179–193, 2023.
https://doi.org/10.1007/978-3-031-40247-0_13

sets of codewords that need not to be generated afresh. Such universal codes are especially useful when encoding integers, for example by means of the Elias' codes known as C_γ and C_δ. Another application may be for the encoding of a large alphabet, when optimal compression is less important and may be traded for increased processing speed and other advantages. The elements of Σ can be ordered by non-increasing probabilities, and the ith element is then assigned the Elias' or other universal codeword for the integer i, with $i \geq 1$. For many distributions, the loss of compression relative to the optimum is small enough to be tolerable.

Some universal codes with the additional advantage of providing robustness against transmission errors have been introduced in [6] and are based on a numeration system using Fibonacci numbers, rather than powers of 2, as their basis elements. The Fibonacci sequence can be defined by

$$F_0 = 0, \quad F_1 = 1,$$
$$\text{and} \quad F_i = F_{i-1} + F_{i-2}, \quad \text{for} \quad i \geq 2.$$

An interesting feature of this sequence, as well as of many other increasing integer sequences, is that every integer x can be expressed as a sum of Fibonacci numbers, so that each element participates in this summation at most once, that is,

$$x = \sum_{i \geq 2} b_i F_i, \quad \text{with} \quad b_i \in \{0, 1\}.$$

This representation will be unique if, at each step, the largest Fibonacci number F_i that is less than or equal to x is chosen, and one continues with the remaining $x - F_i$ value, until $x = 0$. The Fibonacci sequence can therefore be used to yield an alternative binary representation of the integers. While the standard binary encoding refers to powers of 2 as its basis elements, the so-called *Zeckendorf* [15] representation uses Fibonacci numbers instead. As example, consider $20 = 16+4$, which yields the binary encoding 10100, as the bit positions refer, from right to left, to the basis elements $2^0 = 1, \ldots, 2^4 = 16$; using Fibonacci numbers instead, one gets $20 = 13 + 5 + 2$ yielding the bit-string 101010, in which the positions refer, again right to left, to 1, 2, 3, 5, 8, 13. The main property of the Zeckendorf representation is that there is never a need to use consecutive Fibonacci numbers, in other words, there are no adjacent 1-bits in this binary representation.

This property has been exploited to design the *Fibonacci code*: by reversing the codewords and letting the bit positions refer to the sequence 1, 2, 3, 5, ... from left to right, one gets binary strings with a rightmost 1-bit, for example 010101 for the integer 20, so it suffices to append an additional 1-bit at the right end of each codeword to obtain a *prefix code*. Indeed, all the codewords then end in 11, and this string is not found in any internal position, so no codeword can be the prefix of another one. It follows that this code, denoted henceforth by $\mathcal{B}$, is uniquely decipherable and instantaneous [12]. The first few elements of the code can be seen in Table 1. The usefulness of Fibonacci codes has been studied in [9].

There are several possibilities to generalize these codes. One may use other delimiters for the codewords and even multiple ones, as in [1], or pass to higher order. A standard way is to consider recurrences in which each element is the sum of the $d \geq 2$ preceding ones,

$$F_i^{(d)} = F_{i-1}^{(d)} + F_{i-2}^{(d)} + \cdots + F_{i-d}^{(d)} \qquad \text{for} \quad i \geq 2,$$

from which a binary code can be derived in which no more than d consecutive 1s can appear in any codeword, except as suffix. The special case $d = 2$ yields the Fibonacci code above.

Higher order codes of a different kind may be concerned with non-binary extensions. There is an obvious theoretical interest in extending a well established result, in particular when the extension in not straightforward as in the present case. On the practical side, such codes can be motivated by expected technological advances that will enable the storage of more than just two values in an atomic unit. For instance, 3-valued logic has been used in the Soviet Union as early as 1958 to build a ternary computer [13], and even Huffman's original paper [7] dealt with r-ary trees, for general $r \geq 2$.

Consider the following recurrences, depending on a parameter $m \geq 1$:

$$R_{-1}^{(m)} = 1, \quad R_0^{(m)} = 1,$$
$$\text{and} \quad R_i^{(m)} = mR_{i-1}^{(m)} + R_{i-2}^{(m)}, \qquad \text{for} \quad i \geq 1.$$

For $m = 1$, one again gets the standard Fibonacci sequence, but for larger m, the representation of integers is $(m + 1)$-ary and not binary, that is, an integer x can be uniquely represented as

$$x = \sum_{i \geq 1} b_i \, R_i^{(m)}, \qquad \text{with} \quad b_i \in \{0, 1, \ldots, m\}.$$

The generalization to larger m of the non-adjacency of 1-bits in the corresponding encodings of the integers for $m = 1$, is that every occurrence of the digit m has to be followed by an occurrence of the digit 0.

On the basis of this property, an $(m + 1)$-ary code has been proposed in [10]. For example, for $m = 2$, one gets the sequence $\{R_i^{(2)}\}_{i \geq 0} = \{1, 3, 7, 17, 41, \ldots\}$, and the corresponding representation is ternary, i.e., using only the digits 0, 1 and 2. As binary digits are called bits, their ternary equivalent are often referred to as *trits*. The ternary strings representing the first integers are $\{1, 2, 10, 11, 12, 20, 100, 101, 102, 110, 111, 112, 120, 200, 201, 202, \ldots\}$. Note, in particular, the missing strings 21, 22, 121, 122, which are legal ternary strings, yet do not appear in the sequence, because they violate the constraint of having every 2-trit followed by a 0. As there are no leading 0-trits, the code itself is constructed by reversing the strings and adding a trailing 2-trit. In all the codewords of the resulting set, every 2-trit is preceded by a 0-trit, unless it is in the first position, and then it has no preceding trit, or in the last, and then it serves as delimiter. The first codewords of this ternary Fibonacci code, which we shall denote by $\mathcal{R}$, are listed in Table 1.

We here propose an enhanced ternary code in terms of compression efficiency, based on a different generalization of the Fibonacci sequence. We show how

Table 1. *First codewords of the standard binary Fibonacci code, $\mathcal{B}$, the ternary Fibonacci Code $\mathcal{R}$ of [10] and the proposed ternary Fibonacci codes $\mathcal{G}$ and $\mathcal{H}$.*

	$\mathcal{B}$	$\mathcal{R}$	$\mathcal{G}$	$\mathcal{H}$		$\mathcal{B}$	$\mathcal{R}$	$\mathcal{G}$	$\mathcal{H}$
0			22	22	16	0010011	2022	00212	11022
1	11	12	122	022	17	1010011	00012	10212	21022
2	011	22	212	122	18	0001011	10012	20212	02022
3	0011	012	0122	212	19	1001011	20012	01212	12022
4	1011	112	1122	0022	20	0101011	01012	11212	00122
5	00011	212	2112	1022	21	00000011	11012	000122	10122
6	10011	022	0212	2022	22	10000011	21012	100122	20122
7	01011	0012	1212	0122	23	01000011	02012	200122	01122
8	000011	1012	00122	1122	24	00100011	00112	010122	11122
9	100011	2012	10122	2112	25	10100011	10112	110122	21112
10	010011	0112	20122	0212	26	00010011	20112	210122	02112
11	001011	1112	01122	1212	27	10010011	01112	020212	12112
12	101011	2112	11122	00022	28	01010011	11112	120122	00212
13	0000011	0212	21112	10022	29	00001011	21112	001122	10212
14	1000011	0022	02112	20022	30	10001011	02112	101122	20212
15	0100011	1022	12112	01022	31	01001011	12112	201122	01212

to employ the property implied by the numeration system to turn the integer representations into a set of ternary codewords and then study the properties of this new code.

The paper is constructed as follows. Section 2 presents the proposed ternary Fibonacci code and provides proofs for its features. Section 3 provides experimental results about its compression performance as compared to ternary state-of-the-art encoding techniques and concludes.

2   A New Ternary Fibonacci Code

Consider the sub-sequence of every other Fibonacci number, those with the even indices: $1, 3, 8, 21, 55, 144, 377, 987, \ldots$. It turns out that the same sequence can be obtained by the following simple recurrence:

$$G_0 = 0, \quad G_1 = 1,$$
$$\text{and} \quad G_i = 3\,G_{i-1} - G_{i-2}, \quad \text{for} \quad i \geq 2,$$

a fact already noticed in [8]. Indeed,

$$F_{2i} = F_{2i-1} + F_{2i-2} = 2F_{2i-2} + F_{2i-3} = 3F_{2i-2} - F_{2i-4},$$

and since, for every $i > 0$, G_i is strictly smaller than three times the preceding element, the resulting numeration system is ternary.

This recurrence can be the basis of a ternary numeration system, and the first integers expressed according to this system are {1, 2, 10, 11, 12, 20, 21, 100, 101, 102, 110, 111, 112, 120, 121, 200, 201, ...}. The string 1201 for instance, would represent the integer $21 + 2 \times 8 + 1 = 38$. As for the previously mentioned codes $\mathcal{B}, \mathcal{R}$ and their variants, the recurrence relation implies an inherent property of the corresponding ternary integer representations, that can be used to define a code. For the current sequence G, the property is that there must be a zero between any two 2s. Though this property is a special case of a more general theorem shown in [5], we provide a proof for our special case in the following lemma.

Lemma 1. *There must be a 0-trit between any two 2-trits in the representation of every integer expressed according to the ternary numeration system based on the sequence G.*

Proof. Let $T = t_r t_{r-1} \cdots t_2 t_1$ the representation of some integer m in the given numeration system, that is $m = \sum_{i=1}^{r} t_i G_i$ and $t_i \in \{0, 1, 2\}$. An equivalent formulation of the lemma is that T does not contain a substring of the form $21^n 2$ for any $n \geq 0$, that is, all substrings of the form 22, 212, 2112, etc. are prohibited.

Let j be the leftmost (highest) index such that $t_j = 2$. We first show that the j rightmost trits of T cannot be of the form $P = 2111 \cdots 12$. This follows from the fact, based on the well-known formula for the sum of the first Fibonacci numbers, that a string 1^j of j 1-trits at the suffix of T would represent the integer

$$N_j = \sum_{i=1}^{j} G_i = \sum_{i=1}^{j} \left(F_{2i-1} + F_{2i-2} \right) = \sum_{i=0}^{2j-1} F_i = F_{2j+1} - 1.$$

For the numeric value V of the string P, we have to add G_j and G_1 to get

$$V = N_j + G_j + G_1 = (F_{2j+1} - 1) + F_{2j} + 1 = F_{2j+2} = G_{j+1}.$$

But this is a contradiction to the choice of j. If $j = r$, it contradicts the fact that P is the representation of m, since the algorithm, trying to fit in always the largest basis element in what remains to be represented, would prefer the form 10^r to P. If $j < r$, then $t_{j+1} < 2$, so its value could be increased, and the suffix of T would be $(t_{j+1} + 1)0^j$ rather than $t_{j+1} 21^{j-2} 2$.

For the general case, in which the substring P is not a suffix of T, note that since the string $Q = 211 \cdots 1200 \cdots 00$ is lexicographically larger than the string $P = 211 \cdots 111 \cdots 12$ of the same length, the numerical value of the former is larger than that of the latter. For example, 211200 represents the number 380, whereas 211112 is 377. Therefore, if the occurrence of P is impossible, then a fortiori so is that of Q: the representations of both values of P and Q would involve the trit t_{j+1}.

The existence of forbidden substrings 11 and mx, with $x < m$, lead to the definition of the prefix codes $\mathcal{B}$ and $\mathcal{R}$, respectively, which were obtained by reversing the strings representing the integers, and appending a single digit, 1 or m, respectively. The forbidden strings thus appear as suffixes of every codeword and thereby serve as delimiters, and do not appear anywhere else. Extending this idea to our sequence G is not immediate, because no single trit can turn every ternary string representing an integer according to G into one violating the rules. For instance, since the property refers to the existence of a zero between two 2-trits, a string without any 2-trits cannot be disqualified by the addition of a single trit.

A possible solution is to prefix every string by 22, which would yield the strings

$$\{221, 222, 2210, 2211, 2212, 2220, 2221, 22100, 22101, \ldots\},$$

but the code obtained by reversing these strings does not have the prefix property. for example, 122 ($= 1$) is then a prefix of 1222 ($= 7$). What is even worse, the code would not even be uniquely decipherable (UD): the string 1222122 could be interpreted as both 122-2122 ($= 1$-5) and 1222-122 ($= 7$-1).

An alternative is to prefix every string with either 22 or just 2, if this is enough to create an illegal substring. The resulting sequence would then start by

$$\{221, 22, 2210, 2211, 212, 220, 221, 22100, 22101, \ldots\},$$

since all strings with a prefix of the form 1^s2, with $s \geq 0$ need only a single 2-trit as delimiter. The corresponding code, however, is not well-defined, e.g., 221 would represent both the integers 1 and 7. Moreover, by prefixing delimiters of different lengths, the corresponding codewords are not ordered anymore by their lengths, and reordering them will destroy the connection between an index and its representation in the ternary Fibonacci numeration system.

Our suggestion is therefore to use two different delimiters, yet of the same length two trits: 22 or 21. Strings starting with 1^s2, for $s \geq 0$, will be prefixed by 21, all the others, which are those starting with 1^s0, for $s \geq 1$, or strings consisting only of 1s, will be prefixed by 22. This yields

$$\{\underline{221}, \underline{212}, \underline{22}10, \underline{22}11, \underline{2112}, \underline{2120}, \underline{2121}, \underline{22}100, \underline{22}101, \ldots\},$$

where the violating substrings in a right to left scan have been underlined. Reversing the strings, we get a prefix code, denoted by $\mathcal{G}$, the first elements of which appear in Table 1. Each codeword of $\mathcal{G}$ terminates in an illegal string of variable length, the detection of which allows instantaneous decoding.

2.1   Properties of $\mathcal{G}$

The compression capabilities of a fixed universal code depend obviously on the number of different codewords n_ℓ that can be defined for every length ℓ. To evaluate these numbers for our code $\mathcal{G}$, note that for $\ell \geq 1$ the first codeword of length $\ell + 2$ is $0^{\ell-1}122$ and represents the integer G_ℓ. It follows that the number of codewords of length $\ell + 2$ is

$$n_{\ell+2} = G_{\ell+1} - G_\ell = 3G_\ell - G_{\ell-1} - 3G_{\ell-1} + G_{\ell-2} = 3n_{\ell+1} - n_\ell.$$

The sequence n_ℓ thus starts with 2, 5, 13, 34, 89, and follows the same recurrence relation as G, just with different initial values.

The McMillan sum of a d-ary infinite code $\mathcal{C} = \{c_1, c_2, \ldots\}$ is defined as $M = \sum_{i=1}^{\infty} d^{-|c_i|}$, and a necessary condition for a code to be UD is that $M \leq 1$. If $M = 1$, the code is *complete*, which means that no additional codeword can be adjoined without violating the UD property of the code. For $i \geq 1$, there are n_i codewords of length $i+2$ in the ternary code $\mathcal{G}$, so we can evaluate its McMillan sum $M = \sum_{i=1}^{\infty} n_i 3^{-(i+2)}$.

$$M = \sum_{i=1}^{\infty} n_i 3^{-(i+2)} = 2 \cdot 3^{-3} + 5 \cdot 3^{-4} + \sum_{i=3}^{\infty} \left(3n_{i-1} - n_{i-2}\right) 3^{-(i+2)}$$

$$= \frac{11}{81} + \sum_{i=2}^{\infty} n_i 3^{-(i+2)} - \frac{1}{9} \sum_{i=1}^{\infty} n_i 3^{-(i+2)} = \frac{11}{81} + \left(M - 2 \cdot 3^{-3}\right) - \frac{1}{9}M,$$

$$\tag{1}$$

from which one can derive that $M = \frac{5}{9}$. The code is thus not complete, which might indicate compression inefficiency. A first remedy could be to adjoin the codeword 22, that is, one consisting of an empty prefix and only of the terminating string 22 itself. The extended code $\mathcal{G}^* = \mathcal{G} \cup \{22\}$ would still be UD, and its McMillan sum would be improved to $\frac{2}{3}$.

The fact that even $\mathcal{G}^*$ is incomplete can be explained by the missing codewords having suffixes of the form $0^r 22$ with $r \geq 1$, that is, an integer is represented in the numeration system based on G_i without leading zeros. Indeed, while the values of the strings 1201, 12010 and 120100 could not be distinguished, all of them representing the integer $1 + 2 \times 3 + 21 = 28$, the addition of a delimiting suffix may turn them into different *codewords* 120122, 1201022 and 12010022, of which only the first appears in $\mathcal{G}^*$. This leads to the idea of extending the definition of this code to include also these missing codewords.

Note that a similar amendment is not feasible for the codes $\mathcal{B}$ or $\mathcal{R}$: though the illegal substrings to be avoided, 11 for $\mathcal{B}$ and 12 or 22 for $\mathcal{R}$ are of length 2 bits or trits, the leftmost bit or trit of the separator is a part of the representation of the index, actually, its leading bit or trit. Therefore, only a single 1-bit or 2-trit is appended, which does not allow it to be preceded by one or more zeros. In the code $\mathcal{G}$, on the other hand, the separators consist of two trits, and they will be recognized even after a sequence of zeros.

2.2   Extended Definition

The suggested extended code is most easily understood by the following recursive construction. We shall omit the two terminating trits 12 or 22 of every codeword and calculate the number m_ℓ of legal ternary strings of all possible lengths $\ell \geq 1$. Denote by B_ℓ the block of the m_ℓ ternary strings of length ℓ, arranged in backward lexicographic order, that is, scanning the trits of each codeword right to left. It will be convenient to refer to Table 2, showing the blocks B_ℓ for $\ell \leq 5$.

The first block B_1 consists just of the single trits $\{0,1,2\}$. The block B_ℓ is obtained by taking three copies of the preceding block $B_{\ell-1}$, and extending each string to the right, respectively for the three copies, by 0, 1 or 2. The appended 2-trit in the third block of this extension may, however, produce illegal strings, and these appear boldfaced at the end of the blocks, and are assumed not be a part of them.

The following lemma evaluates the number d_ℓ of these illegal strings.

Table 2. *New ternary code construction.*

1	2	3	4	5
0	00	000	0000 0001 0002	00000 00010 00020 00001 00011 00021 00002 00012 00022
1	10	100	1000 1001 1002	10000 10010 10020 10001 10011 10021 10002 10012 10022
2	20	200	2000 2001 2002	20000 20010 20020 20001 20011 20021 20002 20012 20022
	01	010	0100 0101 0102	01000 01010 01020 01001 01011 01021 01002 01012 01022
	11	110	1100 1101 1102	11000 11010 11020 11001 11011 11021 11002 11012 11022
	21	210	2100 2101 2102	21000 21010 21020 21001 21011 21021 21002 21012 21022
	02	020	0200 0201 0202	02000 02010 02020 02001 02011 02021 02002 02012 02022
	12	120	1200 1201 1202	12000 12010 12020 12001 12011 12021 12002 12012 12022
	22	001	0010 0011 0012	00100 00110 00120 00101 00111 00121 00102 00112 00122
		101	1010 1011 1012	10100 10110 10120 10101 10111 10121 10102 10112 10122
		201	2010 2011 2012	20100 20110 20120 20101 20111 20121 20102 20112 20122
		011	0110 0111 0112	01100 01110 01120 01101 01111 01121 01102 01112 01122
		111	1110 1111 1112	11100 11110 11120 11101 11111 11121 11102 11112 11122
		211	2110 2111 **2112**	21100 21110 21101 21111 21102 **21112**
		021	0210 0211 **0212**	02100 02110 02101 02111 02102 **02112**
		121	1210 1211 **1212**	12100 12110 12101 12111 12102 **12112**
		002	0020 0021 **0022**	00200 00210 00201 00211 00202 **00212**
		102	1020 1021 **1022**	10200 10210 10201 10211 10202 **10212**
		202	2020 2021 **2022**	20200 20210 20201 20211 20202 **20212**
		012	0120 0121 **0122**	01200 01210 01201 01211 01202 **01212**
		112	1120 1121 **1122**	11200 11210 11201 11211 11202 **11212**
		212		
		022		
		122		

Lemma 2. *The number d_ℓ of illegal strings following the block B_ℓ is equal to $m_{\ell-2}$, the size of block $B_{\ell-2}$ of legal strings preceding the previous one.*

This fact is highlighted by matching colors for $\ell = 3, 4, 5$ in Table 2.

Proof. Recall that the d_ℓ illegal strings are extensions of different legal strings of length $\ell-1$ that have been extended to their right by adding a 2-trit. Therefore, these legal strings of length $\ell-1$ cannot end with a 0-trit, and if we remove the ending pair 12 or 22 from each of these strings, we are left with legal strings of length $\ell-2$. All these string must be different, because there is no legal string that can be extended by both 12 and 22 to its right and thereby create an illegal string. It follows that $m_{\ell-2} \geq d_\ell$.

On the other hand, each string in $B_{\ell-2}$ can be extended, by adding either 12 or 22 as suffix, into an illegal string of length ℓ, thus $m_{\ell-2} \leq d_\ell$.

The conclusion is that $m_\ell = 3\,m_{\ell-1} - m_{\ell-2}$ for $\ell \geq 3$. The resulting code, $\mathcal{H}$, obtained by adding the suffixes 12 or 22 to each codeword, appears in Table 1.

Here too, the codeword 22 can be added, so that the number m_ℓ of codewords of length $\ell + 2$ satisfies

$$m_0 = 1, \qquad m_1 = 3, \qquad m_i = 3m_{i-1} - m_{i-2} \quad \text{for } i \geq 2,$$

in other words, $m_i = G_{i+1}$. The corresponding McMillan sum can be evaluated as in Eq. (1) to get $\sum_{i=0}^{\infty} m_i 3^{-(i+2)} = 1$, which shows that the code $\mathcal{H}$ is complete.

Note that the completeness of $\mathcal{H}$ proves that the addition of two additional trits, to transform the ternary representation of an integer into a codeword, is optimal in the sense that none of the infinitely many codewords may be shortened by even a single trit. Indeed, the corresponding McMillan sum would then be strictly larger than 1, contradicting unique decipherability.

There is a priori a drawback to the extended code $\mathcal{H}$ versus $\mathcal{G}$, as the obvious connection between the represented integer and the index of the corresponding codeword seems to be broken. The problem is that leading zeros do not affect the integer, but influence the length of the codeword, and to get reasonable compression efficiency, the codewords have to be assigned in order of non-decreasing length. A new connection can, however, be established, in the following way.

Every codeword will be considered as consisting of three parts when scanned from left to right: the first is the prefix P, consisting of any legal sequence of p trits and ending with 1 or 2; the second, Z, is a sequence of $z \geq 0$ zeros; the third is a delimiter 22 or 12. As example,

$$\underbrace{1\,2\,1\,1\,0\,2\,1\,0\,2}_{\text{P}}\,\underbrace{0\,0\,0\,0\,0}_{\text{Z}}\,\underbrace{2\,2}_{\text{D}}.$$

Once the delimiter 22 is detected, we know that the codeword is of length $\ell = p + z + 2$ trits, $9 + 5 + 2 = 16$ in our example. The number of codewords $U_{\ell-1}$ of lengths up to and including $\ell - 1$ is

$$U_{\ell-1} = \sum_{i=0}^{\ell-3} m_i = \sum_{i=1}^{\ell-2} G_i = \sum_{i=1}^{\ell-2} F_{2i} = F_{2\ell-3} - 1, \tag{2}$$

and correspond to the numbers 0 to $U_{\ell-1} - 1$. The m_ℓ codewords of length ℓ can thus be indexed by the integers $U_{\ell-1}$ to $U_{\ell-1} + m_\ell - 1$, and the relative index within this set of a given codeword is encoded by P. For our example, $\ell = 16$, $U_{15} = F_{29} - 1 = 514,228$, and P represents the relative index 5869, so the given codeword represents the number 520,097. The formal encoding and decoding procedures are deferred to the Appendix.

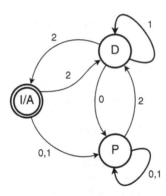

Fig. 1. *Decoding automaton.*

The decoding process for the extended code $\mathcal{H}$, just parsing the encoded string into a sequence of codewords, without evaluation of the corresponding indices, can be represented by the automaton of Fig. 1. The initial state is also the accepting state I/A, in which the process finds itself at the beginning and end of every codeword. If a 0 or 1-trit is scanned, we pass into a Prefix state P and stay there while seeing more 0- or 1-trits. When encountering a 2-trit in either I/A or P, the process enters state D, having detected the potential start of a Delimiter, and stays in D while scanning 1-trits. If such a sequence $S = 21^s$, with $s \geq 0$, is followed by a 0-trit, we return to the P state, as the single 2-trit is not a part of the delimiter; if S is followed by a 2-trit, the end of a codeword is detected and we accept it in I/A. Note that since the code $\mathcal{H}$ is complete, any binary string can be parsed into a sequence of codewords or fragments thereof, and the automaton correctly distinguishes between a legal sequence of codewords, which terminates in the accepting state, and the other strings.

2.3 Densities

The main motivation of devising a new ternary code was to improve the compression efficiency of the ternary code $\mathcal{R}$. Let us first compare the densities of the codes, i.e., the number of codewords of a given length. The characteristic polynomial of the recurrence defining the sequence $R^{(2)}$ is $x^2 - 2x - 1$, with roots $1 \pm \sqrt{2}$ of which only the first, $\rho = 2.414$, is larger than 1. For the sequence G, the polynomial is $x^2 - 3x + 1$, with roots $(3 \pm \sqrt{5})/2$, again with only one of them, $\psi = 2.618$, greater than 1. It follows that the general elements $R_n^{(2)}$ and G_n of the sequences grow proportionally to $\frac{1}{2}\rho^n$ and $\frac{1}{\sqrt{5}}\psi^n$, where the constants $\frac{1}{2}$ and $\frac{1}{\sqrt{5}}$ have been determined by the initial values. It follows that the number of necessary trits to represent n is about $\log_\rho 2n = 1.246 \log_3 n + 0.786$ and $\log_\psi \sqrt{5}n = 1.142 \log_3 n + 0.836$, respectively.

In other words, while the penalty paid for using $\mathcal{R}$ is an increase of about 25% in the number of necessary trits, relative to using the standard ternary system, this is reduced to a waste of only about 14% for $\mathcal{G}$. This does, however, not take the fact into account, that the codewords of $\mathcal{R}$ need only a single trit as delimiters, while those of $\mathcal{G}$ need two. As comparison, recall that in the binary case, the use of the Fibonacci code incurred a loss of 44% over the length of the standard binary encoding [9].

We could thus ask from what values of n onwards does the length of the encoding by $\mathcal{R}$ already exceed that of $\mathcal{G}$ by at least 1 trit, and get the equation

$$1.2465 \log_3 n + 1.7864 = 1.1415 \log_3 n + 2.8361,$$

from which one gets $n \simeq 3^{10.5} = 102,000$. As example, consider the integer 120,000: in $\mathcal{R}$, its representation would be 110021111100012, using 15 trits, in $\mathcal{G}$ it would be 21001100111212, using only 14 trits.

On the other hand, one has to remember that while the density of $\mathcal{G}$ exceeds that of $\mathcal{R}$ because $\psi > \rho$, the compression efficiency depends on the given probability distribution at hand, and the shorter codewords are assigned to the higher probabilities, yet for the shorter codewords, $\mathcal{R}$ uses less trits than $\mathcal{G}$. This lead to the definition of the code $\mathcal{H}$.

To evaluate the density of $\mathcal{H}$, recall that the first index needing ℓ trits (without the 2 suffix trits 12 or 22) is $F_{2\ell+1} \simeq \frac{\phi}{\sqrt{5}}(\phi^2)^\ell$, so that the number of trits needed for n is $1.1415 \log_3 n + 2.3361$, exactly half a trit less than for $\mathcal{G}$.

3 Experimental Results

Comparing the densities of the codes relates only to the number of codewords of the different lengths and does not take any probability distribution into account. This is equivalent to encoding a large alphabet with a uniform distribution of its elements. To compare the codes on some distributions that are more realistic than the uniform one, consider *Zipf's law* [16], defined by the weights $p_i = 1/(i\, H_n)$ for $1 \leq i \leq n$. Here $H_n = \sum_{j=1}^{n} \frac{1}{j}$ is the n-th harmonic number, which is about $\ln n$. The frequency of the different words in a large natural language corpus, and many other natural phenomena, are believed to be governed by this law [4]. We used one million elements, for which the first few probabilities are 0.069, 0.034, 0.023, 0.017, ...

To check the performance also on real-life data, we considered the files *eng*, *sources* and *dblp*, all of size 50MB, downloaded from the Pizza & Chili Corpus[1]: Since the new codes $\mathcal{G}$ and $\mathcal{H}$ have shorter codewords than $\mathcal{R}$ only for larger indices, their performance is expected to improve over that of $\mathcal{R}$ only for quite large alphabets. In particular, for many of the first codewords, those of $\mathcal{R}$ are shorter by 1 trit than those of $\mathcal{G}$ or $\mathcal{H}$, and since precisely these codewords correspond to the highest probabilities, there are certain distributions, and certainly those for smaller alphabets, for which $\mathcal{R}$ will compress better.

To show that this is not always the case, the alphabet elements on our tests were chosen as the set of *words*, separated by one or more blanks, within a given

---

[1] http://pizzachili.dcc.uchile.cl/.

text. Table 3 presents the alphabet size and the average codeword lengths, in trits, for each of the discussed methods, including a ternary Huffman coding variant Huf3. As can be seen, $\mathcal{H}$ slightly outperforms $\mathcal{R}$ on all the examples, except for the *eng* alphabet. The optimal ternary Huffman coding is of course even better, if the compression efficiency is the sole criterion. As all the codewords of $\mathcal{G}$ are at least as long as the corresponding ones of $\mathcal{H}$, its compression is necessarily inferior; the only advantage of using then $\mathcal{G}$ rather than $\mathcal{H}$ may be the simplified encoding and decoding procedures.

One should, however, remember that the encoding of entire texts, using all the elements of an alphabet, is only one of the possible applications of universal codes. Another one could be the transmission of a long list of integers whose sizes are not bounded. One of the advantages of universal codes is precisely the ability of encoding systematically any integer, even so large ones that they may not represent any physical size such as the size of an alphabet. Encoding such large numbers can be done by C_δ or even better by recursively prefixing a string representing a number by its length, until getting to some constant length, as suggested in [3].

Examples for such lists of integers could be astronomical data of very high precision, where each of the many data elements is a large number defined by hundreds of digits. Another application domain could be a list of large prime numbers as those used in cryptography. For instance, the RSA scheme uses as part of its public key the product $n = p \times q$ of two primes p and q, which are kept secret. The security of RSA relies on the fact that factoring n is (still) a difficult task, and therefore the current recommendation is to choose very large primes, of about 1000 bits each.

There are clearly no alphabets of size 2^{1000}, but it may be needed to encode the prime p itself. A 1000-bit number can be written in 631 trits when a standard ternary encoding is used, and 643 trits would be required using C_δ^3. The more robust variants $\mathcal{R}$ and $\mathcal{H}$ would need 790 and 723 trits, respectively.

We conclude that beside the theoretical interest in the derivation of the new ternary codes, they can also be of significant practical value.

Table 3. *Compression performance on the used datasets.*

File	$|\Sigma|$ words	$\mathcal{R}$	$\mathcal{G}$	$\mathcal{H}$	Huf3
ZipfM	1,000,000	8.87	9.18	8.76	8.51
eng	303,385	7.30	7.74	7.32	7.11
sources	867,121	6.93	7.19	6.88	6.62
dblp	559,695	8.95	9.29	8.84	8.51

Appendix: Encoding and decoding

Algorithm 1. *Encoding*

TERNARY-FIB-ENCODE (integer sequence)

---

while *not EOF* **do**

 $x \leftarrow$ read next integer from input

 $k, j \leftarrow \lfloor \frac{1}{2}\log_\phi(\sqrt{5}(x+1))+3 \rfloor$ ▷ k and j are both length of x, including 2 terminating trits

 $x \leftarrow x - F_{2j-3} + 1$

 $B[1..j-2] \leftarrow (0,...,0)$ ▷ $j-2$ is length of relative index, after removing 2 terminating trits

 for $i \leftarrow j - 2$ **to** 1 **by** -1 **do**

  repeat twice **if** $x \geq G_i$ **then**

   $B[i] \leftarrow B[i] + 1$

   $x \leftarrow x - G_i$

  ▷ now adjoin the terminating 22 or 12 trits

 while $B[j] = 1$ **do**

  $j \leftarrow j - 1$ ▷ check whether the relative index has a string of 1-trits as suffix

 if $j = 0$ **or** $B[j] = 0$ **then**

  $B[k-1] \leftarrow 2$ ▷ if suffix of 1's is the whole string or preceded by 0

 else

  $B[k-1] \leftarrow 1$

 $B[k] \leftarrow 2$

 output $B[1..k]$

For the encoding, Algorithm 1 assumes that a sequence of integers is given, and an array B is used to temporarily store the trits representing the current integer. The first step is to calculate the length of the representation, including the terminating delimiter 12 or 22, of a given integer x. According to Eq. (2), this will be the smallest ℓ for which $x \geq F_{2\ell-3} - 1$, so that

$$\ell = \left\lfloor \tfrac{1}{2} \log_\phi \left(\sqrt{5}(x+1) \right) + 3 \right\rfloor,$$

where $\phi = \frac{1+\sqrt{5}}{2} = 1.618$ is the golden ratio, and we have used the fact that Fibonacci numbers are given by $F_n = \frac{1}{\sqrt{5}}\phi^n$, rounded to the nearest integer.

Once the length j is known, the relative index within the block of legal strings of length $j - 2$ is iteratively evaluated. Finally, the last two trits are set to 22 or 12 accordingly. Specifically, if $B[r..j-2] = 21^{j-2-r}$ with $r \geq 1$ and $j-2-r \geq 0$, then the suffix is 12, otherwise, it is 22.

The decoding procedure of Algorithm 2 works on an array A of trits assumed to contain the concatenated ternary representations of a sequence of integers. It accumulates the value of the current integer in a variable *val*. A variable *status* maintains the number of 2-trits, encountered while scanning the current ternary string, that could possibly belong to the suffix of the form 21^s2, with $s \geq 0$, serving as delimiter. Once the end of the current string has been detected, we know the length of the encoding and *val* can be updated to account for the shorter strings, according to Eq. (2). Finally, *val* has to be adjusted because the last two trits, 12 or 22, do not belong to the representation.

Algorithm 2. *Decoding*

TERNARY-FIB-DECODE (A)

$i \leftarrow 1$
while *not EOF* **do**
    $j \leftarrow 0$ *status* $\leftarrow 0$ *val* $\leftarrow 0$
    while *status* < 2 **do**
        $val \leftarrow val + A[i] \times G_j$ ▷ build relative index
        if $A[i] = 2$ **then**
            $\llcorner$ *status* $\leftarrow$ *status* $+ 1$
        else if *status* $= 1$ **and** $A[i] = 0$ **then**
            $\llcorner$ *status* $\leftarrow 0$
        $i \leftarrow i + 1$
        $\llcorner j \leftarrow j + 1$ ▷ length of current codeword, including terminating 12 or 22
    $val \leftarrow val + F_{2j-3} - 1$ ▷ add sum of sizes of shorter codewords
    $\llcorner$ **output** $val - 2 \times G_j - A[i-2] \times G_{j-1}$

References

1. Anisimov, A.V., Zavadskyi, I.O.: Variable-length prefix codes with multiple delimiters. IEEE Trans. Inf. Theory **63**(5), 2885–2895 (2017)
2. Elias, P.: Universal codeword sets and representations of the integers. IEEE Trans. Information Theory **21**(2), 194–203 (1975)
3. Even, S., Rodeh, M.: Economical encoding of commas between strings. Commun. ACM **21**(4), 315–317 (1978)
4. Fagan, S., Gençay, R.: An introduction to textual econometrics. Handbook of Empirical Economics and Finance CRC Press, pp. 133–153 (2010)
5. Fraenkel, A.S.: Systems of numeration. In: 6th IEEE Symposium on Computer Arithmetic, ARITH 1983, Aarhus, Denmark, 20–22 June 1983, pp. 37–42 (1983)
6. Fraenkel, A.S., Klein, S.T.: Robust universal complete codes for transmission and compression. Discrete Appl. Math. **64**(1), 31–55 (1996)
7. Huffman, D.A.: A method for the construction of minimum-redundancy codes. Proc. IRE **40**(9), 1098–1101 (1952)
8. Klein, S.T.: Combinatorial representation of generalized Fibonacci numbers. Fibonacci Quart. **29**(2), 124–131 (1991)
9. Klein, S.T., Kopel Ben-Nissan, M.: On the usefulness of Fibonacci compression codes. Comput. J. **53**, 701–716 (2010)
10. Klein, S.T., Serebro, T.C., Shapira, D.: Generalization of Fibonacci codes to the non-binary case. IEEE Access **10**, 112043–112052 (2022)
11. Moffat, A.: Word-based text compression. Softw. Pract. Exp. **19**(2), 185–198 (1989)
12. Moffat, A., Turpin, A.: Compression and coding algorithms, the international series in engineering and computer science, vol. 669. Kluwer (2002)

13. Trogemann, G., Nitussov, A.Y., Ernst, W.: Computing in Russia: the history of computer devices and information technology revealed. Vieweg Braunschweig (2001)
14. Witten, I.H., Neal, R.M., Cleary, J.G.: Arithmetic coding for data compression. Commun. ACM **30**(6), 520–540 (1987)
15. Zeckendorf, E.: Représentation des nombres naturels par une somme de nombres de Fibonacci ou de nombres de Lucas. Bull. Soc. R. Sci. Liège **41**, 179–182 (1972)
16. Zipf, G.K.: Human behavior and the principle of least effort. Ravenio Books (1949)

Sweeping Input-Driven Pushdown Automata

Martin Kutrib[✉]

Institut für Informatik, Universität Giessen, Arndtstr. 2, 35392 Giessen, Germany
kutrib@informatik.uni-giessen.de

Abstract. Input-driven pushdown automata (IDPDA) are pushdown automata where the next action on the pushdown store (push, pop, nothing) is solely governed by the input symbol. Here, we introduce sweeping input-driven pushdown automata that process the input in multiple passes (also sweeps). That is, a sweeping input-driven pushdown automaton is a two-way device that may change the input head direction only at the endmarkers. First we show that, given an arbitrary SIDPDA, an equivalent SIDPDA that halts on any input can effectively be constructed. From this result further properties follow. Then we address the determinization of SIDPDAs and its descriptional complexity. Furthermore, the computational capacity of SIDPDAs is studied. To this end, we compare the family $\mathscr{L}(\text{SIDPDA})$ with other well-known language families. In particular, we are interested in families that have strong relations to some kind of pushdown machines.

1 Introduction

Input-driven pushdown automata (IDPDAs) have been introduced in [25] and their motivation stems from the search for an upper bound for the space needed for the recognition of deterministic context-free languages. IDPDAs are a subclass of pushdown automata where the actions on the pushdown store are dictated by the input symbols. To this end, the input alphabet is partitioned into three subsets, where one subset contains symbols on which the automaton pushes a symbol onto the pushdown store, one subset contains symbols on which the automaton pops a symbol, and one subset contains symbols on which the automaton leaves the pushdown unchanged and makes a state change only. The results in [25] and the follow-up papers [7,15] comprise the equivalence of nondeterministic and deterministic models and the proof that the membership problem is solvable in logarithmic space.

The investigation of input-driven pushdown automata has been revisited in [2,3], where such devices are called visibly pushdown automata or nested word automata. Some of the results are descriptional complexity aspects for the determinization as well as closure properties and decidability questions which turned out to be similar to those of finite automata. Further aspects such as the minimization of IDPDAs and a comparison with other subfamilies of deterministic

B. Nagy (Ed.): CIAA 2023, LNCS 14151, pp. 194–205, 2023.
https://doi.org/10.1007/978-3-031-40247-0_14

context-free languages have been studied in [11,12]. A recent survey with many valuable references on complexity aspects of input-driven pushdown automata may be found in [26]. An extension of input-driven pushdown automata towards pushdown automata synchronized by a finite transducer has been discussed in [10]. The properties and features of IDPDAs revived the research on input-driven pushdown languages and triggered the study of further input-driven automata types, such as input-driven variants of, for example, (ordered) multi-pushdown automata [9], scope-bounded multi-pushdown automata [22], stack automata [4], queue automata [20], etc. While in the early papers IDPDAs are defined as ordinary pushdown automata whose behavior on the pushdown is solely driven by the input symbols, the definition in [2] requires certain normal forms. These normal forms do not change the computational capacity of pushdown automata but have an impact on the power of IDPDAs [19,21].

Here, we consider input-driven pushdown automata that may read their input in a restricted two-way fashion. We introduce sweeping input-driven pushdown automata that process the input in multiple passes (also sweeps). To this end, it receives the input between two endmarkers, and reads it in alternating sweeps starting by a sweep from left to right. So, a sweeping input-driven pushdown automaton is a two-way device that may change the input head direction only at the endmarkers.

General two-way pushdown automata have been studied in detail in [17] for the first time. They accept non-context-free languages as $\{\, ww \mid w \in \{a,b\}^* \,\}$. It is not known whether deterministic two-way pushdown automata capture all context-free languages. The induced language family is a subset of the family of context-sensitive languages. This is an open problem for the nondeterministic variants (cf. [1]). Descriptional complexity aspects of two-way pushdown automata with restricted head reversals are studied in [24]. Before we turn to our main results and the organization of the paper, we briefly mention a different approach to introduce two-way head motion to input-driven automata and transducers. They are introduced in [14] with general two-way head motion, but with the following change in the way of how input symbols dictate the pushdown behavior: in a step that moves the input head to the left, the role played by push and pop symbols is interchanged. This immediately implies that at any position in the input the height of the pushdown is always the same.

The paper is organized as follows. In the next section, we compile the necessary definitions and give an example of how a non-context-free language is accepted by a sweeping input-driven pushdown automaton (SIDPDA). In Sect. 3, we turn to a crucial lemma for the further considerations. It deals with halting computations. Clearly, a SIDPDA can run into an infinite loop. We are going to show that, given an arbitrary SIDPDA, an equivalent SIDPDA that halts on any input can effectively be constructed. Moreover, the proof reveals that only a constant number of sweeps had to be performed, where the constant depends on the given automaton. So, for example, the closure of the family of induced languages under complementation can be derived. To our knowledge this closure is still an open problem for general deterministic two-way pushdown automata [17].

Section 4 considers the determinization of SIDPDAs and its descriptional complexity, where the upper bound follows from the determinization process and an exponential lower bound is shown. In Sect. 5, the computational capacity of SIDPDAs is studied. To this end, we compare the family $\mathscr{L}(\text{SIDPDA})$ with other well-known language families. In particular, we are interested in families that have strong relations to some kind of pushdown machines. It turns out that SIDPDAs are quite weak on the one hand, but are quite powerful on the other.

2 Definitions and Preliminaries

Let Σ^* denote the set of all words over the finite alphabet Σ. The *empty word* is denoted by λ, and $\Sigma^+ = \Sigma^* \setminus \{\lambda\}$. The set of words of length at most $n \geq 0$ is denoted by $\Sigma^{\leq n}$. The *reversal* of a word w is denoted by w^R. For the *length* of w we write $|w|$. We use $\subseteq$ for *inclusions* and $\subset$ for *strict inclusions*. We write $|S|$ for the cardinality of a set S. We say that two language families $\mathscr{L}_1$ and $\mathscr{L}_2$ are *incomparable* if $\mathscr{L}_1$ is not a subset of $\mathscr{L}_2$ and vice versa.

A classical pushdown automaton is called input-driven if the next input symbol defines the next action on the pushdown store, that is, pushing a symbol onto the pushdown store, popping a symbol from the pushdown store, or changing the state without modifying the pushdown store. To this end, the input alphabet Σ is partitioned into the sets Σ_D, Σ_R, and Σ_N, that control the actions push (D), pop (R), and state change only (N). A sweeping input-driven pushdown automaton processes the input in multiple passes (also sweeps). To this end, it receives the input between two endmarkers, and reads it in alternating sweeps starting by a sweep from left to right. So, a sweeping input-driven pushdown automaton is a two-way device that may change the input head direction only at the endmarkers.

Formally, a *deterministic sweeping input-driven pushdown automaton* (SIDPDA) is a system $M = \langle Q, \Sigma, \Gamma, q_0, F, \rhd, \lhd, \perp, \delta_D, \delta_R, \delta_N \rangle$, where Q is the finite set of *internal states* partitioned into the sets Q_l and Q_r, Σ is the finite set of *input symbols* partitioned into the sets Σ_D, Σ_R, and Σ_N, Γ is the finite set of *pushdown symbols*, $q_0 \in Q_l$ is the *initial state*, $F \subseteq Q$ is the set of *accepting states*, $\rhd \notin \Sigma$ and $\lhd \notin \Sigma$ are the *left and right endmarkers*, $\perp \notin \Gamma$ is the *empty-pushdown symbol*, δ_D is the partial transition function mapping $Q_x \times \Sigma_D \times (\Gamma \cup \{\perp\})$ to $Q_x \times \Gamma$ for $x \in \{l, r\}$, δ_R is the partial transition function mapping $Q_x \times \Sigma_R \times (\Gamma \cup \{\perp\})$ to Q_x for $x \in \{l, r\}$, δ_N is the partial transition function mapping $Q_x \times \Sigma_N \times (\Gamma \cup \{\perp\})$ to Q_x for $x \in \{l, r\}$, $Q \times \{\rhd\} \times (\Gamma \cup \{\perp\})$ to Q_r, and $Q \times \{\lhd\} \times (\Gamma \cup \{\perp\})$ to Q_l.

A *configuration* of a SIDPDA $M = \langle Q, \Sigma, \Gamma, q_0, F, \rhd, \lhd, \perp, \delta_D, \delta_R, \delta_N \rangle$ is a quadruple $(q, \rhd w \lhd, i, s)$, where $q \in Q$ is the current state, $w \in \Sigma^*$ is the input, $0 \leq i \leq |w| + 1$ is the current position of the input head, and $s \in \Gamma^*$ denotes the current pushdown content, where the leftmost symbol is at the top of the pushdown store. The *initial configuration* for an input string w is set to $(q_0, \rhd w \lhd, 0, \lambda)$. During the course of its computation, M runs through a sequence of configurations. One step from a configuration to its successor configuration is denoted by $\vdash$. Let $q, q' \in Q$, $n \geq 0$, $w = a_1 a_2 \cdots a_n \in \Sigma^*$, $z, z' \in \Gamma$, $s \in \Gamma^*$, and $1 \leq i \leq n$. We set

1. $(q, \triangleright w \triangleleft, i, zs) \vdash (q', \triangleright w \triangleleft, j, z'zs)$, if $a_i \in \Sigma_D$ and $(q', z') = \delta_D(q, a_i, z)$,
2. $(q, \triangleright w \triangleleft, i, \lambda) \vdash (q', \triangleright w \triangleleft, j, z')$, if $a_i \in \Sigma_D$ and $(q', z') = \delta_D(q, a_i, \bot)$,
3. $(q, \triangleright w \triangleleft, i, zs) \vdash (q', \triangleright w \triangleleft, j, s)$, if $a_i \in \Sigma_R$ and $q' = \delta_R(q, a_i, z)$,
4. $(q, \triangleright w \triangleleft, i, \lambda) \vdash (q', \triangleright w \triangleleft, j, \lambda)$, if $a_i \in \Sigma_R$ and $q' = \delta_R(q, a_i, \bot)$,
5. $(q, \triangleright w \triangleleft, i, zs) \vdash (q', \triangleright w \triangleleft, j, zs)$, if $a_i \in \Sigma_N$ and $q' = \delta_N(q, a_i, z)$,
6. $(q, \triangleright w \triangleleft, i, \lambda) \vdash (q', \triangleright w \triangleleft, j, \lambda)$, if $a_i \in \Sigma_N$ and $q' = \delta_N(q, a_i, \bot)$,
7. $(q, \triangleright w \triangleleft, 0, zs) \vdash (q', \triangleright w \triangleleft, 1, zs)$, if $q' = \delta_N(q, \triangleright, z)$,
8. $(q, \triangleright w \triangleleft, 0, \lambda) \vdash (q', \triangleright w \triangleleft, 1, \lambda)$, if $q' = \delta_N(q, \triangleright, \bot)$,
9. $(q, \triangleright w \triangleleft, n+1, zs) \vdash (q', \triangleright w \triangleleft, n, zs)$, if $q' = \delta_N(q, \triangleleft, z)$,
10. $(q, \triangleright w \triangleleft, n+1, \lambda) \vdash (q', \triangleright w \triangleleft, n, \lambda)$, if $q' = \delta_N(q, \triangleleft, \bot)$,

where $j = i + 1$ if $q' \in Q_r$, and $j = i - 1$ if $q' \in Q_l$.

So, on the endmarkers only δ_N is defined, and these are the only symbols on which a state from Q_l can be followed by a state from Q_r and vice versa. This implements the sweeping of the automaton. Whenever the pushdown store is empty, the successor configuration is computed by the transition functions with the special empty-pushdown symbol $\bot$. As usual, we define the reflexive and transitive closure of $\vdash$ by $\vdash^*$.

A SIDPDA M starts in state q_0 with its head on the left endmarker. It halts when the transition functions are not defined for the current situation. A computation can also enter an infinite loop. However, an input $w \in \Sigma^*$ *is accepted if and only if M halts in an accepting state on* $\triangleright w \triangleleft$. The *language accepted* by M is $L(M) = \{ w \in \Sigma^* \mid w$ is accepted by $M \}$.

In general, the family of all languages accepted by automata of some type X will be denoted by $\mathscr{L}(X)$.

Some properties of language families implied by classes of input-driven pushdown automata may depend on whether all automata involved share the same partition of the input alphabet. For easier writing, we call the partition of an input alphabet a *signature*.

In order to clarify these notions, we continue with an example.

Example 1. The language $L = \{ a^n b^n a^n \mid n \geq 1 \}$ is accepted by the SIDPDA $M = \langle Q, \Sigma, \Gamma, q_0, F, \triangleright, \triangleleft, \bot, \delta_D, \delta_R, \delta_N \rangle$ with $Q = Q_r \cup Q_l$, $Q_r = \{q_0, q_1, q_2, q_+\}$, $Q_l = \{r_0, r_1, r_2\}$, $\Sigma_D = \{b\}$, $\Sigma_R = \{a\}$, $\Sigma_N = \emptyset$, $\Gamma = \{\bullet\}$, $F = \{q_+\}$, and the transition functions specified as:

(1) $\delta_N(q_0, \triangleright, \bot) = \{q_0\}$ (6) $\delta_D(r_0, b, \bot) = \{(r_1, \bullet)\}$ (10) $\delta_R(q_2, a, \bullet) = q_2$
(2) $\delta_N(q_2, \triangleleft, \bot) = \{r_0\}$ (7) $\delta_D(r_1, b, \bullet) = \{(r_1, \bullet)\}$ (11) $\delta_R(r_0, a, \bot) = r_0$
(3) $\delta_N(r_2, \triangleright, \bot) = \{q_+\}$ (8) $\delta_R(q_0, a, \bot) = q_0$ (12) $\delta_R(r_1, a, \bullet) = r_2$
(4) $\delta_D(q_0, b, \bot) = \{(q_1, \bullet)\}$ (9) $\delta_R(q_1, a, \bullet) = q_2$ (13) $\delta_R(r_2, a, \bullet) = r_2$
(5) $\delta_D(q_1, b, \bullet) = \{(q_1, \bullet)\}$

The basic idea of the construction is as follows. Essentially, M performs two sweeps, except for a possible last accepting step on the left endmarker, where

the head moves to the right. Automaton M pops on a's and pushes a symbol •
on b's. So, in the first left-to-right sweep, the number of b's could be compared
with the number of a's in the suffix. If both are identical, the pushdown is empty
at the end of the sweep. Then, in the second right-to-left sweep, symmetrically,
the number of b's is compared with the number of a's in the prefix.   ∎

3   Halting, Time Complexity, Basic Closure Properties

Here, we first turn to a crucial lemma for the further considerations. It deals with
halting computations. Clearly, a SIDPDA can run into an infinite loop. This is
a big difference compared with one-way input-driven pushdown automata. For
classical deterministic one-way pushdown automata, it is well known that any
deterministic one-way pushdown automaton can effectively be converted into an
equivalent one that halts on any input. The situation for classical deterministic
two-way pushdown automata is again different. To our knowledge it is unknown
whether the class of induced languages is closed under complementation [17].
We are going to show that the problem can be solved for SIDPDAs. A problem
we have to cope with is as follows. Assume, for example, that the number of
input symbols that cause a push operation equals the number of input symbols
that cause a pop operation. Then one can neither conclude that the pushdown
is empty at the end of a sweep nor that the same number of symbols is pushed
or popped in a left-to-right sweep and in a right-to-left sweep. Let the input
be $a^n b^n$ for some $n \geq 1$. If $a \in \Sigma_R$ and $b \in \Sigma_D$ then the pushdown contains n
symbols after the first (left-to-right) sweep. After the second (right-to-left) sweep
it contains again n symbols. Now let $a \in \Sigma_D$ and $b \in \Sigma_R$. Then the pushdown
is empty after the first sweep and contains n symbols after the second sweep. In
preparation we use the following functions. For a signature $\Sigma = \Sigma_D \cup \Sigma_R \cup \Sigma_N$
and a word $w = a_1 a_2 \cdots a_n \in \Sigma^*$, we define $dr_w^\Sigma, dl_w^\Sigma : \{0, 1, \ldots, n+1\} \to \mathbb{N}$ by

$$dr_w^\Sigma(i) = \begin{cases} 0 & \text{if } i \in \{0,1\} \\ dr_w^\Sigma(i-1) + 1 & \text{if } i \in \{2,3,\ldots,n+1\} \text{ and } a_{i-1} \in \Sigma_D \\ dr_w^\Sigma(i-1) & \text{if } i \in \{2,3,\ldots,n+1\} \text{ and } a_{i-1} \in \Sigma_N \\ dr_w^\Sigma(i-1) - 1 & \text{if } i \in \{2,3,\ldots,n+1\} \text{ and } a_{i-1} \in \Sigma_R \end{cases}$$

and

$$dl_w^\Sigma(i) = \begin{cases} dr_w^\Sigma(n+1) & \text{if } i \in \{n+1, n\} \\ dl_w^\Sigma(i+1) + 1 & \text{if } i \in \{0,1,\ldots,n-1\} \text{ and } a_{i+1} \in \Sigma_D \\ dl_w^\Sigma(i+1) & \text{if } i \in \{0,1,\ldots,n-1\} \text{ and } a_{i+1} \in \Sigma_N \\ dl_w^\Sigma(i+1) - 1 & \text{if } i \in \{0,1,\ldots,n-1\} \text{ and } a_{i+1} \in \Sigma_R \end{cases}.$$

Essentially, these functions give the difference of the numbers of symbols
from Σ_D and from Σ_R that have been seen when the input head is on the ith
position. Function dr_w^Σ yields the difference for the first (right-to-left) sweep and
function dl_w^Σ continues the calculation for the subsequent second (left-to-right)
sweep.

Theorem 2. *Given an arbitrary SIDPDA, an equivalent SIDPDA that halts on any input can effectively be constructed.*

Proof. Let $M = \langle Q, \Sigma, \Gamma, q_0, F, \triangleright, \triangleleft, \perp, \delta_D, \delta_R, \delta_N \rangle$ be an arbitrary SIDPDA with signature $\Sigma = \Sigma_D \cup \Sigma_R \cup \Sigma_N$. For any input $w = a_1 a_2 \cdots a_n \in \Sigma^*$, we consider the number $sh_w = |\{ i \mid a_i \in \Sigma_D \}| - |\{ i \mid a_i \in \Sigma_R \}| = dr_w^\Sigma(n+1)$. Note, this number does not depend on the ordering of the symbols in w. However, the functions dr_w^Σ and dl_w^Σ do not give the heights of the pushdown as they appear in the first two sweeps of a computation. The reason is that popping from the empty pushdown does not compensate for some push. So, a pop may or may not compensate for a push. We denote each pop from the empty pushdown as *lost decrease*.

How many lost decreases appear during the first two sweeps? This number is the absolute value of the minimum $\min_{0 \le i \le n+1}\{\min\{dr_w^\Sigma(i), dl_w^\Sigma(i)\}\}$. Since $dr_w^\Sigma(0) = 0$, the minimum cannot be greater than 0.

Next, we can calculate the height of the pushdown after the second sweep when the input head is back on the left endmarker by considering the number $2 \cdot sh_w$ and adding the number of lost decreases to it, that is, the height is $2 \cdot sh_w - \min_{0 \le i \le n+1}\{\min\{dr_w^\Sigma(i), dl_w^\Sigma(i)\}\}$. For the heights of the pushdown after further double sweeps from left to right and back, we distinguish two cases. In the first case we have $sh_w \ge 0$. Due to the pushdown height after the first two sweeps, now no further lost decreases occur. Therefore, the pushdown height increases by $2 \cdot sh_w$ after each double sweep. In the second case we have $sh_w < 0$. Due to the pushdown height after the first two sweeps, now $|2 \cdot sh_w|$ further lost decreases occur. On the other hand, the pushdown height changes by $2 \cdot sh_w$, that is, should decrease by $|2 \cdot sh_w|$. Together, the pushdown height does not change after another double sweep.

Next we are interested at which positions during a double sweep the pushdown height becomes minimal. These are exactly the positions i at which the values $\min\{dr_w^\Sigma(i), dl_w^\Sigma(i)\}$ are minimal. Let us fix one of these positions if there are more than one, say position i_0 with $dl_w^\Sigma(i_0) \le dr_w^\Sigma(i_0)$. From above we know that either the pushdown height increases by $2 \cdot sh_w$ after each further double sweep or the pushdown height remains as it is after each further double sweep. But this implies that in each further double sweep the pushdown height becomes minimal when the input head enters position i_0 from right. That is, the contents of the pushdown below that minimum will never be seen again during the computation on w. There are at most $|Q| \cdot (|\Gamma| + 1)$ possibilities to continue the computation. So, we conclude that M is in an infinite loop if one of these possibilities appears twice. In other words, M will not halt if and only if it performs more than $|Q| \cdot (|\Gamma| + 1)$ double sweeps.

Finally, from M an equivalent SIDPDA M' is constructed as follows. First, automaton M' simulates M in such a way that M' halts only after a number of double sweeps on the left endmarker. This is always possible by adding states that are used to remember the decision of M and to complete a double sweep should M halt anywhere else. Since $|Q| \cdot (|\Gamma| + 1)$ is a constant that does not depend on the input, M' can maintain a finite counter in its state set that counts the double sweeps performed. Now, M' can always halt rejecting if M tries to perform more than $|Q| \cdot (|\Gamma| + 1)$ many double sweeps.   □

Since the proof of Theorem 2 revealed that any SIDPDA can be simulated by a halting SIDPDA that makes at most a constant number of sweeps, we have the following corollary. Recall that the upper bound for the family of languages accepted by deterministic two-way pushdown automata is DTIME$(n^2 \log(n))$ [1].

Corollary 3. *The family* $\mathscr{L}(SIDPDA)$ *is contained in the time complexity class* DTIME(n).

For input-driven pushdown automata, strong closure properties are shown in [2] *provided that* all automata involved share the same partition of the input alphabet. Now, we derive the closure under the Boolean operations from Theorem 2 as follows.

Proposition 4. *Let M and M' be two SIDPDAs with compatible signatures. Then SIDPDAs accepting the complement $\overline{L(M)}$, the intersection $L(M) \cap L(M')$, and the union $L(M) \cup L(M')$ can effectively be constructed.*

4   Determinization

It is well known that nondeterministic one-way IDPDAs can be determinized. Okhotin and Salomaa [26] traced this result back to [7]. They give a clear proof that shows that 2^{n^2} states are sufficient to simulate an n-state nondeterministic one-way IDPDA by a deterministic one. However, in [26] and [2] IDPDAs are considered in a certain normal form. That is, neither the push nor the state change only operations depend on the topmost pushdown symbol. Since any deterministic one-way pushdown automaton and any deterministic one-way input-driven pushdown automaton can be converted to this normal form, the general computational capacity does not change. But by this conversion the number of states changes since the topmost pushdown symbol has to be remembered in the states. Thus, when we compare such automata with the original definition of one-way IDPDAs in [7], that is based on the usual definition of pushdown automata and does not require a normal form, we obtain that the state complexity bound of 2^{n^2} achieved for the determinization of one-way IDPDAs in normal form is lower than in general. Since here we are closer to the original definition, the size of a two-way input-driven pushdown automaton is measured not only by its states but as the product of the number of states and the number of pushdown symbols plus one. Accordingly, we define the function $\mathsf{size}(M) = |Q|(|\Gamma| + 1)$ that maps a two-way input-driven pushdown automaton to its size.

A *nondeterministic sweeping input-driven pushdown automaton* (SNIDPDA) is a system $M = \langle Q, \Sigma, \Gamma, q_0, F, \triangleright, \triangleleft, \bot, \delta_D, \delta_R, \delta_N \rangle$, where all components but the transition functions are defined as for SIDPDAs. The transition functions are now nondeterministic, that is, their codomains are now subsets of the codomains of the deterministic functions. As usual, an input is accepted if there is an accepting computation on it.

The next theorem shows the determinization of nondeterministic sweeping input-driven pushdown automata. Basically, the idea of the proof is along the lines of [26] and [19].

Theorem 5. *Let $n \geq 1$ and M be an n-state nondeterministic sweeping input-driven pushdown automaton with input alphabet $\Sigma = \Sigma_D \cup \Sigma_R \cup \Sigma_N$ and m pushdown symbols. Then, an equivalent SIDPDA with at most $2^{(n+2)^2 m + n + 2}$ states and $2^{(n+2)^2 m + n + 2} \cdot |\Sigma_D|$ pushdown symbols can effectively be constructed.*

One may ask why the determinization of nondeterministic sweeping input-driven pushdown automata is possible at all, when the head movement is limited two way. Intuitively, a big point is that the devices are sweeping, that is, the head movement of all nondeterministic choices is the same, respectively.

So far, Theorem 5 provides an upper bound on the size for determinization. Next, we turn to a lower bound. To this end, we first relate the size trade-off between nondeterministic finite automata (NFA) and sweeping deterministic finite automata (2DFA) to our problem. Here, the size of finite automata without further storage is measured by their number of states. So, let N be a minimal n-state NFA accepting a language over an alphabet with at least two letters, say $\{a, b\} \subseteq \Sigma$. We take a new letter d and define

$$L_N = \{ d^x w \mid x \geq 0, w \in \{a, b\}^*, x = |w| \text{ or } w \in L(N) \}.$$

Lemma 6. *For $n \geq 1$, let N be a minimal n-state NFA accepting a language over an alphabet $\Sigma \supseteq \{a, b\}$. Then, any SIDPDA M accepting language L_N has $\mathsf{size}(M) \geq \mathsf{size}(N')$, where N' is the smallest 2DFA accepting $L(N)$.*

Proof. Let $M = \langle Q, \Sigma, \Gamma, q_0, F, \triangleright, \triangleleft, \bot, \delta_D, \delta_R, \delta_N \rangle$ be a SIDPDA accepting L_N. We consider the signature of M.

If $\{a, b\} \subseteq \Sigma_D \cup \Sigma_N$ or $\{a, b\} \subseteq \Sigma_R \cup \Sigma_N$, we consider inputs of the form $w \in \{a, b\}^*$. For any such input only the topmost pushdown symbol is accessible by M. Therefore, from M, we can construct a 2DFA accepting the language $L(N)$, essentially by storing the topmost pushdown symbol into the states and removing all transitions on d, where the size of the 2DFA is at most $|Q| \cdot (|\Gamma| + 1) \leq \mathsf{size}(M)$. Therefore, $\mathsf{size}(M) \geq \mathsf{size}(N')$.

Next, we consider the remaining cases that either $a \in \Sigma_D$ and $b \in \Sigma_R$ or vice versa. Assume $a \in \Sigma_D$ and $b \in \Sigma_R$, the alternative case is treated symmetrically. Then we consider inputs of the form $d^* a^*$ if $d \in \Sigma_D \cup \Sigma_N$, or of the form $d^* b^*$ if $d \in \Sigma_R$. Again, by storing the topmost pushdown symbol into the states and removing all transitions on the symbol not appearing in the word, we can construct a 2DFA from M accepting either the language $\{ d^x a^x \mid x \geq 0 \}$ or $\{ d^x b^x \mid x \geq 0 \}$. However, both languages are non-regular but any 2DFA accepts a regular language. From the contradiction, we conclude that these cases are impossible for the signature. Therefore, the only possible signatures imply $\mathsf{size}(M) \geq \mathsf{size}(N')$.   □

In order to apply Lemma 6 to obtain a lower bound on the size for determinization, we have to plug in appropriate NFAs. The tight bound $2^n - 1$ for the conversion of an NFA to a 2DFA has been shown in [23] where, for every $n \geq 1$, an n-state NFA was exhibited whose smallest equivalent 2DFA cannot do better in the amount of size than the smallest equivalent incomplete DFA. We use these NFAs $\hat{N}$ depicted in Fig. 1 as witness automata.

Lemma 7. *For all $n \geq 1$, let $\hat{N}$ be the minimal n-state NFA depicted in Fig. 1. Then language $L_{\hat{N}}$ is accepted by some SNIDPDA with $n + 3$ states and one pushdown symbol.*

Applying the determinization of Theorem 5 to the SNIDPDA M of Lemma 7 gives an equivalent SIDPDA with at most $2^{(n+5)^2+n+5}$ states and $2^{(n+5)^2+n+5}$ pushdown symbols. So, the size of M is $2^{(n+5)^2+n+5} \cdot (2^{(n+5)^2+n+5} + 1) \in 2^{O(n^2)}$.

The next lemma shows that the lower bound for the determinization is at least exponential.

Theorem 8. *For all $n \geq 1$, let $\hat{N}$ be the minimal n-state NFA depicted in Fig. 1. Then any SIDPDA M accepting language $L_{\hat{N}}$ has $\mathsf{size}(M) \geq 2^n - 1$.*

Proof. We apply Lemma 6 with the minimal NFAs depicted in Fig. 1 and obtain $\mathsf{size}(M) \geq \mathsf{size}(\hat{N}') = 2^n - 1$.   □

5   Computational Capacity

Here we are going to compare the family $\mathscr{L}(\mathrm{SIDPDA})$ with other well-known language families. In particular, we are interested in families that have strong relations to some kind of pushdown machines. It turns out that SIDPDAs are quite weak on the one hand, but are quite powerful on the other.

From Corollary 3, we know already that the family $\mathscr{L}(\mathrm{SIDPDA})$ shares the attractive time complexity $\mathrm{DTIME}(n)$ with the family of one-way input-driven pushdown automata. Clearly, all regular languages belong to $\mathscr{L}(\mathrm{SIDPDA})$. However, this somehow seems to be a tight lower bound, since there are simple deterministic context-free languages not belonging to $\mathscr{L}(\mathrm{SIDPDA})$.

Proposition 9. *The linear deterministic context-free language $\{\, a^n \# a^n \mid n \geq 0 \,\}$ is not accepted by any SIDPDA.*

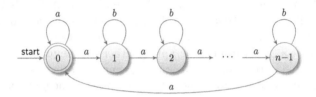

Fig. 1. The n-state NFA $\hat{N}$ where each equivalent 2DFA has at least $2^n - 1$ states.

Let us now turn to a hierarchy of larger families. As is very well known, the family DCFL of deterministic context-free languages is characterized by deterministic one-way pushdown automata, and the family CFL of context-free languages is characterized by nondeterministic one-way pushdown automata. Growing context-sensitive grammars, that is, context-sensitive grammars for which each production rule is strictly length-increasing, have been introduced in [13]. The induced language family GCSL is a strict superset of CFL. It is interesting in our context, since each growing context-sensitive language is accepted in polynomial time by some one-way auxiliary pushdown automaton with a logarithmic space bound [8]. In general, an auxiliary pushdown automaton (auxPDA) with space bound s is a nondeterministic Turing machine M with a read-only input tape, a pushdown store, and an auxiliary work tape which is initialized to have exactly $s(n)$ squares, limited by endmarkers, if M is started on an input of length n. Extending the hierarchy of language families we consider context-sensitive grammars with an additional bound on the length of derivations [5, 16]. With a linear bound these grammars generate the languages in the family CSL_{lin}, which, by definition, is a superset of GCSL. The inclusion is strict. Moreover, the family CSL_{lin} is strictly included in the family of deterministic context-sensitive languages (DCSL) but still contains NP-complete languages [6]. So, we obtain the hierarchy $REG \subset DCFL \subset CFL \subset GCSL \subset CSL_{lin} \subset DCSL$. Since any SIDPDA works in linear time, the family $\mathscr{L}(SIDPDA)$ is strictly included in DCSL. The relation to the remaining families in the hierarchy are given by the next proposition.

Theorem 10. *$\mathscr{L}(SIDPDA)$ is incomparable with each of the families DCFL, CFL, GCSL, and CSL_{lin}.*

Proof. Due to the hierarchy and Proposition 9 it is sufficient to show that there is a language in $\mathscr{L}(SIDPDA)$ that does not belong to CSL_{lin}. It follows from a more general result in [16] that the language $L' = \{\, w\#w^R\#w \mid w \in \{a, b\}^* \,\}$ is not generated by any context-sensitive grammar with $o(n^2)$ derivation steps. In particular, L' does not belong to CSL_{lin}. Since the family CSL_{lin} is closed under non-erasing homomorphisms [5], the language $L = \{\, w\#\bar{w}^R\#w \mid w \in \{a, b\}^* \,\}$ does not belong to CSL_{lin} either, where $\bar{w}^R$ is essentially w^R but with all letters barred.

It remains to be shown that L is accepted by some SIDPDA.   □

So, we have, in particular, an upper bound for the computational capacity of SIDPDA by CSL. Can we do better? As mentioned, each growing context-sensitive language is accepted in polynomial time by some one-way auxiliary pushdown automaton with a logarithmic space bound (OW-auxPDA(log, poly)). If we allow such machines to operate on the input tape in two-way fashion, but restrict the mode of computation to be deterministic, then we obtain auxDPDA(log, poly). The language family $\mathscr{L}(auxDPDA(log, poly))$ is identical to the family LOGDCFL of languages reducible in logarithmic space to a deterministic context-free language [27]. From results in [18] it can be derived that

LOGDCFL $\subseteq$ DSPACE($\log^2(n)$) and, thus, $\mathscr{L}$(auxDPDA(log, poly)) $\subseteq$ DCSL. On the other hand, any SIDPDA is a special case of an auxDPDA(log, poly).

Corollary 11. *The family $\mathscr{L}(SIDPDA)$ is strictly included in LOGDCFL.*

References

1. Aho, A.V., Hopcroft, J.E., Ullman, J.D.: Time and tape complexity of pushdown automaton languages. Inform. Control **13**, 186–206 (1968)
2. Alur, R., Madhusudan, P.: Visibly pushdown languages. In: Symposium on Theory of Computing (STOC 2004), pp. 202–211. ACM (2004)
3. Alur, R., Madhusudan, P.: Adding nesting structure to words. J. ACM **56** (2009)
4. Bensch, S., Holzer, M., Kutrib, M., Malcher, A.: Input-driven stack automata. In: Baeten, J.C.M., Ball, T., de Boer, F.S. (eds.) TCS 2012. LNCS, vol. 7604, pp. 28–42. Springer, Heidelberg (2012). https://doi.org/10.1007/978-3-642-33475-7_3
5. Book, R.V.: Time-bounded grammars and their languages. J. Comput. Syst. Sci. **5**, 397–429 (1971)
6. Book, R.V.: On the complexity of formal grammars. Acta Inform. **9**, 171–181 (1978)
7. von Braunmühl, B., Verbeek, R.: Input-driven languages are recognized in $\log n$ space. In: Topics in the Theory of Computation, Mathematics Studies, vol. 102, pp. 1–19. North-Holland (1985)
8. Buntrock, G., Otto, F.: Growing context-sensitive languages and Church-Rosser languages. Inform. Comput. **141**, 1–36 (1998)
9. Carotenuto, D., Murano, A., Peron, A.: Ordered multi-stack visibly pushdown automata. Theor. Comput. Sci. **656**, 1–26 (2016)
10. Caucal, D.: Synchronization of pushdown automata. In: Ibarra, O.H., Dang, Z. (eds.) DLT 2006. LNCS, vol. 4036, pp. 120–132. Springer, Heidelberg (2006). https://doi.org/10.1007/11779148_12
11. Chervet, P., Walukiewicz, I.: Minimizing variants of visibly pushdown automata. In: Kučera, L., Kučera, A. (eds.) MFCS 2007. LNCS, vol. 4708, pp. 135–146. Springer, Heidelberg (2007). https://doi.org/10.1007/978-3-540-74456-6_14
12. Crespi-Reghizzi, S., Mandrioli, D.: Operator precedence and the visibly pushdown property. J. Comput. Syst. Sci. **78**, 1837–1867 (2012)
13. Dahlhaus, E., Warmuth, M.K.: Membership for growing context-sensitive grammars is polynomial. J. Comput. Syst. Sci. **33**, 456–472 (1986)
14. Dartois, L., Filiot, E., Reynier, P., Talbot, J.: Two-way visibly pushdown automata and transducers. In: Logic in Computer Science (LICS 2016). pp. 217–226. ACM (2016)
15. Dymond, P.W.: Input-driven languages are in $\log n$ depth. Inform. Process. Lett. **26**, 247–250 (1988)
16. Gladkii, A.V.: On complexity of inference in phase-structure grammars. Algebra i Logika. Sem. **3**, 29–44 (1964). (in Russian)
17. Gray, J., Harrison, M.A., Ibarra, O.H.: Two-way pushdown automata. Inform. Control **11**, 30–70 (1967)
18. Harju, T.: A simulation result for the auxiliary pushdown automata. J. Comput. Syst. Sci. **19**, 119–132 (1979)
19. Kutrib, M., Malcher, A.: Digging input-driven pushdown automata. RAIRO Inform. Théor. **55**, Art. 6 (2021)

20. Kutrib, M., Malcher, A., Mereghetti, C., Palano, B., Wendlandt, M.: Deterministic input-driven queue automata: finite turns, decidability, and closure properties. Theor. Comput. Sci. **578**, 58–71 (2015)
21. Kutrib, M., Malcher, A., Wendlandt, M.: On the power of pushing or stationary moves for input-driven pushdown automata. In: Implementation and Application of Automata (CIAA 2022). LNCS, vol. 13266, pp. 140–151. Springer (2022). https:// doi.org/10.1007/978-3-031-07469-1_11
22. La Torre, S., Napoli, M., Parlato, G.: Scope-bounded pushdown languages. Int. J. Found. Comput. Sci. **27**, 215–234 (2016)
23. Leung, H.: Tight lower bounds on the size of sweeping automata. J. Comput. Syst. Sci. **63**, 384–393 (2001)
24. Malcher, A., Mereghetti, C., Palano, B.: Descriptional complexity of two-way pushdown automata with restricted head reversals. Theor. Comput. Sci. **449**, 119–133 (2012)
25. Mehlhorn, Kurt: Pebbling mountain ranges and its application to DCFL-recognition. In: de Bakker, Jaco, van Leeuwen, Jan (eds.) ICALP 1980. LNCS, vol. 85, pp. 422–435. Springer, Heidelberg (1980). https://doi.org/10.1007/3-540-10003-2_89
26. Okhotin, A., Salomaa, K.: Complexity of input-driven pushdown automata. SIGACT News **45**, 47–67 (2014)
27. Sudborough, I.H.: On the tape complexity of deterministic context-free languages. J. ACM **25**, 405–414 (1978)

Verified Verifying: SMT-LIB for Strings in Isabelle

Kevin Lotz[1]([✉])(ID), Mitja Kulczynski[1](ID), Dirk Nowotka[1],
Danny Bøgsted Poulsen[2](ID), and Anders Schlichtkrull[2](ID)

[1] Department of Computer Science, Kiel University, Kiel, Germany
{kel,mku,dn}@informatik.uni-kiel.de
[2] Department of Computer Science, Aalborg University, Aalborg, Denmark
{dannybpoulsen,andsch}@cs.aau.dk

Abstract. The prevalence of string solvers in formal program analysis
has led to an increasing demand for more effective and dependable solving techniques. However, solving the satisfiability problem of string constraints, which is a generally undecidable problem, requires a deep understanding of the structure of the constraints. To address this challenge, the
community has relied on SMT solvers to tackle the quantifier-free first-
order logic fragment of string constraints, usually stated in SMT-LIB
format. In 2020, the SMT-LIB Initiative issued the first official standard
for string constraints. However, SMT-LIB states the semantics in a semi-
formal manner, lacking a level of formality that is desirable for validating
SMT solvers. In response, we formalize the SMT-LIB theory of strings
using Isabelle, an interactive theorem prover known for its ability to formalize and verify mathematical and logical theorems. We demonstrate
the usefulness of having a formally defined theory by deriving, to the
best of our knowledge, the first automated verified model verification
method for SMT-LIB string constraints and highlight potential future
applications.

1 Introduction

Satisfiability Modulo Theories (SMT) solvers [6] have been instrumental in determining the satisfiability of first-order logic formulae within a given theory, driving
much of the development of software verification [18,24] and formal verification
applications. They have also seen widespread use in areas such as security [41,48]
and program analysis [9,39].

A particular area where SMT solving is frequently applied is the verification
of string-heavy programs. This can be partly attributed to the fact that strings
are the most commonly used data type for processing user data and, consequently, mishandling of strings can pose significant security risks. Indeed, third-
ranked security risk on the Open Web Application Security Project's (OWASP)
top ten security risks in 2023 are injection attacks, which are fundamentally
caused by inadequate string handling. While string reasoning is most commonly
associated with web security applications [41,47], it is also applied in other areas

B. Nagy (Ed.): CIAA 2023, LNCS 14151, pp. 206–217, 2023.
https://doi.org/10.1007/978-3-031-40247-0_15

such as model checking [25] and cloud security [2, 40]. Verification and automated reasoning tools typically discharge the heavy lifting of string constraint solving to dedicated SMT solvers. Over the past years, this led to the development of numerous solvers specialized in string reasoning (e.g., [1, 15, 16, 30, 36]) and to widely adapted SMT solvers, such as CVC5 [3] and Z3 [17], adding support for strings. In 2020, efforts converged and were incorporated into the SMT-LIB 2.6 standard [44]. With the addition of the theory of strings, the SMT-LIB 2.6 standard strives to establish a common language and consistent semantics for solving problems related to strings.

Given their extensive usage, it is crucial to ensure that SMT solvers behave correctly and that implementation errors are detected as early as possible. A central question in that regard: *Can we trust an SMT solver's result?* If a solver determines that an input formula is satisfiable, it usually provides evidence of its decisions in the form of a *model*, i.e., an assignment of constants to the variables which satisfies the formula. A common practice to assess the soundness of a solver is to use another solver as an oracle to check whether the produced assignment indeed satisfies the formula at hand. However, that shifts the trust problem from one solver to another and poses a high risk of implementation errors carrying over to new solvers.

To address this problem, we present a novel approach for validating models produced by SMT solvers using Isabelle/HOL, an interactive theorem prover that provides a high-level specification language for expressing logical problems and powerful automation capabilities for proving properties [45]. In particular, our contributions are the following. We formalise the semantics of the SMT-LIB 2.6 theory of strings in Isabelle/HOL and provide an implementation of the standard model that is provably correct. The formalisation proved itself useful as we found inconsistencies in the standard, e.g., in the str. indexof operator, as we highlight in Sect. 3. The formalisation enables us to assess the soundness of SMT solvers by proving that an assignment produced by a solver is indeed a model of the input formula. Unlike using existing SMT solvers as test oracles, this provides a very strong guarantee that a model is indeed correct. We outline our efforts in building a model verification framework in Sect. 4 and show its usefulness on soundness issues reported in the literature.

Related Work. If we look at the Boolean satisfiability problem (SAT) rather than SMT then there are a number of works that use interactive theorem provers to construct verified SAT solvers. These were developed in Isabelle by Maric et al. [33–35], also in Isabelle by Fleury et al. [10–12, 19–23], in Coq by Lescuyer [32] and in PVC by Shankar and Vaucher [43]. Additionally, there is the verified SAT solver in GURU by Oe et al. [38] which ensures model soundness at run time but is not proven to terminate.

SMT has also been combined with interactive theorem proving. In Isabelle, the *smt* tactic will run an SMT solver on a proof goal given in Isabelle, and then the tactic will try to reconstruct the proof in Isabelle's logic. The tactic supports Z3 with the theories of equality, uninterpreted functions, arrays, linear integer arithmetics and real arithmetics [13]. In order to be able to reconstruct the proofs generated by the SMT solver, the SMT solver needs to be able to return a representation of the proof it generated. A recent work in this direction is by Schurr et al. who are building a common format, Alethe, to be used to reconstruct proofs made by SMT solvers [42]. Another recent work is by Barbosa et al. [4] who generate proofs for cvc5 and are also working on exporting them to the Alethe format. The above work concerns proof production from SMT solvers where the interest is in proving that a formula is satisfied by all models. The focus of our work goes in a different direction, namely that of checking in the String theory whether the model given by an SMT solver actually satisfies the input formula.

In the context of string solving, Kan et al. [27] developed a solver for a fragment of string constraints that is completely verified in Isabelle/HOL. On the theoretical side, implementations of regular expressions [29] and general theorems on combinatorics on words [26] were made in Isabelle/HOL.

2   Preliminaries

SMT. SMT extends SAT solving to many-sorted first-order logic within a given logical background theory T that fixes the domain and the interpretation of the non-logical symbols. Solving an SMT formula φ in the theory T involves determining whether a model M exists in T that satisfies φ. The increasing interest in SMT led to the development of the SMT-LIB standard [5], which provides a modeling language for many-sorted first-order logic and specifies various background theories. SMT solvers ingest formulae expressed in the language of SMT-LIB and check their satisfiability according to a theory or a combination of theories specified in the SMT-LIB standard. Theories in SMT-LIB are described in terms of syntactical elements, i.e., the non-logical symbols. The intended semantics of theories are defined informally in natural language [5]. Examples of available theories include fixed-size bitvectors, integers, reals, and strings.

Theory of Strings. We briefly summarise the theory of strings, T_S. An outline of the syntax is depicted in Fig. 1. The theory deals with Boolean combinations of atomic formula including string equalities and inequalities, regular expression membership, and extended string predicates like containment and prefix relations.

A string term is a finite, ordered sequence of characters drawn from a finite alphabet, like ASCII or Unicode. String concatenation is denoted by $t_{str} \cdot t_{str}$. The length of a string term w, denoted by $str.len(w)$, is the number of characters. We also use $|w|$ to refer to the length of w for readability. An empty string is represented by ϵ and has a length of 0. Operations referring to the index of a character or a sub-string within a string utilise zero-based indexing, that is, the first character has an index of zero. The term `str.to_int` treats a string as a non-negative base-10 integer, possibly with leading zeros. If the string is negative or contains non-digit characters, the value is -1. The term `str.from_int` converts a non-negative integer to the shortest possible string representing it in base 10. If the integer is negative, the value is an empty string. The atoms in A_{re} correspond to regular membership constraints of a string term in a regular

$$F \quad ::= Atom \mid F \wedge F \mid F \vee F \mid \neg F$$
$$Atom ::= t_{str} = t_{str} \mid A_{int} \mid A_{ext} \mid A_{re}$$
$$A_{re} \quad ::= t_{str} \in RE$$
$$A_{int} ::= t_{int} = t_{int} \mid t_{int} < t_{int}$$
$$A_{ext} ::= str.contains(t_{str}, t_{str}) \mid str.prefixof(t_{str}, t_{str}) \mid str.suffixof(t_{str}, t_{str})$$
$$t_{int} \quad ::= m \mid v \mid str.len(t_{str}) \mid t_{int} + t_{int} \mid m \cdot t_{int} \mid str.indexof(t_{str}, t_{str}, t_{int}) \mid$$
$$str.to\_int(t_{str}) \text{ where } m \in Con_{int} \ \& \ v \in Var_{int}$$
$$t_{str} \quad ::= s \mid v \mid t_{str} \cdot t_{str} \mid str.from\_int(t_{int}) \mid str.replace(t_{str}, t_{str}, t_{str}) \mid$$
$$str.replace\_all(t_{str}, t_{str}, t_{str}) \mid str.at(t_{str}, t_{int}) \mid str.substr(t_{str}, t_{int}, t_{int})$$
$$\text{where } s \in Con_{str} \text{ and } v \in Var_{str}$$

Fig. 1. The syntax of the theory of strings T_S.

expression RE, constructed using accustomed regular operators. The satisfiability problem for the quantifier-free theory T_S involves determining whether there exists an assignment of some constant in Con_{str} to every string variable in Var_{str} and some constant in Con_{int} to every integer variable in Var_{int}, such that the formula evaluates to true under the semantics given in the SMT-LIB standard. If such an assignment exists, the formula is satisfiable, and if not, it is unsatisfiable.

For more information on the syntax and semantics of the theory of strings, we recommend referring to the SMT-LIB standard for the theory of strings [44].

Isabelle. Isabelle [37, 45, 46] is a generic proof assistant. Proof assistants are computer programs in which users can define objects from mathematics, logic and computer science, and prove lemmas and theorems about them. The proof assistant checks that the proofs are correct, and it can also perform parts of (or whole) proofs automatically. The value of this is that users get a very strong guarantee that their proofs are correct. Isabelle achieves this by having a small kernel that implements a relatively simple logical derivation system. That Isabelle is generic means that it supports several logics, the most prominent being Isabelle/HOL which is Isabelle's implementation of higher-order logic (HOL). This is also the logic that we are using in the present paper. Popularly speaking, Isabelle/HOL combines typed functional programming and logic.

3 Formalising the SMT Theory of Strings

The SMT-LIB theory of strings allows reasoning over finite sequences of characters and regular languages. The signature of the theory contains three sorts String, RegLan, and Int and consists of the various function symbols shown in Fig. 1, such as str. substr, str. at. The SMT-LIB standard not only specifies syntax but also provides corresponding semantics. That is, the symbols are not intended to be interpreted arbitrarily but rather in a standard model in which the domains of the sorts are fixed and the interpretation of the function symbols is predefined. In particular, the domain of String is fixed to the set of all finite words in UC^*, where UC is the alphabet of 196607 Unicode characters, the domain of `RegLan` is the set of all regular languages over String, i.e., the set 2^{UC^*}, and `Int` refers to the standard set of integers. The intended semantics of the function symbols within the respective domains are expressed semi-formally, in natural language. For example, the standard defines the semantics of the function as follows.

> If $0 \leq m < |w|$ and $0 < n$ then $str.\,substr(w, m, n)$ is the unique word w_2 such that $w = w_1 w_2 w_3$, $|w_3| = m$, and $|w_2| = min(n, |w| - m)$ for some words w_1 and w_3. Otherwise $str.\,substr(w, m, n) = \varepsilon$.

We formalise the standard model in Isabelle/HOL. This formalisation comprises an interpretation of all symbols as functions in Isabelle, a translation of the semantics of the standard model into higher-order logic sentences, and proofs that our interpretation satisfies them.

For example, we implement a function `substr::"uc_string ⇒ int ⇒int ⇒ uc_string"`, where `uc_string` denotes the type of Unicode strings, and `int` is the type of integers, and prove the following two lemmas:

```
assumes "0≤m ∧ m<|w| ∧ 0<n" shows "∃!v. substr w m n = v
    ∧ ∃x y. w=x·v·y ∧ |x|=m ∧ |v|=min n (|w|-m)"
assumes "0>m ∨ m≥|w| ∨ 0≥n" shows "substr w m n = ε"
```

These lemmas formalise the meaning of $str.\,substr$ as detailed above. Our formalisation[1] currently encompasses all function symbols except for str.replace_all, str.replace_re, and str.replace_re_all.

Formalisation. We implement SMT-LIB functions based on a formalisation of the notions of words and regular expressions in Isabelle/HOL. To represent words over an arbitrary alphabet, we introduce the type `'a word` which is a synonym for lists of arbitrary type, i.e., `'a word ≡ 'a list`. The type `'a word` allows instantiation with arbitrary types as the underlying alphabet. The SMT-LIB standard specifies the alphabet to be Unicode characters. Therefore, we introduce a new type `UC` defined as the subset $\{0, \ldots, 196,607\}$ of integers, such that each term of type `UC` corresponds to a unique code point in the Unicode alphabet. By using `UC` as the type of the alphabet, the resulting type `UC word` is inhabited by all words over the Unicode alphabet and is thus a formalisation of the String sort defined by the SMT-LIB standard.

We implement the SMT-LIB string functions in terms of `UC word` and show that the implementation satisfies the properties of the standard model, as exemplified above. In doing so, we rely on functions on lists, as well as associated lemmas, that are already present in the Isabelle core libraries. For instance, the $str.\!+\!+$ function directly corresponds to the Isabelle list `append` function. More complex SMT-LIB functions additionally require reasoning about the factors of a word, which we handle by implementing a function that projects a word w onto the factor $w[i; j]$ between two indices $i, j \in \mathbb{N}$ with $0 \leq i \leq j$. We implement SMT-LIB $str.\,substr$ and $str.\,at$ in terms of this function. Moreover, the projection to factors allows searching and replacing occurrences and factors with other

---

[1] Available at https://github.com/formalsmt/isabelle_smt.

words, as required by the SMT-LIB functions str. indexof, str. replace, respectively. However, while Isabelle uses natural numbers to represent the lengths of lists, the SMT-LIB standard defines the length of a string, including the indices of substrings, as integers. Therefore, we require additional conversions between integers and natural numbers and must handle edge cases. For example, in the str. substr implementation, we first check whether the given indices are valid, i.e., whether they are within the bounds of the string, then convert them to natural numbers, and finally project the string onto the corresponding factor:

```
str_substr w m n = if n ≥ 0 ∧ 0 ≤ m ∧ ((nat m) ≤ |w|-1)
    then w[(nat m);(nat (m+n))] else ε
```

To formalise the regular membership predicate, i.e., str.in_re, which evaluates to true whenever a given string term is a member of a given regular language, we first introduce an algebraic data type `'a regex` characterizing regular expressions. The type contains a constructor for each regular operation defined by the SMT-LIB standard, including `'a word` as one of the base cases. Likewise to the type `'a word`, we instantiate `'a regex` with the type UC to obtain the type UC regex, which is inhabited by all regular expressions over the Unicode alphabet. To establish a connection between regular expressions and regular languages as defined by the standard, we additionally define the function `lang::"'a regex ⇒ 'a word set`" that maps a regular expression to its language using accustomed semantics. Hence, for any UC regex term r, the set lang r is a (regular) subset of the set of all UC word terms, which means that UC regex formalises the sort RegLan. We prove that the regular expression type, equipped with the lang function, satisfies all properties that the SMT-LIB theory of strings requires. For example, we show that regular concatenation, re_concat, expresses the exact language specified by the standard, by proving the following lemma:

```
lang (re_concat r e) = { x·y | x y. x ∈ lang r ∧ y ∈ lang e }
```

Finally, we implement the str.in_re predicate in terms of Brzozowski derivatives [14]. We follow the approach outlined in [28], but adapt it to account for the full set of regular operations defined by the SMT-LIB theory of strings. That is, we define a function `deriv::"'a regex ⇒ 'a ⇒ 'a regex` that computes the derivative of a regular expression w.r.t. a single character, and its extensions to words `derivw::"'a regex ⇒ 'a word ⇒ 'a regex`. We then prove that for a term w of type `'a word` and a term r of type `'a regex`, w is contained in the set lang r if and only if the derivative derivw r w contains the empty word.

Using Brzozowski derivatives, testing regular membership amounts to executing a finite number of deterministic derivations steps. This approach is preferable to testing whether a words is contained in a (possibly infinite) set using the lang function, as Isabelle can perform the finitely many derivation steps automatically. This is especially important for the automated model verification as described in Sect. 4.

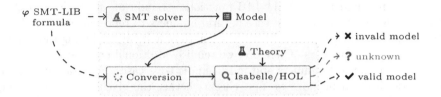

Fig. 2. SMT model verification process overview.

Inconsistencies. During our formalisation, we discovered several inconsistencies in the standard. Foremost, we found that the function str. indexof is not well-defined for all inputs. The standard requires that

> if str. contains(w, w_2) = true and $i \geq 0$ then str. indexof(w, w_2, i) is the smallest n such that $w = w_1 w_2 w_3$ for some words w_1, w_3 with $i \leq n = |w_1|$. Otherwise, str. indexof(w, w_2, i) is -1.

However, if either $i \geq |w|$ or str. contains$(\text{substr}(w, i, |w|), w_2)$ = false, then such an n cannot exist. For instance, str. indexof("ab", ε, 3) is a counterexample concerning the first case. We have $3 \geq 1$ and str. contains("ab", ε) by the definition of str. contains, but there are no words w_1, w_3 with "ab" $= w_1 \cdot \varepsilon \cdot w_3 = w_1 \cdot w_2$ and $3 \leq |w_1|$. For the second case, consider str. indexof("ab", "a", 1). Again str. contains("ab", "a") and $1 \geq 0$, but no words $w_1\ w_2$ with "ab" $= w_1 \cdot$ "a" $\cdot w_2$ and $1 \leq |w_1|$ exist. In order to establish well-defined semantics for str. contains, we suggest modifying the premises such that str. contains$(\text{substr}(w, i, |w|), w_2)$ = true instead of str. contains(w, w_2) = true and additionally ensuring that $i \leq |w|$ is satisfied.

Besides that, we found a minor inconsistency in the signatures of str.replace_re and str.replace_re_all. The standard defines them first as functions of type *String* → *RegLan* → *String* → *String* but later as functions of type *String* → *String* → *String* → *String*. The former was clearly intended.

4   Model Verification Using Isabelle

We present SMTmv[2], an automated tool for SMT model verification that leverages our formalisation to check the accuracy of models generated by SMT solvers. The verification process involves a sequence of steps, which are summarized in Fig. 2. For a satisfiable formula φ, an SMT solver is able to produce a model M, i.e., a variable assignment that satisfies the formula. SMTmv takes this potential model and converts it into Isabelle/HOL by mapping SMT-LIB functions to corresponding counterparts in the presented formalisation, and logical connectives to equivalent Isabelle primitives. The result is a shallow embedding φ_I in Isabelle that is equivalent to φ within the standard model of the theory

---

[2] Available at https://github.com/formalsmt/SMTmv.

Table 1. Results using SMTMV to replicate and verify soundness issues highlighted in [8]. Here, "t.o." stands for "timeout" (60 s) and "$\varnothing$-time" measures the average time it took SMTMV to verify a model, not including timeouts.

Solver	Tested	Solver Result			SMTMV					
		sat	unsat	t.o.	**invalid**	valid	unknown	error	t.o.	$\varnothing$-time
Z3TRAU	5325	2940	2126	259	1846	0	1	0	1093	33.36 s
OSTRICH	28	21	7	0	20	0	0	1	0	4.88 s
Z3STR3	13	0	13	0	0	0	0	0	0	–

of strings. Afterwards, SMTMV represents the assignment M equivalently as a conjunction of equalities M_I. Thus, within the SMT-LIB theory of strings, M is a model to φ if and only if $M_I \models \varphi_I$. SMTMV expresses $M_I \models \varphi_I$ as a lemma in Isabelle/HOL and queries the system to search for a proof. If Isabelle finds a proof, the assignment produced by the SMT solver is provably a model to the formula, and consequently, SMTMV returns *valid*. If Isabelle/HOL instead finds that the model doesn't satisfy the formula, it returns *invalid*. In cases where Isabelle/HOL can neither find a proof nor a counterexample, SMTMV returns *unknown*.

Analysis. To showcase the effectiveness of SMTMV, we utilise it to verify the soundness problems highlighted in [8]. The accompanying artifact offers a database of SMT-LIB instances and their reported outcomes. We re-ran the – at the time – unsound solvers Z3TRAU [1], OSTRICH [15], and Z3STR3 [7] to replicate the soundness issues and used SMTMV to verify the produced models. Note that we intentionally used outdated solver versions on which the errors were reported to demonstrate the effectiveness of SMTMV and that all solvers might have since been fixed. All experiments were run with a timeout of 60 s per instance for each, the solver and SMTMV. Our findings are summarised in Table 1. According to Berzish et al., Z3TRAU had 5,325 soundness issues, out of which SMTMV was able to identify 1846 provably invalid models. On 1,093 instances SMTMV timed out. However, Z3TRAU persistently produced models with a total length of well over 40,000 characters, which Isabelle was unable to handle within the set time limit. OSTRICH supposedly had 28 soundness issues. SMTMV found that 20 of them are due to invalid models. In one case OSTRICH returned *sat* on a formula that contains a unary disjunction, which is not valid syntax according to the SMT-LIB standard. Isabelle rejected this formula and SMTMV reported *error*. For Z3str3 [7], the authors reported 13 soundness issues, none of which were due to an invalid model.

5 Conclusion and Further Work

We presented a formalisation of the SMT-LIB theory of strings in Isabelle/HOL. Through this formalisation, we have identified inconsistencies in the SMT-LIB theory of strings and proposed rectifications for them. Additionally, we have introduced a tool, named SMTMV, that automates the validation of SMT models and successfully identified invalid model production as the cause of known soundness issues in several solvers. We believe SMTMV will be valuable for both SMT solver developers and practitioners in identifying and rectifying soundness errors, e.g. integrated into the benchmarking tool ZALIGVINDER [31]. Our formalisation in Isabelle/HOL lays the groundwork for future research, such as extending the expressiveness to support other SMT-LIB theories beyond strings or providing a deep embedding of the SMT-LIB logic into Isabelle.

References

1. Abdulla, P.A., et al.: TRAU: SMT solver for string constraints. In: 2018 Formal Methods in Computer Aided Design (FMCAD), pp. 1–5. IEEE (2018)
2. Backes, J., et al.: Semantic-based automated reasoning for AWS access policies using SMT. In: 2018 Formal Methods in Computer Aided Design (FMCAD), pp. 1–9 (2018). https://doi.org/10.23919/FMCAD.2018.8602994
3. Barbosa, H., et al.: cvc5: a versatile and industrial-strength SMT solver. In: TACAS 2022, Part I. LNCS, vol. 13243, pp. 415–442. Springer, Cham (2022). https://doi.org/10.1007/978-3-030-99524-9_24
4. Barbosa, H., et al.: Flexible proof production in an industrial-strength SMT solver. In: Blanchette, J., Kovács, L., Pattinson, D. (eds.) IJCAR 2022. LNCS, vol. 13385, pp. 15–35. Springer, Cham (2022). https://doi.org/10.1007/978-3-031-10769-6_3
5. Barrett, C., Fontaine, P., Tinelli, C.: The SMT-LIB Standard: Version 2.6. Technical report, Department of Computer Science, The University of Iowa (2017). http://www.SMT-LIB.org
6. Barrett, C., Tinelli, C.: Satisfiability Modulo Theories. Springer, Cham (2018). https://doi.org/10.1007/978-3-319-10575-8_11
7. Berzish, M., Ganesh, V., Zheng, Y.: Z3str3: a string solver with theory-aware heuristics. In: 2017 Formal Methods in Computer Aided Design (FMCAD), pp. 55–59. IEEE (2017)
8. Berzish, M., et al.: An SMT solver for regular expressions and linear arithmetic over string length. In: Silva, A., Leino, K.R.M. (eds.) CAV 2021. LNCS, vol. 12760, pp. 289–312. Springer, Cham (2021). https://doi.org/10.1007/978-3-030-81688-9_14
9. Bjørner, N., Tillmann, N., Voronkov, A.: Path feasibility analysis for string-manipulating programs. In: Kowalewski, S., Philippou, A. (eds.) TACAS 2009. LNCS, vol. 5505, pp. 307–321. Springer, Heidelberg (2009). https://doi.org/10.1007/978-3-642-00768-2_27
10. Blanchette, J.C., Fleury, M., Lammich, P., Weidenbach, C.: A verified SAT solver framework with learn, forget, restart, and incrementality. J. Autom. Reason. **61**(1-4), 333–365 (2018). https://doi.org/10.1007/s10817-018-9455-7
11. Blanchette, J.C., Fleury, M., Weidenbach, C.: A verified SAT solver framework with learn, forget, restart, and incrementality. In: Olivetti, N., Tiwari, A. (eds.) IJCAR 2016. LNCS (LNAI), vol. 9706, pp. 25–44. Springer, Cham (2016). https://doi.org/10.1007/978-3-319-40229-1_4
12. Blanchette, J.C., Fleury, M., Weidenbach, C.: A verified SAT solver framework with learn, forget, restart, and incrementality. In: Sierra, C. (ed.) Proceedings of the Twenty-Sixth International Joint Conference on Artificial Intelligence, IJCAI 2017, Melbourne, Australia, 19–25 August 2017, pp. 4786–4790. ijcai.org (2017). https://doi.org/10.24963/ijcai.2017/667

13. Böhme, S., Weber, T.: Fast LCF-style proof reconstruction for Z3. In: Kaufmann, M., Paulson, L.C. (eds.) ITP 2010. LNCS, vol. 6172, pp. 179–194. Springer, Heidelberg (2010). https://doi.org/10.1007/978-3-642-14052-5_14

14. Brzozowski, J.A.: Derivatives of regular expressions. J. ACM (JACM) **11**(4), 481–494 (1964)

15. Chen, T., Hague, M., Lin, A.W., Rümmer, P., Wu, Z.: Decision procedures for path feasibility of string-manipulating programs with complex operations. In: Proceedings of the ACM on Programming Languages 3(POPL), pp. 1–30 (2019)

16. Day, J.D., Ehlers, T., Kulczynski, M., Manea, F., Nowotka, D., Poulsen, D.B.: On solving word equations using SAT. In: Filiot, E., Jungers, R., Potapov, I. (eds.) RP 2019. LNCS, vol. 11674, pp. 93–106. Springer, Cham (2019). https://doi.org/10.1007/978-3-030-30806-3_8

17. de Moura, L., Bjørner, N.: Z3: an efficient SMT solver. In: Ramakrishnan, C.R., Rehof, J. (eds.) TACAS 2008. LNCS, vol. 4963, pp. 337–340. Springer, Heidelberg (2008). https://doi.org/10.1007/978-3-540-78800-3_24

18. Eldib, H., Wang, C., Schaumont, P.: Formal verification of software countermeasures against side-channel attacks. ACM Trans. Softw. Eng. Methodol. (TOSEM) **24**(2), 1–24 (2014)

19. Fleury, M.: Optimizing a verified SAT solver. In: Badger, J.M., Rozier, K.Y. (eds.) NFM 2019. LNCS, vol. 11460, pp. 148–165. Springer, Cham (2019). https://doi.org/10.1007/978-3-030-20652-9_10

20. Fleury, M.: Formalization of logical calculi in Isabelle/HOL. Ph.D. thesis, Saarland University, Saarbrücken, Germany (2020). https://tel.archives-ouvertes.fr/tel-02963301

21. Fleury, M., Blanchette, J.C., Lammich, P.: A verified SAT solver with watched literals using imperative HOL. In: Andronick, J., Felty, A.P. (eds.) Proceedings of the 7th ACM SIGPLAN International Conference on Certified Programs and Proofs, CPP 2018, Los Angeles, CA, USA, 8–9 January 2018, pp. 158–171. ACM (2018). https://doi.org/10.1145/3167080

22. Fleury, M., Schurr, H.: Reconstructing veriT proofs in Isabelle/HOL. In: Reis, G., Barbosa, H. (eds.) Proceedings Sixth Workshop on Proof eXchange for Theorem Proving, PxTP 2019, Natal, Brazil, 26 August 2019. EPTCS, vol. 301, pp. 36–50 (2019). https://doi.org/10.4204/EPTCS.301.6

23. Fleury, M., Weidenbach, C.: A verified SAT solver framework including optimization and partial valuations. In: Albert, E., Kovács, L. (eds.) LPAR 2020: 23rd International Conference on Logic for Programming, Artificial Intelligence and Reasoning, Alicante, Spain, 22–27 May 2020. EPiC Series in Computing, vol. 73, pp. 212–229. EasyChair (2020). https://doi.org/10.29007/96wb

24. Grimm, T., Lettnin, D., Hübner, M.: A survey on formal verification techniques for safety-critical systems-on-chip. Electronics **7**(6), 81 (2018)

25. Hojjat, H., Rümmer, P., Shamakhi, A.: On strings in software model checking. In: Lin, A.W. (ed.) APLAS 2019. LNCS, vol. 11893, pp. 19–30. Springer, Cham (2019). https://doi.org/10.1007/978-3-030-34175-6_2

26. Holub, V., Starosta, V.: Formalization of basic combinatorics on words. In: Cohen, L., Kaliszyk, C. (eds.) 12th International Conference on Interactive Theorem Proving (ITP 2021). Leibniz International Proceedings in Informatics (LIPIcs), vol. 193, pp. 22:1–22:17. Schloss Dagstuhl - Leibniz-Zentrum für Informatik, Dagstuhl, Germany (2021). https://doi.org/10.4230/LIPIcs.ITP.2021.22. https://drops.dagstuhl.de/opus/volltexte/2021/13917

27. Kan, S., Lin, A.W., Rümmer, P., Schrader, M.: CertiStr: a certified string solver. In: Proceedings of the 11th ACM SIGPLAN International Conference on Certified Programs and Proofs, pp. 210–224 (2022)

28. Krauss, A., Nipkow, T.: Regular sets and expressions. Archive of Formal Proofs, May 2010. https://isa-afp.org/entries/Regular-Sets.html, Formal proof development

29. Krauss, A., Nipkow, T.: Proof pearl: regular expression equivalence and relation algebra. J. Autom. Reason. 49(1), 95–106 (2011). https://doi.org/10.1007/s10817-011-9223-4

30. Kulczynski, M., Lotz, K., Nowotka, D., Poulsen, D.B.: Solving string theories involving regular membership predicates using SAT. In: Legunsen, O., Rosu, G. (eds.) Model Checking Software. LNCS, vol. 13255, pp. 134–151. Springer, Cham (2022). https://doi.org/10.1007/978-3-031-15077-7_8

31. Kulczynski, M., Manea, F., Nowotka, D., Poulsen, D.B.: The power of string solving: simplicity of comparison. In: Proceedings of the IEEE/ACM 1st International Conference on Automation of Software Test, pp. 85–88 (2020)

32. Lescuyer, S.: Formalizing and implementing a reflexive tactic for automated deduction in Coq. (Formalisation et developpement d'une tactique reflexive pour la demonstration automatique en coq). Ph.D. thesis, University of Paris-Sud, Orsay, France (2011). https://tel.archives-ouvertes.fr/tel-00713668

33. Maric, F.: Formal verification of a modern SAT solver by shallow embedding into Isabelle/HOL. Theor. Comput. Sci. 411(50), 4333–4356 (2010). https://doi.org/10.1016/j.tcs.2010.09.014

34. Maric, F., Janicic, P.: Formalization of abstract state transition systems for SAT. Log. Methods Comput. Sci. 7(3) (2011). https://doi.org/10.2168/LMCS-7(3:19)2011

35. Marić, F.: Formal verification of modern sat solvers. Archive of Formal Proofs, July 2008. https://isa-afp.org/entries/SATSolverVerification.html, Formal proof development

36. Mora, F., Berzish, M., Kulczynski, M., Nowotka, D., Ganesh, V.: Z3str4: a multi-armed string solver. In: Huisman, M., Păsăreanu, C., Zhan, N. (eds.) FM 2021. LNCS, vol. 13047, pp. 389–406. Springer, Cham (2021). https://doi.org/10.1007/978-3-030-90870-6_21

37. Nipkow, T., Paulson, L.C., Wenzel, M.: Isabelle/HOL — A Proof Assistant for Higher-Order Logic. LNCS, vol. 2283. Springer, Heidelberg (2002). https://doi.org/10.1007/3-540-45949-9

38. Oe, D., Stump, A., Oliver, C., Clancy, K.: versat: a verified modern SAT solver. In: Kuncak, V., Rybalchenko, A. (eds.) VMCAI 2012. LNCS, vol. 7148, pp. 363–378. Springer, Heidelberg (2012). https://doi.org/10.1007/978-3-642-27940-9_24

39. Redelinghuys, G., Visser, W., Geldenhuys, J.: Symbolic execution of programs with strings. In: Proceedings of the South African Institute for Computer Scientists and Information Technologists Conference, SAICSIT 2012, pp. 139–148 (2012)

40. Rungta, N.: A billion SMT queries a day (invited paper). In: Shoham, S., Vizel, Y. (eds.) Computer Aided Verification. LNCS, vol. 13371, pp. 3–18. Springer, Cham (2022). https://doi.org/10.1007/978-3-031-13185-1_1

41. Saxena, P., Akhawe, D., Hanna, S., Mao, F., McCamant, S., Song, D.: A symbolic execution framework for JavaScript. In: 2010 IEEE Symposium on Security and Privacy, pp. 513–528. IEEE (2010)

42. Schurr, H., Fleury, M., Barbosa, H., Fontaine, P.: Alethe: towards a generic SMT proof format (extended abstract). In: Keller, C., Fleury, M. (eds.) Proceedings Seventh Workshop on Proof eXchange for Theorem Proving, PxTP 2021, Pittsburg, PA, USA, 11 July 2021. EPTCS, vol. 336, pp. 49–54 (2021). https://doi.org/10.4204/EPTCS.336.6

43. Shankar, N., Vaucher, M.: The mechanical verification of a DPLL-based satisfiability solver. In: Haeusler, E.H., del Cerro, L.F. (eds.) Proceedings of the Fifth Logical and Semantic Frameworks, with Applications Workshop, LSFA 2010, Natal, Brazil, 31 August 2010. Electronic Notes in Theoretical Computer Science, vol. 269, pp. 3–17. Elsevier (2010). https://doi.org/10.1016/j.entcs.2011.03.002

44. Tinelli, C., Barrett, C., Fontaine, P.: SMT: theory of strings. http://smtlib.cs.uiowa.edu/theories-UnicodeStrings.shtml. Accessed 03 Mar 2022

45. Wenzel, M., Paulson, L.C., Nipkow, T.: The Isabelle framework. In: Mohamed, O.A., Muñoz, C., Tahar, S. (eds.) TPHOLs 2008. LNCS, vol. 5170, pp. 33–38. Springer, Heidelberg (2008). https://doi.org/10.1007/978-3-540-71067-7_7

46. Wenzel, M., et al.: The Isabelle/Isar reference manual (2004)

47. Yu, F., Alkhalaf, M., Bultan, T., Ibarra, O.H.: Automata-based symbolic string analysis for vulnerability detection. Form. Methods Syst. Des. **44**(1), 44–70 (2013). https://doi.org/10.1007/s10703-013-0189-1

48. Zbrzezny, A.M., Szymoniak, S., Kurkowski, M.: Practical approach in verification of security systems using satisfiability modulo theories. Log. J. IGPL **30**(2), 289–300 (2022)

Weighted Bottom-Up and Top-Down Tree Transformations Are Incomparable

Andreas Maletti[(✉)] and Andreea-Teodora Nász

Faculty of Mathematics and Computer Science, Universität Leipzig,
PO box 100 920, 04009 Leipzig, Germany
{maletti,nasz}@informatik.uni-leipzig.de

Abstract. Weighted bottom-up and top-down tree transformations over commutative semirings are investigated. It is demonstrated that if the range of a weighted bottom-up tree transformation is well-defined (i.e., for every output tree there are only finitely many input trees that can transform to the output tree with nonzero weight, so that all involved sums remain finite), then the range is hom-regular, which means that it is the image of a regular weighted tree language under a tree homomorphism. Additionally, the strictness of the first level of the weighted bottom-up and top-down tree transformation hierarchy is proved, which was open for any ring.

1 Introduction

Tree transducers [1,3] are a formal model for relations on trees. They extend the classical tree automata [2,16] with the ability to generate output trees. In several application areas like natural language processing [12] a purely qualitative evaluation is insufficient, and weighted versions of tree automata and tree transducers have been proposed and investigated (see [8] for an excellent survey).

In this contribution we consider weighted bottom-up and top-down tree transducers [5,13]. The weights of the transducers are taken from a commutative semiring [10,11]. It is an open question whether the composition hierarchy [6,14] of their computed weighted tree transformations is strict in all commutative semirings; the unweighted version of this result was proved in [4] and later extended to all non-rings [14] (see Fig. 1). The existing extension simply faithfully lifts the unweighted result to the weighted setting. However, this approach does not work in a ring because it does not permit a homomorphism into the BOOLEAN semiring (i.e., the unweighted case).

We prove the necessary strictness and incomparability results for the first level of the hierarchy in any commutative semiring. Essentially this requires to prove that there exists a weighted tree transformation that can be computed by a weighted top-down tree transducer (TOP), but not by any weighted bottom-up tree transducer (BOT), and a weighted tree transformation that can be computed by a BOT, but not by any TOP. While the utilized transformations that

B. Nagy (Ed.): CIAA 2023, LNCS 14151, pp. 218–229, 2023.
https://doi.org/10.1007/978-3-031-40247-0_16

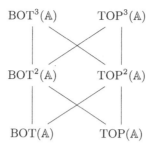

Fig. 1. First levels of the BOT and TOP hierarchy for a commutative semiring $\mathbb{A}$.

separate these classes are strongly inspired by the classical unweighted examples, we do not directly lift the existing unweighted results, but rather provide explicit proofs that work in every commutative semiring.

Since explicit examples for the strictness of the hierarchy are generally unknown even in the unweighted case, the authors doubt that this approach will successfully establish the strictness of the full hierarchy. Instead the seminal paper [4] investigates ranges of the transformations computed by BOTs and TOPs and separates the levels of the hierarchy purely based on the achievable range tree languages. In the same spirit we provide another interesting result. Whenever the range of a weighted tree transformation computed by a (single) BOT is well-defined (i.e., for every output tree there are only finitely many input trees that can transform to the output tree with nonzero weight, so that all involved sums remain finite), it is actually hom-regular (i.e., the homomorphic image of a regular weighted tree language), which is a recently studied class [15]. As a side effect this property of the range can be utilized to apply several recently obtained results (like a pumping lemma [15, Lemma 4] and decidability results [15, Theorem 6] for support emptiness and finiteness) to the range of a weighted tree transformation computed by a BOT.

2 Preliminaries

Let $\mathbb{N}$ denote the set $\{0, 1, 2, \dots\}$ of all nonnegative integers, and for every $k \in \mathbb{N}$ let $[k] = \{i \in \mathbb{N} \mid 1 \le i \le k\}$. For every set A we denote its cardinality by $|A|$ and the set of all strings over A (i.e., finite sequences over A) by A^*. We write ε for the empty string and $|w|$ for the length of a string $w \in A^*$.

Let A, B, and C be sets. A *relation* is any subset $R \subseteq A \times B$. Given relations $R \subseteq A \times B$ and $S \subseteq B \times C$, their *composition* $R \,;\, S$ is given by

$$R \,;\, S = \big\{ (a, c) \in A \times C \mid \exists b \in B \colon (a, b) \in R, (b, c) \in S \big\} \ .$$

We let $\mathrm{id}_A = \{(a, a) \mid a \in A\}$ be the *identity relation* on A, which we also denote by simply 'id' if A is clear from the context. The *inverse of* R is the relation $R^{-1} = \{(b, a) \mid (a, b) \in R\}$. Let $E \subseteq A \times A$ be a relation. Its *transitive* *closure* E^* is $E^* = \bigcup_{n \in \mathbb{N}} E^n$, where $E^0 = \mathrm{id}_A$ and $E^{n+1} = E^n \,; E$ for every $n \in \mathbb{N}$. The relation E is an *equivalence relation* if E is reflexive (i.e., $\mathrm{id}_A \subseteq E$), symmetric (i.e., $E^{-1} = E$), and transitive (i.e., $E^* \subseteq E$). The *equivalence class* of an element $a \in A$ under the equivalence relation E is $[a]_E = \{a' \in A \mid (a, a') \in E\}$. The *domain* and *range* of R are the sets $\mathrm{dom}_R = \{a \in A \mid \exists b \in B \colon (a, b) \in R\}$ and $\mathrm{ran}_R = \mathrm{dom}_{R^{-1}}$, respectively. A relation R is a *(partial) mapping* if $b = b'$ for all $(a, b), (a, b') \in R$. Similarly, R is *injective* if $a = a'$ for all $(a, b), (a', b) \in R$. For a mapping $f \colon A \to B$ and $S \subseteq B$ we denote the *inverse image of S* under f by $f^{-1}(S)$, which equates to $\mathrm{dom}_{f \cap (A \times S)}$. Additionally, we write $f^{-1}(b)$ instead of $f^{-1}(\{b\})$ for every $b \in B$.

A *ranked alphabet* is a pair (Σ, rk) that consists of a finite set Σ and a rank mapping $\mathrm{rk} \colon \Sigma \to \mathbb{N}$. For every $k \in \mathbb{N}$, we let $\Sigma_k = \mathrm{rk}^{-1}(k)$, and we sometimes write $\sigma^{(k)}$ to express that $\sigma \in \Sigma_k$. We often abbreviate (Σ, rk) to Σ leaving 'rk' implicit. For every set S we let

$$\Sigma(S) = \bigcup_{k \in \mathbb{N}} \{\sigma(s_1, \ldots, s_k) \mid \sigma \in \Sigma_k, s_1, \ldots, s_k \in S\} \ .$$

Let Z be a set disjoint with Σ. The set of *Σ" trees over Z*, denoted by $T_\Sigma(Z)$, is the smallest set T such that $Z \subseteq T$ and $\Sigma(T) \subseteq T$. We abbreviate $T_\Sigma(\emptyset)$ simply to T_Σ, and the tree $\alpha()$ to α for every $\alpha \in \Sigma_0$. Any subset $L \subseteq T_\Sigma$ is called a *tree language*.

Let Σ be a ranked alphabet, Z a set and $t \in T_\Sigma(Z)$. The set $\mathrm{pos}(t) \subseteq \mathbb{N}^*$ of *positions* of t is defined inductively by $\mathrm{pos}(t) = \{\varepsilon\}$ for every $t \in \Sigma_0 \cup Z$, and by $\mathrm{pos}(\sigma(t_1, \ldots, t_k)) = \{\varepsilon\} \cup \bigcup_{i \in [k]} \{iw \mid w \in \mathrm{pos}(t_i)\}$ for all $k \in \mathbb{N}$, $\sigma \in \Sigma_k$, and $t_1, \ldots, t_k \in T_\Sigma(Z)$. The *size* $\mathrm{size}(t)$ and *height* $\mathrm{ht}(t)$ of t are defined by $\mathrm{size}(t) = |\mathrm{pos}(t)|$ and $\mathrm{ht}(t) = \max_{w \in \mathrm{pos}(t)} |w|$, respectively. For $w \in \mathrm{pos}(t)$, the *label* $t(w)$ of t at w, the *subtree* $t|_w$ of t at w and the *substitution* $t[t']_w$ of t' into t at w are defined by $t(\varepsilon) = t|_\varepsilon = t$ and $t[t']_\varepsilon = t'$ for $t \in \Sigma_0 \cup Z$, and for $t = \sigma(t_1, \ldots, t_k)$ by $t(\varepsilon) = \sigma$, $t(iw') = t_i(w')$, $t|_\varepsilon = t$, $t|_{iw'} = t_i|_{w'}$, $t[t']_\varepsilon = t'$, and $t[t']_{iw'} = \sigma(t_1, \ldots, t_{i-1}, t_i[t']_{w'}, t_{i+1}, \ldots, t_k)$ for all $k \in \mathbb{N}$, $\sigma \in \Sigma_k$, $t_1, \ldots, t_k \in T_\Sigma(Z)$, $i \in [k]$, and $w' \in \mathrm{pos}(t_i)$. For every subset $S \subseteq \Sigma \cup Z$, we let $\mathrm{pos}_S(t) = \{w \in \mathrm{pos}(t) \mid t(w) \in S\}$, and we abbreviate $\mathrm{pos}_{\{s\}}(t)$ to $\mathrm{pos}_s(t)$ for every $s \in \Sigma \cup Z$. Let $X = \{x_1, x_2, \ldots\}$ be a fixed, countable set of formal variables. For each $k \in \mathbb{N}$ we let $X_k = \{x_i \mid i \in [k]\}$. For any $t \in T_\Sigma(X)$ we let $\mathrm{var}(t) = \{x \in X \mid \mathrm{pos}_x(t) \neq \emptyset\}$. Finally, for $t \in T_\Sigma(Z)$, a subset $V \subseteq Z$, and a mapping $\theta \colon V \to T_\Sigma(Z)$, the *substitution $t\theta$ applied to t* is defined by $v\theta = \theta(v)$ for $v \in V$, $z\theta = z$ for $z \in Z \setminus V$, and

$$\sigma(t_1, \ldots, t_k)\theta = \sigma(t_1\theta, \ldots, t_k\theta)$$

for all $k \in \mathbb{N}$, $\sigma \in \Sigma_k$, and $t_1, \ldots, t_k \in T_\Sigma(Z)$. If $V = \{v_1, \ldots, v_n\}$ is finite, then we write the mapping θ also as $[v_1 \leftarrow \theta(v_1), \ldots, v_n \leftarrow \theta(v_n)]$, or simply as $[\theta(x_1), \ldots, \theta(x_n)]$ if $V = X_n$. Let Σ be a ranked alphabet, $\gamma \in \Sigma_1$, and $n \in \mathbb{N}$. For every tree $s \in T_\Sigma$ we let $\gamma^0(s) = s$ and $\gamma^{n+1}(s) = \gamma^n(\gamma(s))$.

A *(commutative) semiring* [10,11] is an algebraic structure $(\mathbb{A}, +, \cdot, 0, 1)$ such that $(\mathbb{A}, +, 0)$ and $(\mathbb{A}, \cdot, 1)$ are commutative monoids, $\cdot$ distributes over $+$, and

$0 \cdot a = 0$ for all $a \in \mathbb{A}$. Typical examples include the BOOLEAN semiring $\mathbb{B} = (\{0,1\}, \max, \min, 0, 1)$, the semiring $\mathbb{N} = (\mathbb{N}, +, \cdot, 0, 1)$ of nonnegative integers, and the tropical semiring $\mathbb{T} = (\mathbb{N} \cup \{\infty\}, \min, +, \infty, 0)$. When there is no risk of confusion, we refer to a semiring $(\mathbb{A}, +, \cdot, 0, 1)$ simply by its carrier set $\mathbb{A}$, and for the rest of the contribution, let $\mathbb{A}$ be an arbitrary commutative semiring with $0 \neq 1$.

Let S be a set. Any mapping $\varphi \colon S \to \mathbb{A}$ is called a *weighted set* (over $\mathbb{A}$), and its *support* is $\mathrm{supp}(\varphi) = \{s \in S \mid \varphi(s) \neq 0\}$. For every subset $T \subseteq S$ we let $1_T \colon S \to \mathbb{A}$ be the weighted set given by $1_T(s) = 1$ if $s \in T$ and $1_T(s) = 0$ otherwise. If $S = T_\Sigma(Z)$ for some ranked alphabet Σ and set Z, then we also call φ a *weighted tree language*. Similarly, if $S = T_\Sigma \times T_\Delta$ for ranked alphabets Σ and Δ, then we call φ a *weighted tree transformation*. Let $\tau \colon T_\Sigma \times T_\Delta \to \mathbb{A}$ be a weighted tree transformation. Its *inverse* $\tau^{-1} \colon T_\Delta \times T_\Sigma \to \mathbb{A}$ is the weighted tree transformation given by $\tau^{-1}(t, s) = \tau(s, t)$ for every $s \in T_\Sigma$ and $t \in T_\Delta$. For a set $S \subseteq T_\Delta$, the transformation τ is *output-finitary on* S if the set $\{t \in S \mid (s, t) \in \mathrm{supp}(\tau)\}$ of output trees in S is finite for every input tree $s \in T_\Sigma$, and it is *input-finitary on* $S' \subseteq T_\Sigma$ if τ^{-1} is output-finitary on S'. For $S = T_\Delta$ (respectively, $S' = T_\Sigma$) we simply say that τ is *output-finitary* (respectively, *input-finitary*) if τ is output-finitary on S (respectively, input-finitary on S'). If τ is output-finitary, then the *domain of* τ is the weighted tree language $\mathrm{dom}_\tau \colon T_\Sigma \to \mathbb{A}$ given by $\mathrm{dom}_\tau(s) = \sum_{t \in T_\Delta} \tau(s, t)$ for every $s \in T_\Sigma$. As before, if τ is input-finitary, then we let $\mathrm{ran}_\tau = \mathrm{dom}_{\tau^{-1}}$ and call it the *range of* τ. Let τ be output-finitary, and let $\vartheta \colon T_\Delta \times T_\Gamma \to \mathbb{A}$ be another weighted tree transformation. We define their *composition* $(\tau \, ; \vartheta) \colon T_\Sigma \times T_\Gamma \to \mathbb{A}$ to be the weighted tree transformation such that

$$(\tau \, ; \vartheta)(s, u) = \sum_{t \in T_\Delta} \tau(s, t) \cdot \vartheta(t, u)$$

for every $s \in T_\Sigma$ and $u \in T_\Gamma$. If both transformations are output-finitary, then their composition is also output-finitary. Moreover, since composition ; is a form of matrix multiplication, it is associative.

Let Σ and Δ be ranked alphabets. A *(tree) homomorphism* is a mapping $h' \colon \Sigma \to T_\Delta(X)$ such that $h'(\sigma) \in T_\Delta(X_k)$ for every $k \in \mathbb{N}$ and symbol $\sigma \in \Sigma_k$. This mapping h' then extends to a mapping $h \colon T_\Sigma \to T_\Delta$ given for every $k \in \mathbb{N}$, $\sigma \in \Sigma_k$, and $t_1, \ldots, t_k \in T_\Sigma$ by $h(\sigma(t_1, \ldots, t_k)) = h'(\sigma)[h(t_1), \ldots, h(t_k)]$. In the following, we will not distinguish between h' and h and call both homomorphism.

Every homomorphism $h \colon T_\Sigma \to T_\Delta$ induces an output-finitary weighted tree transformation $1_h \colon T_\Sigma \times T_\Delta \to \mathbb{A}$. Additionally, let $\varphi \colon T_\Sigma \to \mathbb{A}$ be a weighted tree language. It induces the weighted tree transformation $\mathrm{id}_\varphi \colon T_\Sigma \times T_\Sigma \to \mathbb{A}$ given for every $s \in T_\Sigma$ and $t \in T_\Sigma$ by $\mathrm{id}_\varphi(s, t) = \varphi(s)$ if $s = t$ and $\mathrm{id}_\varphi(s, t) = 0$ otherwise. Clearly, id_φ is input- and output-finitary, so the composition $\tau = \mathrm{id}_\varphi \, ; 1_h$ is well-defined, and τ is input-finitary if and only if 1_h is input-finitary on $\mathrm{supp}(\varphi)$. For ease of notation, if 1_h is input-finitary on $\mathrm{supp}(\varphi)$, we let $h_\varphi = \mathrm{ran}_\tau$ (also denoted $h(\varphi)$ occasionally) and call it the *image of* φ *under* h. Finally, we say that h is *input-finitary* (on S) if 1_h is input-finitary (on S).

Next we recall the weighted tree automata with homomorphism-constraints of [15, Section 3], where they are called positive classic WTGc. Tree automata with homomorphism-constraints were first introduced in the unweighted case [9].

Definition 1. *A* weighted tree automaton with homomorphism-constraints *(for short: WTAh) is a tuple* $\mathcal{A} = (Q, \Sigma, F, R, \mathrm{wt})$ *such that* Q *is a finite set of states,* Σ *is a ranked alphabet,* $F \subseteq Q$ *is a set of final states,* $\mathrm{wt} \colon R \to \mathbb{A}$ *is a weight assignment, and* R *is a finite set of rules of the form* $\ell \xrightarrow{E} q$ *with* $\ell \in T_\Sigma(Q) \backslash Q, q \in Q$, *and* $E \subseteq \mathrm{pos}_Q(\ell) \times \mathrm{pos}_Q(\ell)$. *The automaton* $\mathcal{A}$ *is a* weighted tree automaton *(for short: WTA) if* $\ell \in \Sigma(Q)$ *and* $E = \emptyset$ *for every rule* $\ell \xrightarrow{E} q \in R$. ☐

We also depict a rule $\ell \xrightarrow{E} q \in R$ as $\ell \xrightarrow{E}_a q$, where $a = \mathrm{wt}(\ell \xrightarrow{E} q)$. The components of this rule are its *left-hand side* ℓ, its *target state* q, its set E of *equality constraints*, and its *weight* a. An equality constraint $(w, w') \in E$ is listed as $w = w'$, of which w and w' are called the *constrained positions*. If $E = \emptyset$, then we simply omit it and just write $\ell \to_a q$. The equivalence relation induced by E is $\equiv_E = (E \cup E^{-1})^*$, and the equivalence class $[w]_{\equiv_E}$ of a position $w \in \mathrm{pos}_Q(\ell)$ is also denoted by $[w]^E$.

In this contribution we are particularly interested in a specific subclass of WTAh, namely the *eq-restricted* WTAh of [15, Definition 5].

Definition 2. *The WTAh* $\mathcal{A} = (Q, \Sigma, F, R, \mathrm{wt})$ *is* eq-restricted *if it has a sink state* $\bot \in Q \setminus F$ *such that*

1. $\rho_\sigma = \sigma(\bot, \ldots, \bot) \to_1 \bot \in R$ *for every* $\sigma \in \Sigma$,
2. $\ell \in \Sigma(\{\bot\})$ *and* $E = \emptyset$ *for every* $\ell \xrightarrow{E} \bot \in R$ *(i.e., the rules of Item 1 are the only rules with target state* $\bot$*), and*
3. $|\{w' \in [w]^E \mid \ell(w') \neq \bot\}| \leq 1$ *for every* $\ell \xrightarrow{E} q \in R$ *and* $w \in \mathrm{pos}_Q(\ell)$ *(i.e., at most one position of every* $\equiv_E$*-equivalence class is labeled by a state other than* $\bot$*).* ☐

In other words, in a rule of an eq-restricted WTAh the constrained positions of the same equivalence class (via $\equiv_E$) are all labeled by $\bot$ except for potentially one position w. In addition, the sink state $\bot$ is able to process each input tree at unit cost 1. Thus all restrictions on this subtree that is shared across the constrained positions need to be implemented by the state labeling w. Similarly, the weight for processing this subtree is only charged in the processing via the state labeling w as all copies are processed via $\bot$ at unit cost. For an eq-restricted WTAh we denote its state set by $Q \, \dot\cup \, \{\bot\}$ to point out the sink-state, which we universally denote by $\bot$.

Let us finally recall the semantics of WTAh from [15, Section 3] in a slightly adjusted, albeit equivalent manner. Let $\mathcal{A} = (Q, \Sigma, F, R, \mathrm{wt})$ be a WTAh. A *run of* $\mathcal{A}$ *is a tree over the ranked alphabet* $\Sigma \cup R$ with $\mathrm{rk}(\ell \xrightarrow{E}_a q) = |\mathrm{pos}_Q(\ell)|$ and it is defined inductively. Let $t_1, \ldots, t_n \in T_\Sigma$, $q_1, \ldots, q_n \in Q$, and suppose that ϱ_i is a run of $\mathcal{A}$ for t_i to q_i with weight $\mathrm{wt}(\varrho_i)$ for each $i \in [n]$. Suppose that there is a rule of the form $\ell \xrightarrow{E}_a q \in R$ such that $\ell = \sigma(\ell_1, \ldots, \ell_m)$, $\mathrm{pos}_Q(\ell) = \{w_1, \ldots, w_n\}$ with $\ell(w_i) = q_i$ and $t_i = t_j$ for all $w_i = w_j \in E$. Then

$$\varrho = (\ell \xrightarrow{E}_a q)(\ell_1, \ldots, \ell_m)[\varrho_1]_{w_1} \cdots [\varrho_n]_{w_n}$$

is a run of $\mathcal{A}$ for $t = \sigma(\ell_1, \ldots, \ell_m)[t_1]_{w_1} \cdots [t_n]_{w_n}$ to q. If $q \in F$, then ϱ is *accepting*. Its *weight* $\mathrm{wt}(\varrho)$ is computed simply as $\mathrm{wt}(\varrho) = a \cdot \prod_{i \in [n]} \mathrm{wt}(\varrho_i)$. The weight $\mathrm{wt}^q(t)$ is the sum of all weights $\mathrm{wt}(\varrho)$ of runs of $\mathcal{A}$ for t to q. Finally, the weighted tree language recognized by $\mathcal{A}$ is the mapping $\|\mathcal{A}\| \colon T_\Sigma \to \mathbb{A}$ that is given by $\|\mathcal{A}\|(t) = \sum_{q \in F} \mathrm{wt}^q(t)$ for every $t \in T_\Sigma$.

If a weighted tree language φ is recognized by some WTA, then it is called *regular*, if it is recognized by some eq-restricted WTAh, then it is called *hom-regular*, and if it is recognized by some arbitrary WTAh, then it is called *constraint-regular*. This choice of naming already hints at the fact that eq-restricted WTAh are tailored to represent homomorphic images of regular weighted tree languages. For an illustration of this feature, we consider the following example.

Example 3. Let $\Sigma = \{\nu^{(1)}, \gamma^{(1)}, \alpha^{(0)}\}$ be a ranked alphabet and $\varphi \colon T_\Sigma \to \mathbb{N}$ be the regular weighted tree language with

$$\varphi(s) = \begin{cases} 2^n & \text{if } s = \nu(\gamma^n(\alpha)) \\ 0 & \text{otherwise} \end{cases}$$

for every $s \in T_\Sigma$. A simple WTA that recognizes this weighted tree language φ is given by $\mathcal{A} = (\{q, q_f\}, \Sigma, \{q_f\}, R, \mathrm{wt})$ with the following set of rules and weights.

$$R = \{\alpha \to_1 q, \ \gamma(q) \to_2 q, \ \nu(q) \to_1 q_f\}$$

Consider the ranked alphabet $\Delta = \{\sigma^{(2)}, \gamma^{(1)}, \alpha^{(0)}\}$ and the tree homomorphism h defined by $h(\alpha) = \alpha$, $h(\gamma) = \gamma(x_1)$, and $h(\nu) = \sigma(x_1, x_1)$. The image h_φ of φ via h is the weighted tree language $h_\varphi \colon T_\Delta \to \mathbb{N}$ given for every $t \in T_\Delta$ by

$$h_\varphi(t) = \begin{cases} 2^n & \text{if } t = \sigma(\gamma^n(\alpha), \gamma^n(\alpha)) \\ 0 & \text{otherwise.} \end{cases}$$

The natural eq-restricted WTAh that represents the weighted tree language h_φ is $\mathcal{A}' = (\{q, q_f, \bot\}, \Delta, \{q_f\}, R', \mathrm{wt}')$ with the following rules and weights.

$$R' = \{ \quad \alpha \to_1 q, \qquad \gamma(q) \to_2 q, \qquad \sigma(q, \bot) \xrightarrow{1=2}_1 q_f \}$$
$$\cup \{ \quad \alpha \to_1 \bot, \qquad \gamma(\bot) \to_1 \bot, \qquad \sigma(\bot, \bot) \to_1 \bot \}.$$

The new rules of $\mathcal{A}'$ are obtained from the old rules of $\mathcal{A}$ by applying the homomorphism to the left-hand sides. The duplicated subtree below σ targets the sink state $\bot$ instead of q to avoid distorting the weight by an additional factor 2^n. □

The following result was shown in [15] for nondeleting and nonerasing tree homomorphisms, but the presented generalization to input-finitary tree homomorphisms presented here is straightforward.

Lemma 4 ([15, **Theorem 5**]). *Let* $\varphi\colon T_\Sigma \to \mathbb{A}$ *be a regular weighted tree language and* $h\colon T_\Sigma \to T_\Delta$ *a tree homomorphism that is input-finitary on* $\mathrm{supp}(\varphi)$. *Then* h_φ *is hom-regular.*   □

3   Weighted Bottom-up Tree Transducers

Let us start by recalling weighted bottom-up tree transducers from [5, Definitions 3.4 and 3.6] in a notation closer to [3, Section 1].

Definition 5. *A* weighted bottom-up tree transducer *(for short: BOT) is a tuple* $\mathcal{B} = (Q, \Sigma, \Delta, F, R, \mathrm{wt})$ *with a finite set* Q *of states, ranked alphabets* Σ *and* Δ *of input and output symbols, respectively, a set* $F \subseteq Q$ *of final states, a finite set* $R \subseteq \bigcup_{k\in\mathbb{N}} \Sigma_k(Q) \times Q \times T_\Delta(X_k)$ *of rules, and a weight map* $\mathrm{wt}\colon R \to \mathbb{A}$.   □

We often depict a rule $\rho = \big(\sigma(q_1,\ldots,q_k), q, t\big) \in R$ as $\sigma(q_1,\ldots,q_k) \to_a q(t)$, where $a = \mathrm{wt}(\rho)$. The BOT $\mathcal{B}$ is a *finite-state relabeling* [5, Definition 3.9] if for every rule $\sigma(q_1,\ldots,q_k) \to_a q(t) \in R$ there is $\delta \in \Delta_k$ such that $t = \delta(x_1,\ldots,x_k)$. We abbreviate finite-state relabeling by QREL.

Next, we present the semantics of BOTs following the derivation semantics of [7, Section 5], which coincides with the original semantics of [5, Definitions 3.2 and 3.5]. In the following, let $\mathcal{B} = (Q, \Sigma, \Delta, F, R, \mathrm{wt})$ be a BOT. We let $\mathrm{SF} = T_\Delta(Q \times T_\Delta)$ be its sentential forms. Given two sentential forms $\xi, \zeta \in \mathrm{SF}$ and a rule $\rho = \sigma(q_1,\ldots,q_k) \to q(t) \in R$, we write $\xi \Rightarrow_\mathcal{B}^\rho \zeta$ if and only if $\xi|_w = \sigma(\langle q_1, t_1\rangle, \ldots, \langle q_k, t_k\rangle)$ and $\zeta = \xi\big[\langle q, t[t_1,\ldots,t_k]\rangle\big]_w$ for some output trees $t_1,\ldots,t_k \in T_\Delta$ and the lexicographically least position $w \in \mathrm{pos}(\xi)$ such that $\xi|_w \in \Sigma(Q \times T_\Delta)$. A *derivation* d from ξ to ζ is a (finite) sequence of rules $d = \rho_1 \cdots \rho_n \in R^*$ such that $\xi(\Rightarrow_\mathcal{B}^{\rho_1} ; \cdots ; \Rightarrow_\mathcal{B}^{\rho_n})\zeta$. The set of all derivations from ξ to ζ is denoted by $\mathrm{Der}_\mathcal{B}(\xi, \zeta)$ and is always finite. The weight $\mathrm{wt}(d)$ of the derivation d is $\mathrm{wt}(d) = \prod_{i=1}^n \mathrm{wt}(\rho_i)$. Now we are ready to define the semantics $\|\mathcal{B}\|\colon T_\Sigma \times T_\Delta \to \mathbb{A}$ of the BOT $\mathcal{B}$. It is defined for every input tree $s \in T_\Sigma$ and output tree $t \in T_\Delta$ by

$$\|\mathcal{B}\|(s,t) = \sum_{\substack{q\in F \\ d\in\mathrm{Der}_\mathcal{B}(s,\langle q,t\rangle)}} \mathrm{wt}(d) \ .$$

Thus for any $(s,t) \in \mathrm{supp}(\|\mathcal{B}\|)$ there exists $q \in F$ and $d \in \mathrm{Der}_\mathcal{B}(s, \langle q, t\rangle)$ such that $\mathrm{wt}(d) \neq 0$.

Let us illustrate weighted bottom-up tree transducers on an example inspired by [3, Property B2], which states that BOTs can check subtrees before deleting them. In our example, this will apply to the second direct subtree s_2 of the input $\sigma(s_1, s_2)$, which causes a different behavior if $s_2 = \alpha$. We first introduce the weighted tree transformation τ_1 followed by a weighted bottom-up tree transducer that computes it. Let $\Sigma = \{\sigma^{(2)}\} \cup \Sigma'$ with $\Sigma' = \{\gamma^{(1)}, \alpha^{(0)}, \beta^{(0)}\}$. Let $\tau_1\colon T_\Sigma \times T_\Sigma \to \mathbb{A}$ be the weighted tree transformation such that

$$\tau_1(s,t) = \begin{cases} 1 & \text{if } s = \sigma(t, \alpha) \text{ and } t \in T_{\Sigma'} \\ 0 & \text{otherwise} \end{cases} \tag{1}$$

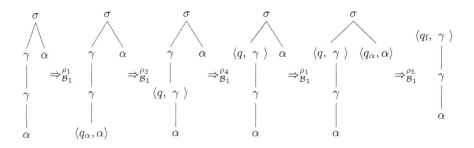

Fig. 2. Sample derivation of the BOT $\mathcal{B}_1$ with weight 1.

for every $s \in T_\Sigma$ and $t \in T_\Sigma$. This transformation can easily be computed by a BOT, as the following example shows.

Example 6. The transformation τ_1 of (1) is computed by the BOT

$$\mathcal{B}_1 = (\{q, q_\alpha, q_\mathrm{f}\}, \Sigma, \Sigma, \{q_\mathrm{f}\}, R, 1_R)$$

with the following set of rules $R = \{\rho_1, \ldots, \rho_6\}$.

$(\rho_1) \qquad \alpha \to_1 q_\alpha(\alpha) \qquad\qquad (\rho_2) \qquad\qquad \beta \to_1 q(\beta)$

$(\rho_3) \qquad \gamma(q_\alpha) \to_1 q\big(\gamma(x_1)\big) \qquad (\rho_4) \qquad \gamma(q) \to_1 q\big(\gamma(x_1)\big)$

$(\rho_5) \qquad \sigma(q, q_\alpha) \to_1 q_\mathrm{f}(x_1) \qquad (\rho_6) \quad \sigma(q_\alpha, q_\alpha) \to_1 q_\mathrm{f}(x_1)$

Clearly, $|\mathrm{Der}_{\mathcal{B}_1}(s, \langle q_\mathrm{f}, t \rangle)| \leq 1$ for every $s \in T_\Sigma$ and $t \in T_\Sigma$. It is straightforward to verify that $\|\mathcal{B}_1\| = \tau_1$, and we present a sample derivation in Fig. 2.   □

Next, let us present another weighted tree transformation τ_2, which is similarly inspired by [3, Property T], for which we will then show that it cannot be computed by any BOT. Property T roughly states that copies of the input tree can be independently processed using nondeterminism. In the next example, the output will be essentially two copies of the input, but the copies need not be exactly equal, but rather may be terminated by different nullary symbols. Let $\Sigma = \{\sigma^{(2)}\} \cup \Sigma'$ with $\Sigma' = \{\gamma^{(1)}, \alpha^{(0)}, \beta^{(0)}\}$. Let $\tau_2 \colon T_\Sigma \times T_\Sigma \to \mathbb{A}$ be the weighted tree transformation such that

$$\tau_2(s, t) = \begin{cases} 1 & \text{if } t(\varepsilon) = \sigma, \text{size}(t|_1) = \text{size}(t|_2) = \text{size}(s) \text{ and } s, t|_1, t|_2 \in T_{\Sigma'} \\ 0 & \text{otherwise.} \end{cases}$$

$$(2)$$

Lemma 7. *The weighted tree transformation τ_2 of (2) cannot be computed by any BOT.*

Proof. Suppose that there exists a BOT $\mathcal{B} = (Q, \Sigma, \Sigma, F, R, \mathrm{wt})$ that computes $\|\mathcal{B}\| = \tau_2$. Select an integer $n \in \mathbb{N}$ such that

$$n > \max \big\{ \text{size}(t) \mid \sigma(q_1, \ldots, q_k) \to_a q(t) \in R \big\} \, ;$$

---

i.e., n is larger than the size of any output tree in any rule of $\mathcal{B}$. Let $s = \gamma^n(\alpha)$ and $t = \gamma^n(\beta)$. Clearly, $\tau_2(s, \sigma(s,t)) = 1$. Thus there exists $d \in \mathrm{Der}_{\mathcal{B}}(s, \langle q_f, \sigma(s,t)\rangle)$ for some final state $q_f \in F$. Let us consider the rule $\ell \to_a q(\sigma(t_1, t_2)) \in R$ of the derivation d that generates the unique occurrence of σ in the output tree $\sigma(s,t)$. By the definition of s, only the variable x_1 may occur in $\sigma(t_1, t_2)$; i.e., $\mathrm{var}(\sigma(t_1, t_2)) \subseteq \{x_1\}$. Since s and t have no subtrees in common, the variable x_1 can only occur once in $\sigma(t_1, t_2)$; i.e., $|\mathrm{pos}_{x_1}(\sigma(t_1, t_2))| \leq 1$. Suppose that $t_2 \in T_\Sigma$ does not contain x_1; the other case is analogous. In this case it is clearly $t_2 = t = \gamma^n(\beta)$. Meanwhile we chose n to be larger than the size of any output tree in any rule of $\mathcal{B}$. This includes the output tree $\sigma(t_1, t_2)$, which in turn is larger than its subtree t_2. Overall we obtain a contradiction via

$$\mathrm{size}(t_2) = \mathrm{size}(t) = n + 1 > \mathrm{size}(\sigma(t_1, t_2)) > \mathrm{size}(t_2) . \qquad \square$$

Finally we show that the range of an input-finitary weighted tree transformation computed by a BOT is hom-regular. This enables the application of several results like, for example, the pumping lemma [15, Lemma 4] or the decidability of support emptiness and finiteness of [15, Theorem 6] for zero-sum free semirings $\mathbb{A}$. To prove that the range is hom-regular we use a well-known decomposition of weighted bottom-up tree transducers, which we recall first. It was lifted from the unweighted case [3, Theorem 3.15] to (essentially) commutative semirings in [5, Theorem 5.7].

Proposition 8 ([5, **Lemma 5.6**])**.** *For every BOT $\mathcal{B}$ there exists a finite-state relabeling $\mathcal{R}$ and a tree homomorphism h such that $\|\mathcal{B}\| = \|\mathcal{R}\|\,; 1_h$.* $\qquad \square$

Theorem 9. *The range $\mathrm{ran}_{\|\mathcal{B}\|}$ is hom-regular for every BOT $\mathcal{B}$ such that $\|\mathcal{B}\|$ is input-finitary.*

Proof. We first apply the decomposition of Proposition 8 to $\mathcal{B}$ and obtain a finite-state relabeling $\mathcal{R}$ and a tree homomorphism h such that $\|\mathcal{B}\| = \|\mathcal{R}\|;1_h$. Since $\mathcal{R}$ is a finite-state relabeling (where input and output trees always have the same set of positions), the transformation $\|\mathcal{R}\|$ is input- and output-finitary, the transformation $\|\mathcal{B}\|$ is input-finitary if and only if h is input-finitary on $\mathrm{supp}(\mathrm{ran}_{\|\mathcal{R}\|})$. Clearly, $\mathrm{ran}_{\|\mathcal{B}\|} = h(\mathrm{ran}_{\|\mathcal{R}\|})$ (i.e., the image of the weighted tree language $\mathrm{ran}_{\|\mathcal{R}\|}$ under h) and $\mathrm{ran}_{\|\mathcal{R}\|}$ is regular by [13, Corollary 14]. Hence $\mathrm{ran}_{\|\mathcal{B}\|}$ is the image of a regular weighted tree language under an input-finitary tree homomorphism, which is hom-regular by Lemma 4. $\qquad \square$

4 Weighted Top-Down Tree Transducers

Following the structure of the previous section, we start by recalling weighted top-down tree transducers from [5, Definitions 3.4 and 3.7] in a notation closer to [3, Section 1].

Definition 10. *A weighted top-down tree transducer (for short: TOP) is a tuple $\mathcal{T} = (Q, \Sigma, \Delta, I, R, \mathrm{wt})$ with a finite set Q of states, ranked alphabets Σ and Δ of input and output symbols, respectively, a set $I \subseteq Q$ of initial states, a finite set of rules $R \subseteq \bigcup_{k \in \mathbb{N}} Q \times \Sigma_k \times T_\Delta(Q \times X_k)$, and a weight map $\mathrm{wt} \colon R \to \mathbb{A}$.* $\qquad \square$

To simplify the presentation, we depict a rule $\rho = (q, \sigma, t) \in R$ as $\langle q, \sigma \rangle \to_a t$, where $a = \text{wt}(\rho)$. Using essentially the same approach as for BOTs, we next present the semantics of TOPs following the derivation semantics of [7, Section 6], which coincides with the semantics of [5, Definitions 3.2 and 3.5]. Let $T = (Q, \Sigma, \Delta, I, R, \text{wt})$ be a TOP, and let $\text{SF} = T_\Delta(Q \times T_\Sigma)$ be its sentential forms. Given two sentential forms $\xi, \zeta \in \text{SF}$ and a rule $\rho = \langle q, \sigma \rangle \to t \in R$, we write $\xi \Rightarrow_T^\rho \zeta$ if $\xi|_w = \langle q, \sigma(s_1, \ldots, s_k) \rangle$ and $\zeta = \xi[t\theta]_w$ for some input trees $s_1, \ldots, s_k \in T_\Sigma$ and the lexicographically least position $w \in \text{pos}(\xi)$ such that $\xi|_w \in Q \times T_\Sigma$, where $\theta \colon Q \times X_k \to Q \times T_\Sigma$ is given by $\theta(\langle q, x_i \rangle) = \langle q, s_i \rangle$ for every $q \in Q$ and $i \in [k]$. A *derivation* d from ξ to ζ is a (finite) sequence of rules $d = \rho_1 \cdots \rho_n \in R^*$ such that $\xi(\Rightarrow_T^{\rho_1} ; \cdots ; \Rightarrow_T^{\rho_n})\zeta$. The set of all derivations from ξ to ζ is denoted by $\text{Der}_T(\xi, \zeta)$ and is always finite. The weight $\text{wt}(d)$ of the derivation d is $\text{wt}(d) = \prod_{i=1}^n \text{wt}(\rho_i)$. Finally, we define the semantics $\|T\| \colon T_\Sigma \times T_\Delta \to \mathbb{A}$ of the TOP T for every input tree $s \in T_\Sigma$ and output tree $t \in T_\Delta$ by

$$\|T\|(s, t) = \sum_{\substack{q \in I \\ d \in \text{Der}_T(\langle q, s \rangle, t)}} \text{wt}(d) .$$

Thus for any $(s, t) \in \text{supp}(\|T\|)$ there exists $q \in I$ and $d \in \text{Der}_T(\langle q, s \rangle, t)$ such that $\text{wt}(d) \neq 0$.

Let us return to the weighted tree transformation τ_2 of (2) and show that it can be computed by a TOP. Recall from Lemma 7 that it cannot be computed by any BOT.

Example 11. The weighted tree transformation τ_2 of (2) is computed by the TOP $T_2 = (\{q_0, q\}, \Sigma, \Sigma, \{q_0\}, R, 1_R)$ with the following rules $R = \{\rho_1, \ldots, \rho_{14}\}$.

$(\rho_1) \quad \langle q_0, \gamma \rangle \to_1 \sigma\big(\gamma(\langle q, x_1 \rangle), \gamma(\langle q, x_1 \rangle)\big)$ $\qquad (\rho_2) \quad \langle q, \gamma \rangle \to_1 \gamma(\langle q, x_1 \rangle)$

$(\rho_3) \quad \langle q, \alpha \rangle \to_1 \alpha$ $\qquad\qquad\qquad\qquad\qquad (\rho_4) \quad \langle q, \alpha \rangle \to_1 \beta$

$(\rho_5) \quad \langle q, \beta \rangle \to_1 \alpha$ $\qquad\qquad\qquad\qquad\qquad (\rho_6) \quad \langle q, \beta \rangle \to_1 \beta$

$(\rho_7) \quad \langle q_0, \alpha \rangle \to_1 \sigma(\alpha, \alpha)$ $\qquad\qquad\qquad (\rho_8) \quad \langle q_0, \alpha \rangle \to_1 \sigma(\alpha, \beta)$

$(\rho_9) \quad \langle q_0, \alpha \rangle \to_1 \sigma(\beta, \alpha)$ $\qquad\qquad\qquad (\rho_{10}) \quad \langle q_0, \alpha \rangle \to_1 \sigma(\beta, \beta)$

$(\rho_{11}) \quad \langle q_0, \beta \rangle \to_1 \sigma(\alpha, \alpha)$ $\qquad\qquad\qquad (\rho_{12}) \quad \langle q_0, \beta \rangle \to_1 \sigma(\alpha, \beta)$

$(\rho_{13}) \quad \langle q_0, \beta \rangle \to_1 \sigma(\beta, \alpha)$ $\qquad\qquad\qquad (\rho_{14}) \quad \langle q_0, \beta \rangle \to_1 \sigma(\beta, \beta)$

It is again straightforward to verify that $\|T_2\| = \tau_2$. We present a sample derivation in Fig. 3. $\qquad\qquad\qquad\qquad\qquad\qquad\qquad\qquad\qquad\qquad\qquad\qquad\qquad\quad \square$

Theorem 12. *There exists a weighted tree transformation that can be computed by a TOP, but not by any BOT.*

Proof. The transformation τ_2 can be computed by a TOP as demonstrated in Example 11, but cannot be computed by any BOT by Lemma 7. $\qquad\qquad\qquad\qquad \square$

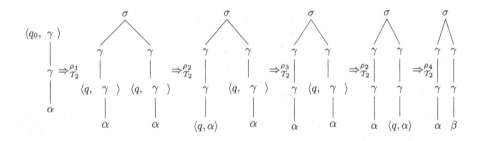

Fig. 3. Sample derivation of the TOP T_2 with weight 1.

Let us now turn to the missing separation with an example from above.

Lemma 13. *The weighted tree transformation τ_1 of (1) cannot be computed by any TOP.*

Proof. Suppose that there exists a TOP $T = (Q, \Sigma, \Sigma, I, R, \mathrm{wt})$ that computes $\|T\| = \tau_1$. We select $n \in \mathbb{N}$ such that $n > \max \{\mathrm{size}(t) \mid \langle q, \sigma \rangle \to_a t \in R\}$, i.e., n is larger than the size of any output tree in any rule of T. Let $s = \sigma(s_1, \alpha)$ and $s' = \sigma(s_1, \beta)$ with $s_1 = \gamma^{2n}(\alpha)$. Clearly,

$$\|T\|(s, s_1) = \tau_1(s, s_1) = 1 \neq 0 = \tau_1(s', s_1) = \|T\|(s', s_1) .$$

However, we will prove that $\|T\|(s, s_1) = \|T\|(s', s_1)$, which yields the desired contradiction. To this end, we will prove that $\mathrm{Der}_T(\langle q, s \rangle, s_1) = \mathrm{Der}_T(\langle q, s' \rangle, s_1)$ for every state $q \in Q$. To this end, let $d \in \mathrm{Der}_T(\langle q, s \rangle, s_1)$ for input s. Our goal is to show that $d \in \mathrm{Der}_T(\langle q, s' \rangle, s_1)$ for input s'. The converse inclusion can be proved in an analogous manner. Let $\langle q, \sigma \rangle \to_a t \in R$ be the rule that is utilized first in d. Obviously, $t \in T_{\Sigma''}(Q \times X_2)$ with $\Sigma'' = \{\gamma^{(1)}, \alpha^{(0)}\}$ because t needs to match s_1. Thus t contains exactly one leaf $w \in \mathrm{pos}_{\{a\} \cup (Q \times X_2)}(t)$. Now we distinguish several cases.

- If $t(w) = \alpha$, then $t \in T_\Sigma$ and we must thus have $s_1 = t$, which is contradictory since $\mathrm{size}(s_1) = 2n + 1 > \mathrm{size}(t)$. Hence this case is impossible.
- If $t(w) = \langle q', x_2 \rangle$, then the next derivation step necessarily uses a rule of the form $\langle q', \alpha \rangle \to_{a'} t' \in R$ with $t' \in T_\Sigma$. Thus we must have $s_1 = t[\langle q', x_2 \rangle \leftarrow t']$, which is again contradictory since

$$\mathrm{size}(s_1) = 2n + 1 > \mathrm{size}(t) + \mathrm{size}(t') > \mathrm{size}(t[\langle q', x_2 \rangle \leftarrow t']) .$$

This renders this case impossible as well.
- Finally, let $t(w) = \langle q', x_1 \rangle$. Thus the derivation does not visit the second direct subtree $s|_2 = \alpha$. Since $s(\varepsilon) = s'(\varepsilon)$ and $s|_1 = s_1 = s'|_1$ the derivation d is also a valid derivation for input s' (with the same output and weight naturally). Thus, $d \in \mathrm{Der}_T(\langle q, s' \rangle, s_1)$ as required.

Since $\mathrm{Der}_T(\langle q, s \rangle, s_1) = \mathrm{Der}_T(\langle q, s' \rangle, s_1)$ for every state $q \in Q$, we must also have $1 = \|T\|(s, s_1) = \|T\|(s', s_1) = 0$, which is the desired contradiction. □

Theorem 14. *There exists a weighted tree transformation that can be computed by a BOT, but not by any TOP.*

Proof. The transformation τ_1 can be computed by a BOT as demonstrated in Example 6, but cannot be computed by any TOP by Lemma 13. $\quad\square$

Corollary 15 (of Theorems 12 **and** 14**).** *The first level of the BOT and TOP hierarchy (see Fig. 1) is strict for every commutative semiring* $\mathbb{A}$*.*

References

1. Baker, B.S.: Tree transductions and families of tree languages. In: Proceedings of the STOC, pp. 200–206 (1973). https://doi.org/10.1145/800125.804051
2. Doner, J.: Tree acceptors and some of their applications. J. Comput. Syst. Sci. **4**(5), 406–451 (1970). https://doi.org/10.1016/S0022-0000(70)80041-1
3. Engelfriet, J.: Bottom-up and top-down tree transformations – a comparison. Math. Syst. Theory **9**(3), 198–231 (1975). https://doi.org/10.1007/BF01704020
4. Engelfriet, J.: Three hierarchies of transducers. Math. Syst. Theory **15**(2), 95–125 (1982). https://doi.org/10.1007/BF01786975
5. Engelfriet, J., Fülöp, Z., Vogler, H.: Bottom-up and top-down tree series transformations. J. Autom. Lang. Comb. **7**(1), 11–70 (2002). https://doi.org/10.25596/jalc-2002-011
6. Fülöp, Z., Gazdag, Z., Vogler, H.: Hierarchies of tree series transformations. Theoret. Comput. Sci. **314**(3), 387–429 (2004). https://doi.org/10.1016/j.tcs.2004.01.001
7. Fülöp, Z., Vogler, H.: Weighted tree transducers. J. Autom. Lang. Comb. **9**(1), 31–54 (2004). https://doi.org/10.25596/jalc-2004-031
8. Fülöp, Z., Vogler, H.: Weighted tree automata and tree transducers. In: Droste, M., Kuich, W., Vogler, H. (eds.) Handbook of Weighted Automata, chap. 9, pp. 313–403. Springer, Heidelberg (2009). https://doi.org/10.1007/978-3-642-01492-5_9
9. Godoy, G., Giménez, O.: The HOM problem is decidable. J. ACM **60**(4) (2013). https://doi.org/10.1145/2508028.2501600
10. Golan, J.S.: Semirings and their Applications. Kluwer Academic, Dordrecht (1999). https://doi.org/10.1007/978-94-015-9333-5
11. Hebisch, U., Weinert, H.J.: Semirings: Algebraic Theory and Applications in Computer Science, Series in Algebra, vol. 5. World Scientific (1998). https://doi.org/10.1142/3903
12. Jurafsky, D., Martin, J.H.: Speech and Language Processing, 2nd edn. Prentice Hall, Hoboken (2008)
13. Kuich, W.: Tree transducers and formal tree series. Acta Cybernet. **14**(1), 135–149 (1999)
14. Maletti, A.: Hierarchies of tree series transformations revisited. In: Ibarra, O.H., Dang, Z. (eds.) DLT 2006. LNCS, vol. 4036, pp. 215–225. Springer, Heidelberg (2006). https://doi.org/10.1007/11779148_20
15. Maletti, A., Nász, A.T.: Weighted tree automata with constraints. arXiv (2023). https://doi.org/10.48550/ARXIV.2302.03434
16. Thatcher, J.W.: Characterizing derivation trees of context-free grammars through a generalization of finite automata theory. J. Comp. Syst. Sci. **1**(4), 317–322 (1967). https://doi.org/10.1016/S0022-0000(67)80022-9

Deciding Whether an Attributed Translation Can Be Realized by a Top-Down Transducer

Sebastian Maneth and Martin Vu$^{(\boxtimes)}$

Universität Bremen, Bremen, Germany
{maneth,martin.vu}@uni-bremen.de

Abstract. We prove that for a given partial functional attributed tree transducer with monadic output, it is decidable whether or not an equivalent top-down transducer (with or without look-ahead) exists. We present a procedure that constructs an equivalent top-down transducer (with or without look-ahead) if it exists.

1 Introduction

It is well known that *two-way (string) transducers* are strictly more expressive than *one-way (string) transducers*. For instance, a two-way transducer can compute the reverse of an input string, which cannot be achieved by any one-way transducer. For a given (functional) two-way transducer, it is a natural question to ask: Can its translation be realized by a one-way transducer? This question was recently shown to be decidable [8], see also [1]. Decision procedures of this kind have several advantages; for instance, the smaller class of transducers may be more efficient to evaluate (i.e., may use less resources), or the smaller class may enjoy better closure properties than the larger class.

One possible pair of respective counterparts of two-way (string) transducers and one-way (string) transducers in the context of trees are *attributed tree transducers* and *top-down tree transducers*. As the name suggests, states of the latter process an input tree strictly in a top-down fashion while the former can, analogously to two-way transducers, change direction as well. As for their string counterparts, attributed tree transducers are strictly more expressive than top-down tree transducers [10]. Hence, for a (functional) attributed tree transducer, it is a natural question to ask: Can its translation be realized by a deterministic top-down transducer?

In this paper, we address this problem for a subclass of attributed tree transducers. In particular, we consider attributed tree transducer with *monadic output* meaning that all output trees that the transducer produces are *monadic*, i.e., "strings". We show that the question whether or not for a given attributed tree transducer A with monadic output an equivalent top-down transducer (with or without look-ahead) T exists can be reduced to the question whether or not a given two-way transducer can be defined by a one-way transducer.

© The Author(s), under exclusive license to Springer Nature Switzerland AG 2023
B. Nagy (Ed.): CIAA 2023, LNCS 14151, pp. 230–241, 2023.
https://doi.org/10.1007/978-3-031-40247-0_17

First, we test whether A can be equipped with look-ahead so that A has the *single-path property*, which means that attributes of A only process a single input path. Intuitively, this means that given an input tree, attributes of A only process nodes occurring in a node sequence $v_1, \ldots, v_n$ where v_i is the parent of v_{i+1} while equipping A with look-ahead means that input trees of A are preprocessed by a deterministic bottom-up relabeling. The single-path property is not surprising given that if A is equivalent to some top-down tree transducer T (with look-ahead) then states of T process the input tree in exactly that fashion. Assume that A extended with look-ahead satisfies this condition. We then show that A can essentially be converted into a two-way (string) transducer. We now apply the procedure of [1]. It can be shown that the procedure of [1] yields a one-way (string) transducer if and only if a (nondeterministic) top-down tree transducer T exists which uses the same look-ahead as A. We show that once we have obtained a one-way transducer T_O equivalent to the two-way transducer converted from A, we can compute T from T_O, thus obtaining our result. It is well-known that for a functional top-down tree transducer (with look-ahead) an equivalent deterministic top-down tree transducer with look-ahead can be constructed [5]. Note that for the latter kind of transducer, it can be decided whether or not an equivalent transducer *without* look-ahead exists [12] (this is because transducers with monadic output are linear by default).

We show that the above results are also obtainable for attributed tree transducers with look-around. This model was introduced by Bloem and Engelfriet [2] due to its better closure properties. For this we generalize the result from [5], and show that every functional partial attributed tree transducer (with look-around) is equivalent to a deterministic attributed tree transducer with look-around.

Note that in the presence of origin, it is well known that even for (nondeterministic) macro tree transducers (which are strictly more expressive than attributed tree transducers) it is decidable whether or not an origin-equivalent deterministic top-down tree transducer with look-ahead exists [9]. In the absence of origin, the only definability result for attributed transducers that we are aware of is, that it is decidable for such transducers (and even for macro tree transducers) whether or not they are of linear size increase [7]; and if so an equivalent single-use restricted attributed tree transducer can be constructed (see [6]).

2   Attributed Tree Transducers

For $k \in \mathbb{N}$, we denote by $[k]$ the set $\{1, \ldots, k\}$. Let $\Sigma = \{e_1^{k_1}, \ldots, e_n^{k_n}\}$ be a *ranked alphabet*, where $e_j^{k_j}$ means that the symbol e_j has *rank* k_j. By Σ_k we denote the set of all symbols of Σ which have rank k. The set T_Σ of *trees over* Σ consists of all strings of the form $a(t_1, \ldots, t_k)$, where $a \in \Sigma_k$, $k \geq 0$, and $t_1, \ldots, t_k \in T_\Sigma$. Instead of $a()$ we simply write a. For a tree $t \in T_\Sigma$, its nodes are referred to as follows: We denote by ϵ the root of t while $u.i$ denotes the i-th child of the node u. Denote by $V(t)$ the set of nodes of t, e.g. for the tree $t = f(a, f(a, b))$ we have $V(t) = \{\epsilon, 1, 2, 2.1, 2.2\}$. For $v \in V(t)$, $t[v]$ is the label of v, t/v is the subtree of t rooted at v, and $t[v \leftarrow t']$ is obtained from t by replacing t/v by t'. For a set Λ disjoint with Σ, we define $T_\Sigma[\Lambda]$ as $T_{\Sigma'}$ where $\Sigma'_0 = \Sigma_0 \cup \Lambda$ and $\Sigma'_k = \Sigma_k$ for $k > 0$.

A *(partial deterministic) attributed tree transducer* (or *att* for short) is a tuple $A = (S, I, \Sigma, \Delta, a_0, R)$ where S and I are disjoint sets of *synthesized attributes* and *inherited attributes*, respectively. The sets Σ and Δ are ranked alphabets of *input* and *output symbols*, respectively. We denote by $a_0 \in S$ the *initial attribute* and define that $R = (R_\sigma \mid \sigma \in \Sigma \cup \{\#\})$ is a collection of finite sets of rules. We implicitly assume *att*'s to include a unique symbol $\# \notin \Sigma$ of rank 1, the so-called *root marker*, that only occurs at the root of a tree. Let $\sigma \in \Sigma \cup \{\#\}$ be of rank $k \geq 0$. Let π be a variable for nodes for which we define $\pi 0 = \pi$. For every $a \in S$ the set R_σ contains at most one rule of the form $a(\pi) \to t$ and for every $b \in I$ and $i \in [k]$, R_σ contains at most one rule of the from $b(\pi i) \to t'$ where $t, t' \in T_\Delta[\{a'(\pi i) \mid a' \in S, i \in [k]\} \cup \{b(\pi) \mid b \in I\}]$. The right-hand sides t, t' are denoted by $\mathrm{rhs}_A(\sigma, a(\pi))$ and $\mathrm{rhs}_A(\sigma, b(\pi i))$, respectively, if they exist. If $I = \emptyset$ then we call A a *deterministic top-down tree transducer* (or simply a *dt*). In this case, we call S a set of *states* instead of attributes.

To define the semantics of the *att* A, we first define the *dependency graph of A for the tree* $s \in T_\Sigma$ as $D_A(s) = (V, E)$ where $V = \{(a_0, \epsilon)\} \cup ((S \cup I) \times (V(\#(s)) \setminus \{\epsilon\}))$ and $E = \{((\gamma', uj), (\gamma, ui)) \mid u \in V(s), \gamma'(\pi j) \text{ occurs in } \mathrm{rhs}_A(s[u], \gamma(\pi i)),$ with $0 \leq i, j$ and $\gamma, \gamma' \in S \cup I\}$. If $D_A(s)$ contains a cycle for some $s \in T_\Sigma$ then A is called *circular*. We define $v0 = v$ for a node v. For a given tree $s \in T_{\Sigma \cup \{\#(s') \mid s' \in T_\Sigma\}}$, let $N = \{a_0(\epsilon)\} \cup \{\alpha(v) \mid \alpha \in S \cup I, v \in V(s) \setminus \{\epsilon\}\}$. For trees $t, t' \in T_\Delta[N]$, $t \Rightarrow_{A,s} t'$ holds if t' is obtained from t by replacing a node labeled by $\gamma(vi)$, where $i = 0$ if $\gamma \in S$ and $i > 0$ if $\gamma \in I$, by $\mathrm{rhs}_A(s[v], \gamma(\pi i))[\gamma'(\pi j) \leftarrow \gamma'(vj) \mid \gamma' \in S \cup I, 0 \leq j]$. If A is non-circular, then every $t \in T_\Delta[N]$ has a unique normal form with respect to $\Rightarrow_{A,s}$ denoted by $\mathrm{nf}(\Rightarrow_{A,s}, t)$. The *translation realized by A*, denoted by τ_A, is the set $\{(s, \mathrm{nf}(\Rightarrow_{A,\#(s)}, a_0(\epsilon))) \in T_\Sigma \times T_\Delta\}$. As A is deterministic, τ_A is a partial function. Thus we also write $\tau_A(s) = t$ if $(s, t) \in \tau_A$ and say that on input s, A produces the tree t. Denote by $\mathrm{dom}(A)$ the *domain of A*, i.e., the set of all $s \in T_\Sigma$ such that $(s, t) \in \tau_A$ for some $t \in T_\Delta$. Similarly, $\mathrm{range}(A)$ denotes the *range of A*, i.e., the set of all $t \in T_\Delta$ such that for some $s \in T_\Sigma$, $(s, t) \in \tau_A$.

Example 1. Consider the *att* $A_1 = (S, I, \Sigma, \Delta, a, R)$ where $\Sigma = \{f^2, e^0\}$ and $\Delta = \{g^1, e^0\}$. Let the set of attributes of A be given by $S = \{a\}$ and $I = \{b\}$. We define $R_f = \{a(\pi) \to a(\pi 1), b(\pi 1) \to a(\pi 2), b(\pi 2) \to b(\pi)\}$. Furthermore we define $R_e = \{a(\pi) \to g(b(\pi))\}$ and $R_\# = \{a(\pi) \to a(\pi 1), b(\pi 1) \to e\}$. The tree transformation realized by A_1 contains all pairs (s, t) such that if s has n leaves, then t is the tree over Δ that contains n occurrences of the symbol g. The domain of A is T_Σ and its range is $T_\Delta \setminus \{e\}$. The dependency graph of A_1 for the tree $f(f(e, e), f(e, e))$ is depicted in Fig. 1. As usual, occurrences of inherited and synthesized attributes are placed to the left and right of nodes, respectively. If clear from context, names of attribute occurrences are omitted. □

We emphasize that we always consider input trees to be trees over Σ. The root marker is a technical necessity. For instance, the translation of A_1 in Example 1 is not possible without it. It is well known that whether or not a given *att*

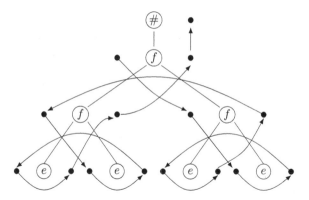

Fig. 1. Dependency Graph of the *att* in Example 1 for $f(f(e,e), f(e,e))$.

A is circular can be tested by computing the *is-dependencies* of A [11]. Informally, the is-dependency of a tree s depicts the dependencies between inherited and synthesized attributes at the root of s. Formally, the is-dependency of s is the set $\mathrm{IS}_A(s) = \{(b,a) \in I \times S \mid a(1)$ is reachable from $b(1)$ in $D_A(s)\}$. As a dependency graph is a directed graph, we say for $v, v_1, v_2 \in V$, that v_1 is *reachable* from v_2 if (a) $v_1 = v_2$ or (b) v is reachable from v_2 and $(v, v_1) \in E$ as usual. If $s = \sigma(s_1, \ldots, s_k)$ and the is-dependency of $s_1, \ldots, s_k$ is known, then the is-dependency of s can be easily computed in a bottom-up fashion using the rules of R_σ. In the the rest of the paper, we only consider non-circular *att*'s.

We define an *attributed tree transducer with look-ahead (or att^R)* as a pair $\hat{A} = (B, A)$ where B is a deterministic bottom-up relabeling which preprocesses input trees for the *att* A. A *(deterministic) bottom-up relabeling* B is a tuple $(P, \Sigma, \Sigma', F, R)$ where P is the set of states, Σ, Σ' are ranked alphabets and $F \subseteq P$ is the set of final states. For $\sigma \in \Sigma$ and $p_1, \ldots, p_k \in P$, the set R contains at most one rule of the form $\sigma(p_1(x_1), \ldots, p_k(x_k)) \to p(\sigma'(x_1, \ldots, x_k))$ where $p \in P$ and $\sigma' \in \Sigma'$. These rules induce a derivation relation $\Rightarrow_B$ in the obvious way. The translation realized by B is given by $\tau_B = \{(s,t) \in T_\Sigma \times T_\Delta \mid s \Rightarrow_B^* p(t), p \in F\}$. As τ_B is a partial function, we also write $\tau_B(s) = t$ if $(s,t) \in \tau_B$. The translation realized by $\hat{A}$ is given by $\tau_{\hat{A}} = \{(s,t) \in T_\Sigma \times T_\Delta \mid t = \tau_A(\tau_B(s))\}$. We write $\tau_{\hat{A}}(s) = t$ if $(s,t) \in \tau_{\hat{A}}$ as usual. If A is a *dt* then $\hat{A}$ is called a deterministic top-down transducer with look-ahead (or *dt^R*).

3   The Single Path Property

In the this section, we show that given an *att* with monadic output, it is decidable whether or not an equivalent *dt^R* exists. Monadic output means that output symbols are at most of rank 1. First consider the following definition. For an *att* A with initial attribute a_0, an input tree s and $v \in V(s)$, we say that on input s, an attribute α of A *processes* the node v if (a_0, ϵ) is reachable from $(\alpha, 1.v)$ in $D_A(s)$. Recall that the dependency graph for s is defined on the tree $\#(s)$. Now, consider an arbitrary $dt^R\ \check{T} = (B', T')$ with monadic output. Then the behavior of T' is limited in a particular way: Let s be an input tree and s' be obtained from s via the relabeling B'. On input s', the states of T' only process the nodes on a single *path* of s'. A path is a sequence of nodes $v_1, \ldots, v_n$ such that v_i is the parent node of v_{i+1}. This property holds obviously as the output is monadic and hence at most one state occurs on the right-hand side of any rule of T'.

Using this property, we prove our claim. In the following, we fix an *att* $A = (S, I, \Sigma, \Delta, a_0, R)$ with monadic output and show that if a $dt^R\ T = (B', T')$ equivalent to A exists then we can equip A with look-ahead such that attributes of A become limited in the same way as states of T': They only process nodes of a single path of the input tree. Our proof makes use of the result of [1]. This result states that for a *functional two-way transducer* it is decidable whether or not an equivalent *one way transducer* exists. Such transducers are essentially attributed transducers and top-down transducers with monadic input and output, respectively. Functional means that the realized translation is a function. We show that A equipped with look-ahead so that attributes of A are limited as described earlier can be converted into a two-way transducer T_W. It can be shown that the procedure of [1] yields a one-way transducer T_O equivalent to T_W if and only if T exists. We then show that we can construct T from T_O.

Subsequently, we define the look-ahead with which we equip A. W.l.o.g. we assume that only right-hand sides of rules in $R_\#$ are ground (i.e., in T_Δ). Clearly any *att* can be converted so that it conforms to this requirement. Let $\alpha(\pi) \to \tau \in R_\sigma$ such that τ is ground. First, we remove this rule from R_σ. Then we introduce a new inherited attribute $\langle\tau\rangle$ and the rules $\alpha(\pi) \to \langle\tau\rangle(\pi) \in R_\sigma$, and $\langle\tau\rangle(\pi 1) \to \tau \in R_\#$. For all $\sigma' \in \Sigma_k$ and $j \in [k]$ we define $\langle\tau\rangle(\pi j) \to \langle\tau\rangle(\pi) \in R_{\sigma'}$.

Let $s \in dom(A)$ and let $v \in V(s)$. We call $\psi \subseteq I \times S$ the *visiting pair set at v on input s* if $(b, a) \in \psi$ if and only if

- on input s, the attribute a processes v and
- $(b, a) \in \mathrm{IS}_A(s/v)$

Let ψ be the visiting pair set at v on input s. In the following, we denote by Ω_ψ the set consisting of all trees $s' \in T_\Sigma$ such that $\psi \subseteq \mathrm{IS}_A(s')$. Thus the set Ω_ψ contains all trees s' such that the visiting pair set at v on input $s[v \leftarrow s']$ is also ψ. If an arbitrary $a \in S$ exists such that $(b, a) \in \psi$ for some $b \in I$ and the range of a when translating trees in Ω_ψ is unbounded, i.e., if the cardinality of $\{\mathrm{nf}(\Rightarrow_{A,s'}, a(\epsilon)) \mid s' \in \Omega_\psi\}$ is unbounded, then we say that the *variation of Ω_ψ is unbounded*. If ψ is the visiting pair set at v on input s and the variation of Ω_ψ is unbounded then we also say that the *variation at v on input s* is unbounded. The variation plays a key role for proving our claim. In particular, the following property is derived from it: We say that A has the *single path property* if for all trees $s \in dom(A)$ a path ρ exists such that the variation at $v \in V(s)$ is bounded whenever v does not occur in ρ. The following lemma states that the single path property is a necessary condition for the *att* A to have an equivalent dt^R.

Lemma 1. *Let $s \in dom(A)$ and $v_1, v_2 \in V(s)$ such that v_1 and v_2 have the same parent node. If a $dt^R\ T$ equivalent to A exists then the variation at either v_1 or v_2 on input s is bounded.*

Proof. For simplicity assume that T is a dt. The proof for dt^R is obtained similarly. Assume that the variations at both v_1 and v_2 on input s is unbounded. As T produces monadic output trees, on input s, only one of the nodes v_1 and v_2 is processed by a state of T. W.l.o.g. let ψ be the visiting pair set at v_1 on input s and assume that only v_2 is processed by a state of T. Then for all $s' \in \Omega_\psi$, T produces the same output on input $s[v_1 \leftarrow s']$. However, A does not produce the same output as the visiting pair set at v_1 on input $s[v_1 \leftarrow s']$ is ψ and the variation of Ω_ψ is unbounded contradicting the equivalence of T and A.   □

Example 2. Consider the att $A_2 = (S, I, \Sigma, \Delta, a, R)$ where $\Sigma = \{f^2, e^0, d^0\}$ and $\Delta = \{f^1, g^1, e^0, d^0\}$. The set of attributes are given by $S = \{a, a_e, a_d\}$ and $I = \{b_e, b_d, \langle e \rangle, \langle d \rangle\}$. In addition to $\langle e \rangle(\pi 1) \to \langle e \rangle(\pi)$ and $\langle d \rangle(\pi 1) \to \langle d \rangle(\pi)$, the set R_f contains the rules

$$a_d(\pi) \to f(a(\pi 1)) \quad b_d(\pi 1) \to a_d(\pi 1) \quad b_d(\pi 2) \to b_d(\pi) \quad a(\pi) \to a(\pi 2)$$
$$a_e(\pi) \to g(a(\pi 1)) \quad b_e(\pi 1) \to a_e(\pi 1) \quad b_e(\pi 2) \to b_e(\pi)$$

while $R_\#$ contains in addition to $\langle e \rangle(\pi 1) \to e$ and $\langle d \rangle(\pi 1) \to d$ the rules

$$a(\pi) \to a(\pi 1) \quad b_e(\pi 1) \to a_e(\pi 1) \quad b_d(\pi 1) \to a_d(\pi 1).$$

Furthermore, we define $R_e = \{a(\pi) \to b_e(\pi), a_e(\pi) \to \langle e \rangle(\pi)\}$ and $R_d = \{a(\pi) \to b_d(\pi), a_d(\pi) \to \langle d \rangle(\pi)\}$. Let $s \in T_\Sigma$ and denote by n the length of the leftmost path of s. On input s, A outputs the tree t of height n such that if $v \in V(t)$ is not a leaf and the rightmost leaf of the subtree s/v is labeled by e then $t[v] = g$, otherwise $t[v] = f$. If v is a leaf then $t[v] = s[v]$.   □

Clearly, the att A_2 in Example 2 is equivalent to a dt^R and A_2 has the single path property. In particular, it can be verified that the variations of all nodes that do not occur on the left-most path of the input tree are bounded. More precisely, if v does not occur on the leftmost path of the input tree then its visiting pair set is either $\psi_e = \{(b_e, a)\}$ or $\psi_d = \{(b_d, a)\}$. Thus, Ω_{ψ_e} consists of all trees in T_Σ whose rightmost leaf is labeled by e. For all such trees the attribute a yields the output $b_e(\epsilon)$. The case ψ_d is analogous.

In contrast, consider the att A_1 in Example 1. Recall that it translates an input tree s into a monadic tree t of height $n + 1$ if s has n leaves. This translation is not realizable by any dt^R. This is reflected in the fact that the att of Example 1 does *not* have the single path property. In particular, consider $s = f(f(e, e), f(e, e))$. The visiting pair set at all nodes of s is $\psi = \{(a, b)\}$ (cf. Fig. 1) and Ω_ψ is T_Σ. It can be verified that the variation of Ω_ψ is unbounded.

Recall that we aim to equip A with look-ahead for obtaining the att^R $\hat{A}$ that has the following property: On input $s \in \text{dom}(A)$, attributes of A only process nodes of a single path of s. Before we show how to test whether or not A has the single path property, we describe how to construct $\hat{A}$. Denote by B the bottom-up relabeling of $\hat{A}$ and let A' be A modified to process output trees produced by B. Let $s \in \text{dom}(A)$ and let s' be obtained from s via the relabeling B. The idea is that on input s', if attributes of A' process $v \in V(s')$ then the variation of v on input s with respect to A is unbounded. Note that obviously $V(s) = V(s')$. Clearly, if A has the single path property then attributes of A' only process nodes of a single path of s'.

Now the question is how precisely do we construct $\hat{A}$? It can be shown that all $\psi \subseteq I \times S$ that are visiting pair sets of A can be computed. Let ψ be a visiting pair set. If the variation of Ω_ψ is bounded, then a minimal integer κ_ψ can be computed such that for all $a \in S$ such that $(b, a) \in \psi$ for some $b \in I$ and for all $s' \in \Omega_\psi$, height(nf($\Rightarrow_{A,s'}, a(\epsilon)$)) $\leq \kappa_\psi$. Whether or not the variation of Ω_ψ is bounded can be decided as finiteness of ranges of att's is decidable [3]. Thus, $\kappa = \max\{\kappa_\psi \mid \psi$ is a visiting pair set of A and the variation of Ω_ψ is bounded$\}$ is computable.

Denote by $T_\Delta^\kappa[I(\{\epsilon\})]$ the set of all trees in $T_\Delta[I(\{\epsilon\})]$ of height at most κ. Informally, the idea is that the bottom-up relabeling of the att^R $\hat{A}$ precomputes output subtrees of height at most κ that contain the inherited attributes of the root of the current input subtree. Hence, the att of $\hat{A}$ does not need to compute those output subtrees itself; the translation is continued immediately with those output subtrees. Formally, the bottom-up relabeling B is constructed as follows. The states of B are sets $\varrho \subseteq \{(a, \xi) \mid a \in S$ and $\xi \in T_\Delta^\kappa[I(\{\epsilon\})]\}$. The idea is that if $s \in \text{dom}_B(\varrho)$ and $(a, \xi) \in \varrho$ then $\xi = \text{nf}(\Rightarrow_{A,s}, a(\epsilon))$. Given $\sigma(s_1, \ldots, s_k)$, B relabels σ by $\sigma_{\varrho_1,\ldots,\varrho_k}$ if for $i \in [k]$, $s_i \in \text{dom}_B(\varrho_i)$. Note that knowing $\varrho_1, \ldots, \varrho_k$ and the rules in the set R_σ of A, we can easily compute ϱ such that $\sigma(s_1, \ldots, s_k) \in \text{dom}_B(\varrho)$ and hence the rules of B. In particular, ϱ contains all pairs (a, ξ) such that $\xi = \text{nf}(\Rightarrow_{A,\sigma(s_1,\ldots,s_k)}, a(\epsilon)) \in T_\Delta^\kappa[I(\{\epsilon\})]$. Therefore, B contains the rule $\sigma(\varrho_1(x_1), \ldots, \varrho_k(x_k)) \to \varrho(\sigma_{\varrho_1,\ldots,\varrho_k}(x_1, \ldots, x_k))$.

Example 3. Consider the att A_2 in Example 2. Recall that all nodes that do not occur on the leftmost path of the input tree s of A_2 have bounded variation. Let v be such a node. Then the visiting pair set at v is either $\psi_e = \{(a, b_e)\}$ or $\psi_d = \{(a, b_d)\}$. Assume the former. Then $\text{nf}(\Rightarrow_{A_2,s/v}, a(\epsilon)) = b_e(\epsilon)$. If we know beforehand that a produces $b_e(\epsilon)$ when translating s/v, then there is no need to process s/v with a anymore. This can be achieved via a bottom-up relabeling B_2 that precomputes all output trees of height at most $\kappa = \kappa_{\psi_e} = \kappa_{\psi_d} = 1$. In particular the idea is that if for instance $v \in V(s)$ is relabeled by $f_{\{(a,b_d(\epsilon))\},\{a,b_e(\epsilon)\}}$ then this means when translating $s/v.1$ and $s/v.2$, a produces $b_d(\epsilon)$ and $b_e(\epsilon)$, respectively. The full definition of B_2 is as follows: The states of B_2 are $\varrho_1 = \{(a_e, \langle e \rangle(\epsilon)), (a, b_e(\epsilon))\}$, $\varrho_2 = \{(a_d, \langle d \rangle(\epsilon)), (a, b_d(\epsilon))\}$, $\varrho_3 = \{(a, b_d(\epsilon))\}$, and $\varrho_4 = \{(a, b_e(\epsilon))\}$. All states are final states. In addition to $e \to \varrho_1(e)$ and $d \to \varrho_2(d)$, B_2 also contains the rules

$$f(\varrho(x_1), \varrho_1(x_2)) \to \varrho_4(f_{\varrho,\varrho_1}(x_1, x_2)) \quad f(\varrho(x_1), \varrho_2(x_2)) \to \varrho_3(f_{\varrho,\varrho_2}(x_1, x_2))$$
$$f(\varrho(x_1), \varrho_3(x_2)) \to \varrho_3(f_{\varrho,\varrho_3}(x_1, x_2)) \quad f(\varrho(x_1), \varrho_4(x_2)) \to \varrho_4(f_{\varrho,\varrho_4}(x_1, x_2)),$$

where $\varrho \in \{\varrho_1, \ldots, \varrho_4\}$. It is easy to see that using B_2, attributes of A_2', that is, A_2 modified to make use of B_2, only process nodes of the leftmost path of s_2.□

With B, we obtain $\hat{A} = (B, A')$ equivalent to A. Note that A is modified into A' such that its input alphabet is the output alphabet of B and its rules make use of B. In particular, the rules of A' for a symbol $\sigma_{\varrho_1,\ldots,\varrho_k}$ are defined as follows. First, we introduce for each state ϱ of B an auxiliary symbol $\langle\varrho\rangle$ of rank 0 to A. We define that $a(\pi) \to t \in R_{\langle\varrho\rangle}$ if $(a, t[\pi \leftarrow \epsilon]) \in \varrho$. Denote by $[\pi \leftarrow v]$ the substitution that substitutes all occurrences of π by the node v, e.g. for $t_1 = f(b(\pi))$ and $t_2 = f(a(\pi2))$ where f is a symbol of rank 1, $a \in S$ and $b \in I$, we have $t_1[\pi \leftarrow v] = f(b(v))$ and $t_2[\pi \leftarrow v] = f(a(v2))$. Let $t = \mathrm{nf}(\Rightarrow_{A,\sigma(\langle\varrho_1\rangle,\ldots,\langle\varrho_k\rangle)}$ $, a(\epsilon)) \in T_\Delta[I(\{\epsilon\}) \cup S([k])]$. Then we define the rule $a(\pi) \to t' \in R_{\sigma_{\varrho_1,\ldots,\varrho_k}}$ for A' where t' is the tree such that $t'[\pi \leftarrow \epsilon] = t$. It should be clear that since all output subtrees of height at most κ are precomputed, attributes of A' only process nodes whose variation with respect to A are unbounded (cf. Example 3).

In parallel with the above construction of $\hat{A}$, we can decide whether or not a given att A has the single path property as follows. Let $s \in \mathrm{dom}(A)$ and let s' be the tree obtained from s via the relabeling B. If nodes $v_1, v_2 \in V(s')$ with the same parent node exist such that on input s', attributes of A' process both v_1 and v_2 then A does not have the single path property. Thus, to test whether A has the single path property, we construct the following att^R $\check{A} = (\check{B}, \check{A}')$ from $\hat{A} = (B, A')$. Input trees of $\check{A}$ are trees $s \in \mathrm{dom}(\hat{A})$ where two nodes v_1, v_2 with the same parent node are annotated by flags f_1 and f_2 respectively. The relabeling $\check{B}$ essentially behaves like B ignoring the flags. Likewise, the att $\check{A}'$ behaves like A' with the restriction that output symbols are only produced if an annotated symbol is processed by a synthesized attribute or if a rule in $R_\#$ is applied. For $i = 1, 2$ we introduce a special symbol g_i which is only outputted if the node with the flag f_i is processed. Hence, we simply need to check whether there is a tree with occurrences of both g_1 and g_2 in the range of $\check{A}$. Obviously, the range of $\check{A}$ is finite. Thus it can be computed.

Lemma 2. *It is decidable whether or not A has the single path property.*

4 From Tree to String Transducers and Back

Assume that A has the single path property. We now convert $\hat{A} = (B, A')$ into a two-way transducer T_W. Recall that two-way transducers are essentially attributed tree transducers with monadic input and output[1]. Informally the idea is as follows: Consider a tree $s \in \mathrm{dom}(A)$ and let s' be obtained from s via B. As on input s', attributes of A' only process nodes occurring on a single path ρ of s, the basic idea is to 'cut off' all nodes from s' not occurring in ρ. This way, we effectively make input trees of A' monadic. More formally, T_W is constructed by converting the input alphabet of A' to symbols of rank 1. Denote by Σ' the set of

---

[1] Note that the two-way transducers in [1] are defined with a *left end marker* $\vdash$ and a *right end marker* $\dashv$. While $\vdash$ corresponds to the root marker of our tree transducers, $\dashv$ has no counterpart. Monadic trees can be considered as strings with specific end symbols, i.e. symbols in Σ_0, that only occur at the end of strings. Thus, $\dashv$ is not required. Two-way transducers can test if exactly one end symbol occurs in the input string and if it is the rightmost symbol. Thus they can simulate tree transducers with monadic input and output.

all output symbols of B. Accordingly, let Σ'_k with $k \geq 0$ be the set of all symbols in Σ' of rank k. Let $\sigma' \in \Sigma'_k$ with $k > 0$. Then the input alphabet Σ^W of T_W contains the symbols $\langle \sigma', 1 \rangle, \ldots, \langle \sigma', k \rangle$ of rank 1. Furthermore, Σ^W contains all symbols in Σ'_0. Informally, a symbol $\langle \sigma', i \rangle$ indicates that the next node is to be interpreted as the i-th child. Thus trees over Σ^W can be considered prefixes of trees over Σ', e.g., let $f \in \Sigma'_2$, $g \in \Sigma'_1$ and $e \in \Sigma'_0$ then $\langle f, 2 \rangle \langle f, 1 \rangle \langle f, 1 \rangle e$ encodes the prefix $f(x_1, f(f(e, x_1), x_1))$ while $\langle f, 1 \rangle \langle g, 1 \rangle e$ encodes $f(g(e), x_1)$. Note that for monadic trees we omit parentheses for better readability. We call $t_1 \in T_{\Sigma'}[\{x_1\}]$ a prefix of $t_2 \in T_{\Sigma'}$ if t_2 can be obtained from t_1 by substituting nodes labeled by x_1 by ground trees. The idea is that as attributes of A' only process nodes occurring on a single path of the input tree, such prefixes are sufficient to simulate A'.

The attributes of T_W are the same ones as those of A'. The rules of T_W are defined as follows: Firstly the rules for $\#$ are taken over from A'. As before, assume that only rules for $\#$ have ground right-hand sides. Let A' contains the rule $a(\pi) \to t \in R_{\sigma'}$ where $\sigma' \in \Sigma'_k$ and $a, \alpha \in S$ and $\alpha(\pi i)$ occurs in t. Then T_W contains the rule $a(\pi) \to t[\pi i \leftarrow \pi 1] \in R_{\langle \sigma', i \rangle}$, where $[\pi i \leftarrow \pi 1]$ denotes the substitution that substitutes occurrences of πi by $\pi 1$. Analogously, if A' contains the rule $b(\pi i) \to t' \in R_{\sigma'}$ where $b \in I$ and for $\alpha \in S$, $\alpha(\pi i)$ occurs in t', then T_W contains the rule $b(\pi 1) \to t'[\pi i \leftarrow \pi 1] \in R_{\langle \sigma', i \rangle}$. We remark that as A has the single path property, A' will never apply a rule $b(\pi i) \to t'$ where $\alpha(\pi j)$ with $j \neq i$ occurs in t'. Thus, we do not need to consider such rules.

For the correctness of subsequent arguments, we require a technical detail: Let $\tilde{s}$ be an input tree of T_W. Then we require that an output tree s of B exists such that $\tilde{s}$ encodes a prefix of s. If such a tree s exists we say $\tilde{s}$ *corresponds to* s. To check whether for a given input tree $\tilde{s}$ of T_W an output tree s of B exists such that $\tilde{s}$ corresponds to s, we proceed as follows. As B is a relabeling, its range is effectively recognizable. Denote by $\bar{B}$ the bottom-up automaton recognizing it. Given $\bar{B}$, we can construct a bottom-up automaton $\bar{B}'$ that accepts exactly those trees $\tilde{s}$ for which an output tree s of B exists such that $\tilde{s}$ corresponds to s. W.l.o.g. assume that for all states l of $\bar{B}$, $\mathrm{dom}_B(l) \neq \emptyset$. If for $\sigma' \in \Sigma'_k$, $\sigma'(l_1(x_1), \ldots, l_k(x_k)) \to l(\sigma'(x_1, \ldots, x_k))$ is a rule of $\bar{B}$ then $\langle \sigma', i \rangle (l_i(x_1)) \to l(\langle \sigma', i \rangle (x_1))$ is a rule of $\bar{B}'$. We define that $\bar{B}'$ has the same accepting states as $\bar{B}$. Using $\bar{B}'$, we check whether for a given input tree $\tilde{s}$ of T_W an output tree s of B exists such that $\tilde{s}$ corresponds to s as follows. Before producing any output symbols, on input $\tilde{s}$, T_W goes to the leaf of $\tilde{s}$ and simulates $\bar{B}'$ in a bottom-up fashion while going back to the root. If $\tilde{s}$ is accepted by $\bar{B}'$ then T_W begins to simulate A', otherwise no output is produced. Thus the domain of T_W only consists of trees $\tilde{s}$ for which an output tree s of B exists such that $\tilde{s}$ corresponds to s. We remark that $\bar{B}'$ may be nondeterministic which in turn means that T_W may be nondeterministic as well, however the translation it realizes is a function. In fact the following holds.

Lemma 3. *Consider the att^R $\hat{A} = (B, A')$ and the two-way transducer T_W constructed from $\hat{A}$. Let $\tilde{s}$ be a tree over Σ^W. If on input $\tilde{s}$, T_W outputs t then for all $s \in range(B)$ such that $\tilde{s}$ corresponds to s, A' also produces t on input s.*

In the following, consider the two-way transducer T_W. Assume that the procedure of [1] yields a one-way transducer T_O that is equivalent to T_W. Given T_O, we construct a top-down transducer T' that produces output trees on the range of B. In particular, T' has the same states as T_O. Furthermore, a rule $q(\langle \sigma', i \rangle(x_1)) \to t$ of T_O where $\sigma' \in \Sigma'_k$ and $i \in [k]$ induces the rule $q(\sigma'(x_1, \ldots, x_k)) \to \hat{t}$ for T' where $\hat{t}$ is obtained from t by substituting occurrences of x_1 by x_i, e.g., if $t = f(g(q'(x_1)))$ then $\hat{t} = f(g(q'(x_i)))$. Recall that the domain of T_W only consists of trees $\tilde{s}$ for which an output tree s of B exists such that $\tilde{s}$ corresponds to s. As T_W and T_O are equivalent, the domain of T_O also consists of such trees. Hence, by construction, the following holds.

Lemma 4. *Consider the top-down transducer T' constructed from the one-way transducer T_O. Let $\tilde{s}$ be a tree over Σ^W. If on input $\tilde{s}$, T_O outputs t then for all $s \in range(B)$ such that $\tilde{s}$ corresponds to s, T' also produces t on input s.*

With Lemmas 3 and 4, it can be shown that the following holds.

Lemma 5. *The top-down transducer T' and the att A' are equivalent on the range of B.*

Therefore it follows that $\hat{A} = (B, A')$ and $N = (B, T')$ are equivalent. Consequently, A and N are equivalent. We remark that there is still a technical detail left. Recall that our aim is to construct a dt^R T equivalent to A. However, the procedure of [1] may yield a nondeterministic T_O. Thus T' and hence N may be nondeterministic (but functional). However, as shown in [5], we can easily compute a dt^R equivalent to N.

The arguments above are based on the assumption that the procedure of [1] yields a one-way transducer equivalent to T_W. Now the question is, does such a one-way transducer always exists if a dt^R equivalent to A exists? The answer to this question is indeed affirmative. In particular the following holds.

Lemma 6. *Consider the att^R $\hat{A} = (B, A')$ equivalent to A. If a dt^R T equivalent to A exists, then a (nondeterministic) top-down transducer with look-ahead $N = (B, N')$ exists such that $\hat{A}$ and N are equivalent.*

Proof. (sketch) Recall that given $\sigma(s_1, \ldots, s_k)$, B relabels σ by $\sigma_{\varrho_1, \ldots, \varrho_k}$ if for $i \in [k]$, $s_i \in dom_B(\varrho_i)$. In the following, denote by s_ϱ a fixed tree in $dom_B(\varrho)$.

Let $T = (B', T')$. We sketch the main idea of the proof, i.e., how to so simulate B' using B. First consider the following property which we call the *substitute-property*. Let $s \in T_\Sigma$ and $\hat{s}$ be obtained from s via the relabeling B. Let v_1 and v_2 be nodes of s with the same parent. As T exists, the single path property holds for A. Thus on input $\hat{s}$, v_1 or v_2 is not processed by attributes of A'. Assume that v_1 is not processed and that $s/v_1 \in dom_B(\varrho)$. Then $\tau_{\hat{A}}(s) = \tau_{\hat{A}}(s[v_1 \leftarrow s_\varrho])$ follows. Informally, this means that s/v_1 can be substituted by s_ϱ without affecting the output of the translation. By definition, $\hat{A}$ and A are equivalent while T and A are equivalent by our premise. Thus, $\tau_T(s) = \tau_T(s[v_1 \leftarrow s_\varrho])$.

Now we show how B' is simulated using B. Let $\hat{q}$ be a state of N'. Each such state $\hat{q}$ is associated with a state q of T' and a state l of B'. Consider the tree $\hat{s}$ obtained from s via B. Assume that the node v labeled by $\sigma_{\varrho_1,\ldots,\varrho_k}$ is processed by $\hat{q}$ in the translation of N' on input $\hat{s}$, i.e., v has k child nodes. It can be shown that $\hat{q}$ can compute which of v's child nodes is processed by attributes in the translation of A' on input $\hat{s}$. W.l.o.g. let $v.1$ be that node and let $s/v.i \in \text{dom}_B(\varrho_i)$ for $i \in [k]$. Due to the substitute-property, N can basically assume that $s/v.i = s_{\varrho_i}$ for $i \neq 1$. For $i \neq 1$, let $s_{\varrho_i} \in \text{dom}_{B'}(l_i)$. Now N' guesses a state l_1 for $s/v.1$ such that $\sigma(l_1(x_1),\ldots,l_k(x_k)) \to l(\hat{\sigma}(x_1,\ldots,x_k))$ is a rule of B'. The state $\hat{q}$ then 'behaves' as q would when processing a node labeled by $\hat{\sigma}$. It can be guaranteed that N' verifies its guess.   □

The dt^R $N = (B, N')$ can be constructed such that on input $s \in \text{range}(B)$ an attribute of A' processes the node v if and only if a state of N' processes v. The existence of such a transducer with look-ahead N implies the existence of a one-way transducer T_O equivalent to T_W. In fact, T_O is obtainable from N similarly to how T_W is obtainable from A_p. Therefore, the procedure of [1] yields a top-down transducer T_O equivalent to T_W if and only if T exists.

5   Final Results

The considerations in Sects. 3 and 4 yield the following theorem.

Theorem 1. *For an att with monadic output, it is decidable whether or not an equivalent dt^R exists and if so then it can be constructed.*

We show that the result of Theorem 1 is also obtainable for nondeterministic functional attributed tree transducers with look-around. Look-around is a formalism similar to look-ahead. An attributed tree transducers with look-around (or att^U) consists of a *top-down relabeling with look-ahead* R and an att A where output trees are computed as $\tau_A(\tau_R(s))$ for input trees s. Before we prove our claim, we prove the following result. Denote by $nATT^U$ and $dATT^U$ the classes of translations realizable by nondeterministic and deterministic att^U's, respectively. Denote by *func* the class all functions.

Lemma 7. $nATT^U \cap func = dATT^U$.

Proof. (sketch) Informally, the basic idea is the same as for functional top-down transducers in [5]. For simplicity, let A be a functional *att* without look-around. Consider a rule set R_σ of A where $\sigma \in \Sigma$. Let all rules in R_σ with the same left-hand side be ordered. Whenever several such rules are applicable, i.e., utilizing these particular rule leads to the production of a ground tree, the first applicable rule in the given order will be executed. Using look-around, we can test whether or not a rule is applicable.   □

We remark that look-around is required; just look-ahead for instance is not sufficient as $ATT \cap func \not\subseteq dATT^R$ can be shown.

Due to Lemma 7, it is sufficient to show that for a deterministic att^U it is decidable whether or not an equivalent dt^R exists. Roughly speaking, this result is obtained as follows. Let a deterministic att^U $\hat{A} = (R, A')$, where R is a top-down relabeling with look-ahead and A' is an att, be given. To show that a dt^R equivalent to $\hat{A}$ exists it is sufficient to show that a dt^R T exists such that A' and T are equivalent on the range of R. This is because R is also a dt^R and dt^R's are closed under composition [4]. The dt^R T can be obtained by slightly modifying the procedure in Sect. 3.

Theorem 2. *For a functional att^U with monadic output, it is decidable whether or not an equivalent dt^R exists and if so then it can be constructed.*

Note that by definition dt^R's with monadic output are by default *linear*. For a linear dt^R it is decidable whether or not an equivalent linear dt exists [12]. If such a dt exists it can be constructed. Hence, we obtain the following corollary.

Corollary 1. *For a functional att^U with monadic output, it is decidable whether or not an equivalent dt exists and if so then it can be constructed.*

6 Conclusion

We have shown how to decide for a given attributed transducer with look-around but restricted to monadic output, whether or not an equivalent deterministic top-down tree transducers (with or without look-ahead) exists. Clearly we would like to extend this result to non-monadic output trees. The latter seems quite challenging, as it is not clear whether or not the result [1] can be applied in this case. Other questions that remain are the exact complexities of our constructions.

References

1. Baschenis, F., Gauwin, O., Muscholl, A., Puppis, G.: One-way definability of two-way word transducers. Log. Methods Comput. Sci. 14(4) (2018)
2. Bloem, R., Engelfriet, J.: A comparison of tree transductions defined by monadic second order logic and by attribute grammars. JCSS **61**(1), 1–50 (2000)
3. Drewes, F., Engelfriet, J.: Decidability of the finiteness of ranges of tree transductions. Inf. Comput. **145**(1), 1–50 (1998)
4. Engelfriet, J.: Top-down tree transducers with regular look-ahead. Math. Syst. Theory **10**, 289–303 (1977)
5. Engelfriet, J.: On tree transducers for partial functions. Inf. Process. Lett. **7**(4), 170–172 (1978)
6. Engelfriet, J., Maneth, S.: Macro tree transducers, attribute grammars, and MSO definable tree translations. Inf. Comput. **154**(1), 34–91 (1999)
7. Engelfriet, J., Maneth, S.: Macro tree translations of linear size increase are MSO definable. SIAM J. Comput. **32**(4), 950–1006 (2003)
8. Filiot, E., Gauwin, O., Reynier, P., Servais, F.: From two-way to one-way finite state transducers. In: LICS, pp. 468–477 (2013)
9. Filiot, E., Maneth, S., Reynier, P., Talbot, J.: Decision problems of tree transducers with origin. Inf. Comput. **261**, 311–335 (2018)
10. Fülöp, Z.: On attributed tree transducers. Acta Cybern. **5**(3), 261–279 (1981)
11. Knuth, D.E.: Semantics of context-free languages. Math. Syst. Theory **2**(2), 127–145 (1968)
12. Maneth, S., Seidl, H.: When is a bottom-up deterministic tree translation top-down deterministic? In: ICALP. LIPIcs, vol. 168, pp. 1–18 (2020)

A Time to Cast Away Stones

Olga Martynova and Alexander Okhotin(✉)

Department of Mathematics and Computer Science, St. Petersburg State University,
7/9 Universitetskaya nab., Saint Petersburg 199034, Russia
st062453@student.spbu.ru, alexander.okhotin@spbu.ru

Abstract. Deterministic graph-walking automata equipped with finitely many "drop-once" pebbles are investigated: these automata explore a graph by following its edges using finitely many states, and they can also leave pebbles at some nodes, but cannot lift these pebbles anymore. It is proved that for every such automaton there is a graph-walking automaton without pebbles that recognizes the same set of graphs. To be precise, for an automaton with m drop-once pebbles and n states, operating on graphs with k labels of edge end-points, an automaton without pebbles with $(2k + 1)^m m! 3^{\frac{m(m+1)}{2}} n^{m+1}$ states accepting the same set of graphs is constructed. The results apply to two-way finite automata (with $k = 2$) and to tree-walking automata as special cases.

1 Introduction

Various families of automata that walk on input objects, such as a string, a tree or a graph, are well-known: these are two-way finite automata (2DFA), tree-walking automata (TWA), graph-walking automata (GWA), etc. Also their variants equipped with pebbles have been studied for a long time. An automaton may drop a pebble at some square or a node, and when it later returns there, it will see the pebble and may lift it to move it to another place.

Automata with pebbles were first investigated by Blum and Hewitt [1], who introduced *two-way finite automata with pebbles*, as well as automata with pebbles on rectangular grids. In particular, Blum and Hewitt [1] proved that 2DFA with a single pebble recognize only regular languages. At the same time, two pebbles allow the automata to recognize some non-regular languages, such as $\{ a^n b^n \mid n \geqslant 0 \}$. Later, Goralčík et al. [10] proved regularity of languages recognized by one-pebble alternating two-way automata. Geffert and Ištoňová [8] studied the state complexity of deterministic and nondeterministic two-way automata with one pebble.

Later, Globerman and Harel [9] investigated pebble 2DFA with m *nested pebbles*. At every moment, such an automaton has pebbles $1, \ldots, i$ placed, and pebbles $i + 1, \ldots, m$ in its hands. Then, the automaton may either lift the i-th pebble, or drop the $(i + 1)$-th pebble, but cannot do any other operations on pebbles. Globerman and Harel [9] proved that such automata, with a certain

This work was supported by the Russian Science Foundation, project 23-11-00133.

extra assumption, also can recognize only regular languages (regularity without the extra assumption was later proved by Engelfriet and Hoogeboom [6]). Also, Globerman and Harel [9] established a lower bound: that simulating an automaton with m nested pebbles and n states by an ordinary 2DFA may require a number of states given by a tower of $m + 1$ exponents.

Tree-walking automata with pebbles were first considered by Engelfriet and Hoogeboom [6]; they proved that every nondeterministic tree-walking automaton with m nested pebbles recognizes a regular tree language. Bojańczyk et al. [3] established a strict hierarchy of such automata by the number of pebbles, both in the deterministic and in the nondeterministic cases. Muscholl et al. [12] proved the closure of deterministic automata with nested pebbles under complementation, using extra pebbles; later, Kunc and Okhotin [11] have shown that the same can be achieved without extra pebbles.

Pebbles are also being considered in the more general case of (deterministic) *graph-walking automata*, that is, automata that walk on graphs of the general form, like tree-walking automata walk on trees and 2DFA walk on strings. The model of graph-walking automata with pebbles was reportedly first proposed by Michael Rabin in his 1967 public lecture [7], who conjectured that for every graph-walking automaton with finitely many pebbles there is a graph that it cannot fully explore. A decade later, Budach [4] proved Rabin's conjecture for graph-walking automata without pebbles. Rollik [13] established an even stronger result than Rabin's conjecture: namely, that if a finite team of automata simultaneously explore the same graph, being able to see each other when they meet and knowing all that the others know, then even in this case there is a graph that they shall not fully explore. Rollik's model is at least as powerful as automata with pebbles, because some of the automata from the team may assume the role of pebbles: that is, stand still when they have been "dropped", and follow their teammate who "lifted" them.

On the other hand, there are some positive results on traversing graphs using pebbles. Blum and Kozen [2] proved that an automaton with 2 pebbles can traverse every plane graph embedded in a rectangular grid with directions of motion marked according to the grid. Disser et al. [5] have shown that every graph with n nodes can be traversed by a single automaton equipped with $\log \log n$ pebbles and $\log \log n$ bits of memory.

What is the class of graph languages recognized by graph-walking automata with pebbles? It follows from the results of Bojańczyk et al. [3] that already deterministic tree-walking automata become stronger with 1 pebble, which can be lifted and moved. Hence, neither for tree-walking nor for graph-walking automata there could be a result similar to the result of Globerman and Harel [9] for 2DFA.

This paper investigates a weaker kind of pebbles, *drop-once pebbles*, which can only be dropped once, and cannot be lifted anymore. This is a special case of nested pebbles. An automaton proverbially *casts away stones*, and never gathers stones. It turns out that no finite number of drop-once pebbles can increase the expressive power of graph-walking automata. To show this, for every graph-walking automaton with m drop-once pebbles, a graph-walking automaton without pebbles recognizing the same set of graphs is constructed.

At first, in Sect. 3, the case of only one pebble is investigated. For any given automaton with 1 drop-once pebble, a new automaton without pebbles simulates its moves step by step, and also remembers whether the original automaton has already dropped its pebble. If the pebble has been dropped already, then at every step the simulating automaton needs to inquire whether this pebble has been dropped precisely at the current node. For this purpose, a separate *pebble-checking automaton* is executed at every step: it checks whether the pebble was dropped at the current node by traversing the tree of the original automaton's computations that end by dropping the pebble at the current node. The general idea of traversing the tree of computations is due to Sipser [14], and it is often used in the literature to simulate various kinds of automata; this paper employs a variant of Sipser's method for graph-walking automata described by Kunc and Okhotin [11]. Using this method, the pebble-checking automaton is constructed.

More or less the same is done in the general case of m pebbles, considered in Sect. 4. The set of pebbles in hands is remembered in the state of the simulating automaton. At each step, the presence of each of the previously dropped pebbles at the current node is determined using a separate pebble-checking automaton for each pebble. This checking also gets more complicated, because the pebble in question could have been dropped after some other pebbles, and then the computations resulting in its being dropped get more complicated to simulate. These pebble-checking automata are actually constructed inductively on the number of pebbles.

2   Definitions

All definitions for graph-walking automata are inherited from the paper by Kunc and Okhotin [11]. For graph-walking automata, the *signature* is analogous to an alphabet for a 2DFA, and it determines the set of all labeled graphs that are admissible as inputs for a graph-walking automaton. The signature defines the possible labels for nodes and for edge end-points in graphs.

Definition 1 ([11]). *A signature S is a quintuple $S = (D, -, \Sigma, \Sigma_0, (D_a)_{a \in \Sigma})$, where:*

- *D is a finite set of directions, which are labels attached to edge end-points;*
- *a bijection $-\colon D \to D$ provides an opposite direction, with $-(-d) = d$ for all $d \in D$;*
- *Σ is a finite set of node labels;*
- *$\Sigma_0 \subseteq \Sigma$ is a non-empty subset of possible labels of the initial node;*
- *$D_a \subseteq D$, for every $a \in \Sigma$, is the set of directions in each node labelled with a.*

Like strings are defined over an alphabet, graphs are defined over a signature.

Definition 2. A graph *over a signature* $S = (D, -, \Sigma, \Sigma_0, (D_a)_{a \in \Sigma})$ *is a quadruple* $(V, v_0, +, \lambda)$, *where:*

- V *is a finite set of nodes;*
- $v_0 \in V$ *is the initial node;*
- *edges are defined by a partial function* $+: V \times D \to V$, *such that if* $v + d$ *is defined, then* $(v+d)+(-d)$ *is defined and equals* v; *also denote* $v - d = v + (-d)$;
- *node labels are assigned by a total mapping* $\lambda: V \to \Sigma$, *such that*
    i. $v + d$ *is defined if and only if* $d \in D_{\lambda(v)}$, *and*
    ii. $\lambda(v) \in \Sigma_0$ *if and only if* $v = v_0$.

The function $+$ defines the edges of the graph. If $u + d = v$, then the nodes u and v in the graph are connected with an edge with its end-points labelled with directions d (on the side of u) and $-d$ (on the side of v). Multiple edges and loops are possible.

A graph-walking automaton is defined similarly to a 2DFA, with an input graph instead of an input string.

Definition 3. A (deterministic) graph-walking automaton (GWA) over a signature $S = (D, -, \Sigma, \Sigma_0, (D_a)_{a \in \Sigma})$ is a quadruple $A = (Q, q_0, F, \delta)$, where

- Q *is a finite set of states;*
- $q_0 \in Q$ *is the initial state;*
- $F \subseteq Q \times \Sigma$ *is a set of acceptance conditions;*
- $\delta: (Q \times \Sigma) \backslash F \to Q \times D$ *is a partial transition function, with* $\delta(q, a) \in Q \times D_a$ *for all* q *and* a *where* δ *is defined.*

An automaton operating on a graph at each moment sees only the label of the current node and knows its current state. Then, according to the transition function, it enters a new state and moves in the given direction to one of the neighbouring nodes. If the automaton enters an accepting pair of a state and a node label (one in F), then it accepts, and if it enters a pair not in F with no transition defined, then it rejects. It may also loop, that is, continue walking indefinitely.

Formally, an automaton's *configuration* on a graph $G = (V, v_0, +, \lambda)$ is a pair (q, v), with $q \in Q$ and $v \in V$. A *computation* of an automaton A on a graph G is the following uniquely defined sequence of configurations. The computation starts in the initial configuration (q_0, v_0). For every configuration (q, v) in the computation, if $\delta(q, \lambda(v))$ is defined and equals (q', d), then the next configuration after (q, v) is $(q', v + d)$. Otherwise, the configuration (q, v) is the last one in the computation; if $(q, \lambda(v)) \in F$, then the automaton *accepts* in the configuration (q, v), otherwise it rejects. If the computation is an infinite sequence, then the automaton is said to *loop*.

A graph walking automaton A defines the language $L(A)$, this is the set of graphs it accepts.

This paper investigates what happens, if a graph-walking automaton is given finitely many distinguishable pebbles $p_1, \ldots, p_m$, which it can drop at the nodes of a graph, but cannot lift. Once a pebble p_i is dropped at some node v, it remains at that node until the end of the computation. Later on it will be proved that this model is not any stronger than graph-walking automata without pebbles.

Definition 4. A graph-walking automaton with drop-once pebbles *over a signature* $S = (D, -, \Sigma, \Sigma_0, (D_a)_{a \in \Sigma})$ *is a quintuple* $A_p = (Q, q_0, P, F, \delta)$*, that consists of the following components.*

- Q *is a finite set of states.*
- $q_0 \in Q$ *is the initial state.*
- P *is a finite set of distinct pebbles.*
  When the automaton works on a graph, at every moment it is in a state $q \in Q$, has a set of pebbles $P_1 \subseteq P$ in its hands, is at a node labelled with $a \in \Sigma$ and sees pebbles from $P_2 \subseteq P$ placed at this node. The quadruple (q, P_1, a, P_2) contains all the information the automaton can use to decide what to do, and it is called the environment.
- $F \subseteq Q \times 2^P \times \Sigma \times 2^P$ *is the set of accepting environments. If the automaton is in an environment (q, P_1, a, P_2) in F, then it accepts.*
- $\delta \colon (Q \times 2^P \times \Sigma \times 2^P) \backslash F \to Q \times D \times 2^P$ *is a partial transition function, which determines the automaton's action in each environment. If $\delta(q, P_1, a, P_2) = (q', d, P_2')$, this means that the automaton enters the state q' and leaves the node in the direction d, leaving behind the pebbles from P_2' (these are both the pebbles that were at this node previously, and the pebbles dropped by this transition). Accordingly, such a transition must satisfy the conditions $d \in D_a$, $P_2 \subseteq P_2'$ and $P_2' \subseteq P_1 \cup P_2$, that is, there should be a direction d in nodes labelled with a, the pebbles which are already at the node must remain there, and in addition the automaton can leave any pebbles it has in its hands. After the transition, the automaton has pebbles from $P_1 \backslash P_2'$ in its hands.*

An automaton with drop-once pebbles starts its computation on a graph at the initial node v_0 in the state q_0, with all pebbles in its hands, and with no pebbles in any nodes of the graph. Then the automaton moves according to the transition function, possibly dropping pebbles at some nodes. A pebble once dropped at a node remains there until the end of the computation. If the automaton gets into an accepting environment, then it accepts the graph; if it gets into an environment with no transition defined, then it rejects, and if it works forever, then it is said to loop.

Formally, a configuration of an automaton with drop-once pebbles on a graph $G = (V, v_0, +, \lambda)$ is a triple (q, v, f), where $q \in Q$ is the automaton's current state, $v \in V$ is the node it is currently observing, and a partial function $f \colon P \to V$ says for each pebble where it currently is. If a pebble is in the automaton's hands, then the value of f is undefined. The initial configuration is (q_0, v_0, f_0), where f_0 is defined nowhere. The next configuration, and accepting and rejecting configurations, are naturally defined according to the explanations above. A computation is defined in the same way as for an ordinary graph-walking automaton.

The language $L(A_p)$ recognized by an automaton A_p with drop-once pebbles is the set of all graphs it accepts.

3   Simulating One Drop-Once Pebble

The main result of this paper is that every graph-walking automaton with m drop-once pebbles can be simulated by a graph-walking automaton without pebbles. This will first be proved in the case of one drop-once pebble.

Theorem 1. *Let $A_p = (Q, q_0, \{p\}, F, \delta)$ be a graph-walking automaton with one drop-once pebble p, defined over a signature $S = (D, -, \Sigma, \Sigma_0, (D_a)_{a \in \Sigma})$. Let $n = |Q|$ be the number of its states, and let $k = |D|$ be the number of directions in the signature.*

Then there exists a graph-walking automaton A_0 without pebbles, with $(2k + 1)n^2 + 2n$ states, which accepts the same set of graphs as A_p. Furthermore, A_0 accepts or rejects at the same node as A_p (and if A_p loops, then so does A_0).

Until A_p drops its pebble, A_0 can just repeat the actions of A_p using the set of states Q, simulating the transitions and the accepting conditions of A_p with an imaginary pebble in its hands (that is, for $P_1 = \{p\}$ and $P_2 = \varnothing$). If A_p never drops the pebble on a graph G, then A_0 simulates it correctly to the end, using only states from Q. But a time may come when A_p decides to drop its pebble, while the automaton A_0 has no pebble to drop. What shall it do? It can remember that the pebble has been dropped and simulate the subsequent transitions of A_p using $P_1 = \varnothing$. But in order to obtain P_2, the set of pebbles at the current node, the automaton should somehow determine whether there is a pebble in it or not.

This is done in special states, in which A_0 remembers the current state of the simulated A_p, and performs a search in the graph to check whether the pebble is at the current node. For convenience, this search procedure is defined as a separate *pebble-checking* automaton B, which can be started from any node to determine whether A_p drops the pebble precisely at that node. This B is defined in the following lemma.

Lemma 1. *For every automaton A_p with 1 drop-once pebble and with n states, defined over a signature with k directions, there exists an automaton without pebbles $B = (Q^{(B)}, q_0^{(B)}, F^{(B)}, \delta^{(B)})$ over the same signature, with $(2k + 1)n + 1$ states, such that, if B is started at some node v of some graph G in its initial state $q_0^{(B)}$, then it eventually stops at the same node v, and*

- *if A_p, working on the graph G from its initial configuration, drops the pebble at the node v, then B accepts at the node v;*
- *and if A_p, working on G from its initial configuration, never drops the pebble, or drops it at a node other than v, then B rejects at v.*

The computation of A_p before it drops the pebble is like a computation of an automaton without pebbles. Therefore, what is done by A_p up to this point can be reformulated without using pebble automata. Consider an automaton A without pebbles with the same set of states as A_p, that operates as A_p with the pebble in its hands. Once A_p decides to drop the pebble, A accepts. If A_p stops (that is, accepts or rejects) without dropping the pebble, then A rejects. Then, instead of solving the problem whether A_p drops the pebble at the current node, B can solve another equivalent problem: does A accept at the current node v? Thus, Lemma 1 has been reduced to the following lemma.

Lemma 2. *Let A be a graph-walking automaton with n states over some signature with k directions. Then there exists a graph-walking automaton B with $(2k+1)n+1$ states, such that, if B is started at some node v of some graph G in its initial state $q_0^{(B)}$, then it eventually stops at the same node v, and, if the automaton A accepts the graph G at the node v, then B also accepts at v, and otherwise B rejects at v.*

Proof (a sketch). Automata answering the question whether another automaton accepts at the current node were already considered in the literature. Kunc and Okhotin [11, Lemma 6] constructed an automaton B_1 that answers a very similar question. This automaton B_1 works by traversing the tree of possible computations of A that converge at a particular configuration at the node v, in a particular state (the general idea of such a traversal is due to Sipser [14]). The automaton B_1 tries to find the initial configuration of A in this tree. If it succeeds, then it stops at the initial node. If it fails, then it completes the traversal and rejects at the same node v.

There are several differences with the desired automaton B: first, B_1 starts at the neighbouring node of v, rather than in v itself; secondly, B_1 checks not for any kind of acceptance at a node, but for acceptance in a particular state; and if the answer is "yes", then it gives it not at the node v, but at the initial node of the graph.

The automaton B is constructed using B_1. It runs B_1 several times starting from all neighbouring nodes of the current node v, and for all states in which A could accept. Every time B_1 fails to find an initial configuration, it returns to v, and B restarts it from another accepting configuration. If B_1 ever gives a positive answer, then it does so at the initial node, and, next, B simulates A to get back to v and accept.

The details are omitted due to space constraints. $\square$

The automaton B is used in the construction of the automaton A_0 as follows. Once the simulation of A_p carried out by A_0 comes to the moment of dropping the pebble, the automaton A_0 continues as follows. Before simulating every subsequent move of A_p, the automaton A_0 will run B first, in order to determine whether there is a pebble at the current node. The automaton B returns to the same node with an answer, and then A_0 can simulate the correct transition of A_p. The simulation of A_p by A_0 is illustrated in Fig. 1.

Proof (of Theorem 1). The set of states of the automaton A_0 is $Q^{(A_0)} = Q \cup (Q \times Q^{(B)})$, where the states from Q are used to simulate A_p up to the moment when it drops the pebble. In the rest of the computation, if A_0 is in a state $(q, q') \in Q \times Q^{(B)}$, this means that the automaton A_0 simulates A_p in the state q and B in the state q'. While B runs, the state q of the automaton A_p

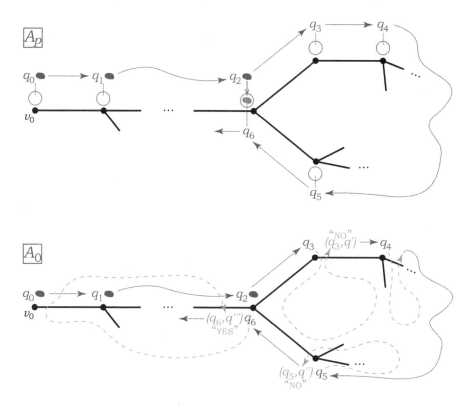

Fig. 1. (top) A pebble automaton A_p drops the pebble and later sees it again; (bottom) an automaton A_0 without pebbles first works as A_p until it drops a pebble, and then continues the simulation using B at each step to check whether the pebble was dropped at the current node.

stays unchanged. Once the computation of B ends, A_0 will have returned to the node in which it is simulating A_p in the state q. The configuration of B will be accepting if there is a pebble at this node, and rejecting otherwise. This is the time to simulate the next step of A_p.

The initial state is $q_0 \in Q$.

The transitions and acceptance conditions of A_0 are defined as follows. Until A_p drops the pebble, the automaton A_0 repeats its transitions verbatim, using the states from Q.

$$\delta^{(A_0)}(q, a) = (r, d), \qquad \text{for } q \in Q, \ a \in \Sigma, \ \delta(q, \{p\}, a, \varnothing) = (r, d, \varnothing)$$

If A_p accepts without having dropped a pebble, then A_0 accepts as well.

$$(q, a) \in F^{(A_0)}, \qquad \text{for } q \in Q, \ a \in \Sigma, \ (q, \{p\}, a, \varnothing) \in F$$

If A_p rejects without having dropped a pebble, then so does A_0.

Let the automaton A_p drop the pebble by a transition $\delta(q, \{p\}, a, \varnothing) = (r, d, \{p\})$. In this case, A_0 moves to the same node, remembers the state r and immediately starts simulating B in order to determine whether there is a pebble at the node at the next step (and it may be there, since the graph may contain loops).

$$\delta^{(A_0)}(q, a) = ((r, q_0^{(B)}), d), \quad \text{for } q \in Q,\ a \in \Sigma,\ \delta(q, \{p\}, a, \varnothing) = (r, d, \{p\})$$

The automaton B is simulated until the end, that is, up to its accepting or rejecting configuration. While it runs, the current state of the automaton A_p at the node where the simulation of B has started and where the simulation of A_p has been suspended, is kept unchanged in the first component of the state.

$$\delta^{(A_0)}((q, q'), a) = ((q, r'), d), \quad \text{for } q \in Q,\ q' \in Q^{(B)},\ a \in \Sigma,\ \delta^{(B)}(q', a) = (r', d)$$

If the simulation of B ends in its accepting configuration, it means that the automaton A_0 has come to the node, in which the simulation of A_p was suspended, and at the same time A_0 has learned that the pebble is at this node. Then A_0 simulates the transition of A_p for the case of a pebble present at the current node, and immediately starts a new simulation of B at the next node.

$$\delta^{(A_0)}((q, q'), a) = ((r, q_0^{(B)}), d), \quad \text{for } q \in Q,\ q' \in Q^{(B)},\ a \in \Sigma,$$
$$(q', a) \in F^{(B)},\ \delta(q, \varnothing, a, \{p\}) = (r, d, \{p\})$$

If A_p accepts with the pebble at the current node, then A_0 accepts as well, and thus concludes the simulation of A_p at the same node, at which A_p accepts the graph.

$$((q, q'), a) \in F^{(A_0)}, \quad \text{for } q \in Q,\ q' \in Q^{(B)},\ a \in \Sigma,$$
$$(q', a) \in F^{(B)},\ (q, \varnothing, a, \{p\}) \in F$$

If A_p rejects with a pebble at the current node, then so does A_0.

If there is no pebble at the node where A_p is being simulated, then the simulation of B ends in a rejecting configuration. In this case, A_0 simulates the transition, acceptance or rejection of A_p for the case of no pebble at the current node.

$$\delta^{(A_0)}((q, q'), a) = ((r, q_0^{(B)}), d), \quad \text{for } q \in Q,\ q' \in Q^{(B)},\ a \in \Sigma,$$
$$\delta^{(B)}(q', a) \text{ is undefined, } \delta(q, \varnothing, a, \varnothing) = (r, d, \varnothing)$$
$$((q, q'), a) \in F^{(A_0)}, \qquad \text{for } q \in Q,\ q' \in Q^{(B)},\ a \in \Sigma,$$
$$\delta^{(B)}(q', a) \text{ is undefined, } (q, \varnothing, a, \varnothing) \in F$$

The automaton A_0 without pebbles has been constructed, and it accepts the same set of graphs as A_p, and accepts or rejects at the same nodes as A_p, and if A_p loops, then A_0 loops as well. The automaton A_0 has exactly $n + ((2k+1)n+1)n = (2k+1)n^2 + 2n$ states, as desired. □

4   When Pebbles Grow in Number

Automata with multiple drop-once pebbles can also be simulated by graph-walking automata without pebbles. This is what the next theorem is about.

Theorem 2. *Let $A_p = (Q, q_0, P, F, \delta)$ be a graph-walking automaton with $n = |Q|$ states and $m = |P|$ drop-once pebbles, defined over a signature $S = (D, -, \Sigma, \Sigma_0, (D_a)_{a \in \Sigma})$ with $k = |D|$ directions. Then there exists a graph-walking automaton A_0 without pebbles, over the same signature, which has at most $(2k+1)^m m! 3^{\frac{m(m+1)}{2}} n^{m+1}$ states and accepts the same set of graphs as the automaton A_p.*

Furthermore, A_0 accepts or rejects at the same node as the pebble automaton A_p, and loops if A_p loops.

Proof (a sketch). The theorem is proved by induction on the number of pebbles.

Base case: $m = 0$. This holds for $A_0 = A_p$.

Induction step: $m - 1 \to m$.

Let $P = \{1, \ldots, m\}$. Similarly to the proof in Sect. 3, there will be a pebble-checking automaton B_i for each $i \in \{1, \ldots, m\}$, which tests whether the i-th pebble is at the current node or not. This is done as follows.

Fix $i \in \{1, \ldots, m\}$. The goal is to define a graph-walking automaton B_i without pebbles, which, if it is started in its initial state $q_0^{(B_i)}$ at any node v of any graph G, then it eventually stops at the node v, and accepts if the automaton A_p operating on G ever drops the i-th pebble at the node v, and rejects otherwise.

If the i-th pebble were always the first to be dropped by A_p, then the computation of A_p up to the moment it is dropped would be the computation of an automaton without pebbles, and then B_i could be obtained directly by Lemma 2. The main difficulty is that the i-th pebble need not be the first to be dropped by A_p. In this case, the computation of A_p leading to the i-th pebble's being dropped is a computation of a certain automaton $A_{i,p}$ with $m - 1$ drop-once pebbles. The solution is to make $A_{i,p}$ accept as it is about to drop the i-th pebble, apply the induction hypothesis to $A_{i,p}$ to make it an automaton without pebbles, and finally use Lemma 2 as before.

The automaton $A_{i,p}$ has the same set of states Q as A_p, and has the set of pebbles $P_i = P \setminus \{i\}$. It works as A_p with the i-th pebble in its hands. If A_p drops the i-th pebble, then $A_{i,p}$ accepts. If A_p accepts or rejects without dropping the i-th pebble, then $A_{i,p}$ rejects.

Since the automaton $A_{i,p}$ uses $m - 1$ pebbles, the induction hypothesis is applicable to it, and it asserts that there exists a graph-walking automaton A_i without pebbles with at most $N_{m-1} = (2k+1)^{m-1}(m-1)! 3^{\frac{(m-1)m}{2}} n^m$ states, which accepts at the node, at which A_p drops the i-th pebble, and not accepts, if A_p never drops the i-th pebble.

Next, Lemma 2 is applied to the graph-walking automaton A_i, and it yields the desired pebble-checking automaton B_i with at most $(2k+1)N_{m-1}+1$ states.

Now, for every pebble i, with $i \in \{1,\ldots,m\}$, there is its own *pebble-checking automaton* B_i without pebbles, which can be started in its initial state $q_0^{(B_i)}$ at any node v of any graph, and then it will return to that node with information on whether or not the automaton A_p drops the i-th pebble at this node.

This allows the simulating automaton A_0 to work by the same principle as for a single pebble, but this time at each step it will have to run multiple automata B_i, for all pebbles previously dropped, to determine which of them lie at the current node. The automaton A_0 always remembers which pebbles it has in its hands, and will not run the corresponding automata B_i for those pebbles (because if an i-th pebble is currently in A_0's hands, then B_i might still report the presence of this pebble at the current node, even though it would be dropped here only in the future).

The desired graph-walking automaton A_0 without pebbles that simulates the pebble automaton A_p has two groups of states. First, there are the states Q used to simulate the automaton A_p until it drops the first pebble. Later A_0 uses more complicated states of the form (q, P_1, P_2, i, q'), where

- $q \in Q$ is the simulated state of A_p;
- the set $P_1 \subseteq P$ is the set of pebbles in its hands;
- $i \in \{1,\ldots,m\}\backslash P_1$ is the pebble that A_0 is currently checking; it must not be in P_1, for otherwise it is known not to lie at the current node;
- $P_2 \subseteq \{1,\ldots,i-1\}\backslash P_1$ is the set of pebbles less than i, which are known to be at the node in which A_p is being simulated (that is, were put there before the current step);
- The state $q' \in Q^{(B_i)}$ is the state of the pebble-checking automaton B_i, executed to determine whether the i-th pebble is at the node where A_p is being simulated.

The construction of the transition function is omitted due to space constraints.

It remains to estimate the number of states in the automaton A_0. It should be noted that, in a quintuple (q, P_1, P_2, i, q'), the sets P_1 and P_2 are disjoint and do not contain i, and hence for each i there are at most $3^{|P|-1}$ pairs (P_1, P_2). Then the number of states in A_0 does not exceed $|Q| + |Q| \cdot 3^{|P|-1} \cdot |P| \cdot \max\{|Q^{(B_i)}| \mid i = 1,\ldots,m\}$, which has the following upper bound.

$$|Q| + |Q| \cdot 3^{|P|-1} \cdot |P| \cdot \max\{|Q^{(B_i)}| \mid i = 1,\ldots,m\}$$
$$\leqslant |Q| + |Q| \cdot 3^{|P|-1} \cdot |P| \cdot ((2k+1)N_{m-1} + 1)$$
$$\leqslant |Q| \cdot 3^{|P|} \cdot |P| \cdot ((2k+1)N_{m-1})$$

This is at most $n3^m m((2k+1)N_{m-1})$. Substituting $N_{m-1} = (2k+1)^{m-1}(m-1)!3^{\frac{(m-1)m}{2}}n^m$ gives that the number of states in A_0 is at most $N_m = (2k+1)^m m! 3^{\frac{(m+1)m}{2}} n^{m+1}$, as claimed. □

5   Conclusion

It has been shown that every graph-walking automaton with finitely many distinct drop-once pebbles can be simulated by a graph-walking automaton without pebbles with a certain number of states. How many states are *necessary* for this construction? Does the simulation of an automaton with one drop-once pebble indeed require a quadratic number of states? Proving any lower bounds is proposed for future research.

Another question is: is there a more efficient construction for the cases of TWA or 2DFA? For the more powerful *nested pebbles*, already one such pebble makes TWA stronger [3], and even though 2DFA with m nested pebbles can be simulated by ordinary 2DFA, this in the worst case requires an $(m + 1)$-fold exponential number of states [9]—much more than for m drop-once pebbles.

References

1. Blum, M., Hewitt, C.: Automata on a 2-dimensional tape. In: 8th Annual Symposium on Switching and Automata Theory, Austin, Texas, USA, 18–20 October 1967, pp. 155–160. IEEE Computer Society (1967)
2. Blum, M., Kozen, D.: On the power of the compass (or, why mazes are easier to search than graphs). In: 19th Annual Symposium on Foundations of Computer Science, Ann Arbor, Michigan, USA, 16–18 October 1978, pp. 132–142. IEEE Computer Society (1978)
3. Bojańczyk, M., Samuelides, M., Schwentick, T., Segoufin, L.: Expressive power of pebble automata. In: Bugliesi, M., Preneel, B., Sassone, V., Wegener, I. (eds.) ICALP 2006. LNCS, vol. 4051, pp. 157–168. Springer, Heidelberg (2006). https://doi.org/10.1007/11786986_15
4. Budach, L.: Automata and labyrinths. Math. Nachr. **86**(1), 195–282 (1978)
5. Disser, Y., Hackfeld, J., Klimm, M.: Tight bounds for undirected graph exploration with pebbles and multiple agents. J. ACM **66**(6), 40:1–40:41 (2019)
6. Engelfriet, J., Hoogeboom, H.J.: Tree-walking pebble automata. In: Karhumäki, J., Maurer, H.A., Paun, G., Rozenberg, G. (eds.) Jewels are Forever, pp. 72–83. Springer, Heidelberg (1999). https://doi.org/10.1007/978-3-642-60207-8_7
7. Fraigniaud, P., Ilcinkas, D., Peer, G., Pelc, A., Peleg, D.: Graph exploration by a finite automaton. Theor. Comput. Sci. **345**(2–3), 331–344 (2005)
8. Geffert, V., Istonová, L.: Translation from classical two-way automata to pebble two-way automata. RAIRO Theor. Inform. Appl. **44**(4), 507–523 (2010)
9. Globerman, N., Harel, D.: Complexity results for two-way and multi-pebble automata and their logics. Theor. Comput. Sci. **169**(2), 161–184 (1996)
10. Goralcik, P., Goralciková, A., Koubek, V.: Alternation with a pebble. Inf. Process. Lett. **38**(1), 7–13 (1991)
11. Kunc, M., Okhotin, A.: Reversibility of computations in graph-walking automata. Inf. Comput. **275**, 104631 (2020)
12. Muscholl, A., Samuelides, M., Segoufin, L.: Complementing deterministic tree-walking automata. Inf. Process. Lett. **99**(1), 33–39 (2006)
13. Rollik, H.: Automaten in planaren Graphen. Acta Informatica **13**, 287–298 (1980). https://doi.org/10.1007/BF00288647
14. Sipser, M.: Halting space-bounded computations. Theor. Comput. Sci. **10**, 335–338 (1980)

Two-Way Machines and de Bruijn Words

Giovanni Pighizzini[ID] and Luca Prigioniero[(✉)][ID]

Dipartimento di Informatica, Università degli Studi di Milano, Milan, Italy
{pighizzini,prigioniero}@di.unimi.it

Abstract. We consider de Bruijn words and their recognition by finite automata. While on one-way nondeterministic automata the recognition of de Bruijn words of order k requires exponentially many states in k, we show a family of de Bruijn words such that the word w_k of order k, for $k > 0$, can be recognized by a deterministic two-way finite automaton with $O\left(k^3\right)$ states. Using this result we are able to obtain an exponential-size separation from deterministic two-way finite automata to equivalent context-free grammars. We also show how w_k can be generated by a 1-limited automaton with $O\left(k^3\right)$ states and a constant-size work alphabet. This allows to obtain small 1-limited automata for certain unary languages and to show an exponential-size separation from unary deterministic 1-limited automata to equivalent deterministic pushdown automata.

1 Introduction

Given an alphabet Δ with m symbols and an integer $k > 0$, a *de Bruijn word* of order k over Δ is a string of length m^k that, considered as a "circular word", contains each string of length k over Δ exactly once [4]. These combinatorial objects have been widely investigated in the literature. In particular, an efficient iterative construction for de Bruijn words over a binary alphabet was discovered some years ago [19]. This construction suggested us to study the connection between the generation of de Bruijn words and the investigation on descriptional complexity of computational and formal models describing regular languages. We now summarize these concepts and the results we present in this paper.

The typical recognizers for the regular languages are finite automata, in both deterministic and nondeterministic variants, operating in one- and two-way fashion [17]. In the one-way case, the nondeterminism can be eliminated with an exponential increase in size, by using the well-known *powerset construction*. This cost cannot be reduced in the worst case, and hence it is optimal [11]. The situation is quite different in the two-way case: while there exist exponential-size transformations to convert two-way nondeterministic automata into deterministic ones, it is not known whether this cost is necessary [18].

In the first part of the paper, we consider the problem of recognizing all the finite prefixes of the word obtained by repeating infinitely many times a fixed de Bruijn word of order k.[1] It is not difficult to show that, to recognize such a language with a one-way automaton, a number of states exponential in k is necessary, even when nondeterministic choices are allowed. A natural question is whether this cost can be reduced for automata that are able to move the head in both directions.

Here we give a positive answer for the de Bruijn word w_k, of any order $k > 0$, that is generated with the method proposed in [19]. Indeed, we show that such a generation method can be turned into a "small" deterministic recognizer. In this way we prove that, for each $k > 0$, the language consisting only of w_k, as well as the language consisting of the finite prefixes of the string w_k^ω which is obtained by concatenating infinitely many occurrences of w_k, are accepted by two-way deterministic finite automata with $O\left(k^3\right)$ states.

Since generating the language consisting of a single de Bruijn word of order k with a context-free grammar in Chomsky normal form requires a number of variables exponential in k [1,5], this gives an exponential-size separation from two-way deterministic finite automata to context-free grammars.

Finite automata are not the only recognizers for regular languages. Providing two-way finite automata with rewriting capabilities restricted to the first time they visit tape cells, the computational power does not change [20], but the resulting model can represent languages in a more concise way. These devices are known in the literature as 1-*limited automata*. They are a special case of a more general model called d-limited automata, which are one-tape Turing machines that, for a fixed d, are allowed to rewrite each tape cell contents only the first d times the cell is visited. For each $d > 1$, this model characterizes the class of context-free languages [7]. For a recent overview on limited automata see [13].

The investigation of 1-limited automata from the descriptional complexity point of view has started recently [14]. It has been proved that each 1-limited automaton with n states can be simulated by a one-way deterministic automaton with a number of states double exponential in a polynomial in n. This cost is optimal and reduces to a single exponential when the 1-limited automaton is deterministic.

The restricted case of unary 1-limited automata, i.e., with a one-letter input alphabet, has been investigated in [15]. Also in this case an exponential lower bound for the conversion of deterministic 1-limited automata into equivalent deterministic finite automata has been proved. The same lower bound holds for the conversion of deterministic 1-limited automata into equivalent two-way nondeterministic finite automata. Due to upper bounds for general alphabets, these lower bounds cannot be improved.

In the second part of the paper we present a deterministic 1-limited automaton with $O\left(k\right)$ states and $O\left(1\right)$ symbols, that given in input any unary string of length $\ell \geq 0$, ends the computation finally leaving on the tape the prefix of length ℓ of w_k^ω. We consider some variants of this 1-limited automaton that,

---

[1] We shortly mention that, for every $k > 2$, the de Bruijn word of order k is not unique. Indeed, the number of de Bruijn words of order k is $2^{2^{k-1}-k}$ [4].

keeping the same costs in terms of states and work alphabet, recognize some interesting unary languages.

The first variant recognizes a language which is known to require exponential size in k to be accepted by deterministic pushdown automata [12]. This gives an exponential-size separation from deterministic 1-limited automata to deterministic pushdown automata. We point out that the definition of such a language is given using de Bruijn words.

Another variant we consider allows to recognize the language $\{a^{2^k}\}$, where k is any fixed integer. This gives an alternative proof of the exponential gap from unary 1-limited automata to finite automata [15]. While in the previous proof the size of the working alphabet of the witness 1-limited automaton is linear in k, here it is constant. This is paid by increasing the number of states.

The results concerning the recognition of the de Bruijn word w_k using two-way deterministic finite automata are presented in Sect. 4, together with the exponential-size separation from these devices to context-free grammar. Section 5 contains the results related to 1-limited automata and the exponential-size separation from deterministic 1-limited automata to deterministic pushdown automata. Before these parts, Sect. 2 contains preliminary notions and Sect. 3 is devoted to show how to recognize some strings, called *necklaces*, using "small" two-way deterministic automata. This tool is fundamental for the implementation of the machines presented in the paper.

2   Preliminaries

In this section we recall some basic definitions useful in the paper. Given a set S, $\#S$ denotes its cardinality and 2^S the family of all its subsets. Given an alphabet Σ and a string $w \in \Sigma^*$, $|w|$ denotes the length of w, and Σ^k the set of all strings on Σ of length k.

We assume the reader familiar with notions from formal languages and automata theory, in particular with the fundamental variants of finite automata (1DFAs, 1NFAs, 2DFAs, 2NFAs, for short, where 1/2 mean *one-way/two-way* and D/N mean *deterministic/nondeterministic*, respectively) and with *context-free grammars* (CFGs, for short) in Chomsky normal form. For any unfamiliar terminology see, e.g., [8].

A 1-*limited automaton* (1-LA, for short) is a tuple $\mathcal{A} = (Q, \Sigma, \Gamma, \delta, q_I, F)$, where Q is a finite *set of states*, Σ is a finite *input alphabet*, Γ is a finite *work alphabet* such that $\Sigma \cup \{\triangleright, \triangleleft\} \subseteq \Gamma$, $\triangleright, \triangleleft \notin \Sigma$ are two special symbols, called the *left* and the *right end-markers*, and $\delta : Q \times \Gamma \to 2^{Q \times (\Gamma \setminus \{\triangleright, \triangleleft\}) \times \{-1, +1\}}$ is the *transition function*.

At the beginning of the computation, the input word $w \in \Sigma^*$ is stored onto the tape surrounded by the two end-markers, the left end-marker being in position zero and the right end-marker being in position $|w| + 1$. The head of the automaton is on cell 1 and the state of the finite control is the *initial state* q_I.

In one move, according to δ and the current state, $\mathcal{A}$ reads a symbol from the tape, changes its state, replaces the symbol just read from the tape by a new symbol, and moves its head to one position forward or backward. Furthermore, the head cannot pass the end-markers, except at the end of computation, to accept the input, as explained below. Replacing symbols is allowed to modify the content of each cell only during the first visit, with the exception of the cells containing the end-markers, which are never modified. For technical details see [14].

The automaton $\mathcal{A}$ accepts an input w if and only if there is a computation path which starts from the initial state q_I with the input tape containing w surrounded by the two end-markers and the head on the first input cell, and which ends in a *final state* $q \in F$ after passing the right end-marker. The device $\mathcal{A}$ is said to be *deterministic* (deterministic 1-LA, for short) whenever $\#\delta(q, \sigma) \leq 1$, for any $q \in Q$ and $\sigma \in \Gamma$.

In this paper we are interested to compare the size of formal systems. The *size* of a model is given by the total number of symbols used to write down its description. Therefore, the size of deterministic 1-LAs is bounded by a polynomial in the number of states and of work symbols, while, in the case of finite automata, since no writings are allowed, the size is linear in the number of instructions and states, which is bounded by a polynomial in the number of states and in the number of input symbols. For recognizing models we will specify explicitly the cardinality of the state set and of the work alphabet, if any.

Following the line of [5], the size of context-free grammars in Chomsky normal form is bounded by a polynomial in the total number of variables.

A *de Bruijn word* β_k of order k is a word of length 2^k having each word of length k over the alphabet $\{0, 1\}$ exactly once as a factor, when β_k is considered as a *circular word*; namely, where the word "wraps around" at the end back to the beginning. So, using this representation, the last $k - 1$ symbols of the "unrolled" word, that are equal to the first symbols of the de Bruijn word, are omitted. For example, the circular word $\beta_3 = 00011101$ is a de Bruijn word for $k = 3$. All the strings of length 3 appear exactly once in β_3, when considered circularly, in the following order: 000, 001, 011, 111, 110, 101, 010, and 100. Instead of seeing de Bruijn as circular words, sometimes it is suitable to present them as standard words. In this case the "unrolled" word of order k, denoted β'_k has length $2^k + k - 1$ and it is obtained by concatenating to β_k its prefix of length $k - 1$. For instance $\beta'_3 = 0001110100$.

3   Recognizing Necklaces with 2DFAs

In this section we develop a tool that will be used later to obtain our result: we study the recognition, by 2DFAs, of special strings, called necklaces, of a given length. A *necklace* is the lexicographically smallest string in an equivalence class of strings under rotation. Let

$$N_k = \{w \in \{0, 1\}^k \mid w \text{ is a necklace}\}$$

be the language of all necklaces of length k on the alphabet $\{0, 1\}$.

Procedure 1: `isNecklace(`k`)`

Returns true if the portion of the tape that starts from the current head position and consists of k consecutive cells contains a necklace of length k; false otherwise (See Footnote 2)

```
1  p ← 1
2  for j ← 2 to k do
3  |    move the head to the cell in position j − p
4  |    σ ← read()
5  |    move the head to the cell in position j
6  |    if read() < σ then return false
7  |    if read() > σ then p ← j
8  if k mod p ≠ 0 then return false
9  return true
```

We are going to show how N_k can be recognized by a 2DFA $\mathcal{A}$ of size polynomial in k. In particular, $\mathcal{A}$ implements Procedure 1. Such a procedure is inspired by an algorithm presented by Gabric et al. [6, Algorithm 2], which is based on the *least circular shift* algorithm [3], a generalization of the *Knuth-Morris-Pratt pattern matching* algorithm [9].

Let us give some details about the implementation of `isNecklace`.

The 2DFA $\mathcal{A}$ uses this procedure to check whether the k consecutive cells starting from the one scanned by the input head contain a necklace of length k. To do so, $\mathcal{A}$ uses a variable, say head, to store the position of the head along the portion of the tape containing the string to recognize. The values that head can get range from 1 to k. Therefore, at the beginning of the execution of `isNecklace`, head is initialized to 1, and then it is incremented or decremented when the head is moved to right or left, respectively. In such a way, it is possible to ensure that the device does not visit any extra cell during the execution of this procedure.

When the procedure is called, $\mathcal{A}$ initializes the variables p and j to 1 and 2, respectively (Lines 1 and 2). The variable j is used to keep track of the position of the next symbol that must be compared with the symbol in position j − p, where p contains the length of the longest period occurring in the necklace, starting from the leftmost symbol. To reach position j − p (Line 3), the machine moves the head to the right, until cell j, and then it moves the head to the left, decrementing p at each step, until p becomes 0. The 2DFA stores the scanned symbol (in position j − p) into the variable σ (Line 4).[2] Then, it moves the head to the right until cell j, incrementing p at each step to restore its correct value (Line 5).

If the two symbols in position j and j − p are not equal, then two cases are possible: either the scanned symbol is less than the symbol in position j − p,

---

[2] Here and in the next procedures, the function **read** returns the input symbol scanned by the device, without moving the input head. All the head movements are explicitly indicated in the instructions.

then the string under consideration can be rotated to obtain a lexicographically smallest one, so it is not a necklace (Line 6); or the scanned symbol is greater than the symbol in position $j - p$, then the prefix analyzed so far must be the period of the necklace, so the variable p is updated accordingly (Line 7). We point out that this is a general procedure for recognizing necklaces and that here comparisons are based on the lexicographic order of the symbols.

After having analyzed the whole string of length k (Lines 2 to 7), the machine checks whether the string is periodic. So the string is a necklace, and hence it is accepted, if and only if the period length p divides k (Lines 8 and 9).

As already mentioned, the correctness of this technique is discussed in [6].

We observe that the automaton $\mathcal{A}$ has to store the variables p, j, and the position of the head. The values of these three variables range from 1 to k. This allows us to prove:

Theorem 1. *For every fixed integer k, the language N_k can be recognized by a* 2DFA *with $O\left(k^3\right)$ states.*

Proof. The 2DFA $\mathcal{A}$ described above can recognize N_k by first checking whether the input string has length k and, if so, by performing one call to isNecklace. Using the analysis above, we easily conclude that this can be done with $O\left(k^3\right)$ states. □

4   de Bruijn Words and Finite Automata

This section is devoted to the recognition of de Bruijn words. We shall present a 2DFA of polynomial size in k, accepting a language related to some de Bruijn words of order k. This automaton, which is interesting *per se*, will allow us to obtain an exponential-size separation from 2DFAs to CFGs.

Given a de Bruijn word β_k, let us consider the language

$$dB_{\beta_k} = \{x \in \{0,1\}^* \mid x \text{ is a finite prefix of } \beta_k^\omega\},$$

where β_k^ω is the infinite word obtained by concatenating infinitely many copies of β_k.

The language dB_{β_k} requires an exponential number of states in k to be accepted by one-way finite automata, even when nondeterministic moves are allowed:

Theorem 2. *For every fixed integer k, every* 1NFA *for dB_{β_k} has at least 2^k states.*

Proof. To prove that each 1NFA accepting dB_{β_k} needs 2^k states, we use the *fooling set technique* [2]. Consider the set

$$X = \{(x_i, y_i) \mid x_i y_i \text{ is the prefix of length } i + k \text{ of } \beta_k^\omega,$$
$$\text{where } |x_i| = i \text{ and } |y_i| = k, \text{ for } i = 0, \ldots, 2^k - 1\}.$$

Notice that, by the definition of de Bruijn words, the y_i's are pairwise different and $\{y_i \mid i = 0,\dots, 2^k - 1\} = \{0,1\}^k$. We are going to show that X is a fooling set for dB_{β_k}.

It is obvious that the word $x_i y_i$ belongs to dB_{β_k}, for $i = 0,\dots, 2^k - 1$. Now, take $j \neq i$, $0 \leq j < 2^k$. Since all strings in dB_{β_k} are prefixes of the same infinite string, $x_i y_j \in dB_{\beta_k}$ would imply $y_i = y_j$, which is a contradiction. Hence, $x_i y_j \notin dB_{\beta_k}$ and X is a fooling set for dB_{β_k}.

This allows to conclude that every 1NFA for dB_{β_k} has at least $\#X = 2^k$ states. $\qquad\square$

As mentioned in the introduction, for $k > 2$ there are many different de Bruijn words of order k. We point out that the proof of Theorem 2 does not depend on the specific de Bruijn word β_k chosen to define the language. From now on, we consider the *fixed* de Bruijn word w_k that is obtained with the efficient iterative construction presented by Sawada, Williams, and Wong [19] (this method is also called *PCR3 (J1)* [6]). Such a construction starts from the string 0^k and, at each step, computes the next symbol of the word by applying, to the suffix $\alpha = a_1 a_2 \cdots a_k$ of length k of the word generated until then, the function $f : \{0,1\}^k \to \{0,1\}$ defined as follows:

$$f(a_1 a_2 \cdots a_k) = \begin{cases} 1 - a_1 & \text{if } a_2 a_3 \cdots a_k 1 \text{ is a necklace} \\ a_1 & \text{otherwise.} \end{cases}$$

For further details and the correctness of this algorithm we address the reader to [19].

We now consider the language dB_{w_k} that, for brevity, in the following will be denoted as dB_k. By Theorem 2, an exponential number of states is necessary to recognize it with a 1NFA. Let us now turn our attention to the recognition by 2DFAs. To this aim, let us now show how to adapt the generation technique mentioned above to obtain a 2DFA $\mathcal{N}$ recognizing dB_k (see Procedure 2). In this case, the automaton uses a "virtual" *relative window* of k consecutive cells. The cells of the relative window are indexed from 1 to k. The position of the window is implicitly shared with the procedure isNecklaceSuccessor called by isDeBruijn, by using the variable head that keeps track of the position of the head along the window (cf., Sect. 3). In fact, the contents of the relative window when such a procedure is called, corresponds to the k cells on which it operates. The window initially contains the first k input cells, and the head is on the cell 1.

At the beginning of the computation, the device has to check whether the first k symbols are equal to 0 (Line 10); this is done by scanning the contents of the relative window entirely. If, during this step, the right end-marker is reached, then $\mathcal{N}$ accepts.

After checking that the prefix of the input word is 0^k, the machine has to iteratively compute the next symbols of w_k and compare them with the following input symbols (Lines 11 to 19). Hence, for each iteration, it computes $f(a_1 a_2 \cdots a_k)$, where $a_1 a_2 \cdots a_k$ is the contents of the relative window, and stores the result in the variable successor (Lines 12 to 16, details are discussed below). After that, the machine shifts the relative window one cell to the

---

Procedure 2: isDeBruijn(k)

Accepts if the string on the tape is in dB_k; rejects otherwise

---

10 **if** *the first cells do not contain 0^k* **then** Reject
11 **while** read() $\neq \lhd$ **do**
12 | move the head to the cell in relative position 1
13 | $\tau \leftarrow$ isNecklaceSuccessor(k)
14 | move the head to the cell in relative position 1
15 | successor $\leftarrow$ read()
16 | **if** τ **then** successor $\leftarrow 1 -$ successor
17 | shift the current window one cell to the right
18 | move the head to the cell in relative position k
19 | **if** read() $\notin \{\lhd,$ successor$\}$ **then** Reject
20 Accept

---

right (Line 17), moves the head to the rightmost cell of the window (Line 18), and compares the input symbol with the symbol stored in the variable successor. If the two symbols do not match and the input is not over, then the input is not a prefix of w_k^ω, hence the machine rejects (Line 19). This process is repeated until the end of the input is reached (Line 11). In this case, if every symbol of the input matches with the content of the variable successor computed at each iteration, then the machine accepts (Line 20).

To check whether the string $a_2 \cdots a_k 1$ is a necklace, a call to the procedure isNecklaceSuccessor is made (Line 13). Such a procedure (Procedure 3) is a slightly-modified version of isNecklace (Procedure 1) that analyzes the symbols of the relative window starting from the second one (in fact, j is initialized with the value 3, cf. Line 22). Moreover, in this case, for analyzing the last symbol, which is always a 1, the last loop is unrolled to prevent the machine entering the cell in position $k + 1$ (Lines 28 and 29). Notice that Line 29 corresponds to Line 27, in which the symbol in position $k + 1 -$ p is not stored (so it is directly scanned using read) and the symbol in position k is always 1, so it is sufficient to test if read() $= 0$.

For implementing the procedure isDeBruijn the 2DFA $\mathcal{N}$ uses a variable successor, which can assume values in $\{0, 1\}$, and the same variables used by the procedure isNecklaceSuccessor. The position of the head along the relative window of size k is shared with the main procedure isDeBruijn. Hence $\mathcal{N}$ can implement isDeBruijn with $O(k^3)$ states, which is analogue to the cost of isNecklace. This analysis leads us to the following result:

Theorem 3. *For every fixed integer k, the language dB_k can be recognized by a 2DFA with $O(k^3)$ states.*

Let us now restrict our attention to the language $W_k = \{w_k'\}$, consisting only of the "unrolled" de Bruijn word of order k.

In order to recognize W_k, we can modify Procedure 2 and the 2DFA implementing it in such a way that, at the end of each iteration of the while-loop

---

Procedure 3: isNecklaceSuccessor(k)

Returns true if the symbols contained in the $k-1$ consecutive cells starting after the one scanned by the input head concatenated with the symbol 1 is a necklace, false otherwise. Notice that the cell in position $k+1$ is never entered during the execution of this procedure

---

```
21  p ← 1
22  for j ← 3 to k do
23  |     move the head to the cell in relative position j − p
24  |     σ ← read()
25  |     move the head to the cell in relative position j
26  |     if read() < σ then return false
27  |     if read() > σ then p ← j
28  move the head to the cell in position k + 1 − p
29  if read() = 0 then p ← k
30  if k mod p ≠ 0 then return false
31  return true
```

(Line 19), it verifies that the string in the current relative window (i.e., the suffix of length k of the input prefix inspected so far) coincides with the suffix of length k of w'_k. In this case the machine accepts, otherwise it continues the execution of the while-loop. To make such a modification, $O(n)$ extra states are enough. Furthermore, since W_k consists only of one string of length $2^k + k - 1$, each 1NFA accepting it cannot contain any loop in accepting paths. Hence, it requires at least $2^k + k$ states. This allows us to extend Theorems 3 and 2 to W_k:

Theorem 4. *For each fixed integer k, the language W_k can be recognized by a 2DFA with $O(k^3)$ states, while every 1NFA accepting it requires at least $2^k + k$ states.*

The exponential-size gap in Theorem 4 can be strengthen by replacing 1NFAs by CFGs. In fact, in [5], using a result from [1], the following was proved:

Lemma 1. *There exists a constant c such that, for all $k \geq 1$, the number of variables in the smallest CFG in Chomsky normal form generating W_k is at least $c\frac{2^k}{k}$.*

As a consequence we obtain the following size gap:

Theorem 5. *Converting a 2DFA in an equivalent CFG can cost exponential in size.*

Concerning the opposite direction, we remind the reader that the cost of the conversion of CFGs generating regular languages into finite automata cannot be bounded by any recursive function [11]. However, in the restricted case of CFGs generating unary languages this cost is exponential, as proved in [16] for the conversion into 1DFAs and 1NFAs. Because of a lower bound presented in [10, Theorem 9], this cost cannot be reduced even if the target machine is a 2NFA.

Procedure 4: deBruijnUnary(k)

A deterministic 1-LA executing this procedure accepts if the input string is in U_k; rejects otherwise

---

32 write 0 in the first k cells
33 while read() $\neq \lhd$ **do**
34 | move the head to the cell in relative position 1
35 | $\tau \leftarrow$ isNecklaceSuccessor(k)
36 | move the head to the cell in relative position 1
37 | successor $\leftarrow$ read()
38 | **if** τ **then** successor $\leftarrow 1 -$ successor
39 | shift the current window one cell to the right
40 | move the head to the cell in relative position k
41 | **if** read() $\neq \lhd$ **then** write(successor)
42 move the head one cell to the left
43 **if** read() $= 1$ **then** Accept
44 **else** Reject

---

5 de Bruijn Words and 1-Limited Automata

In this section we exploit the techniques illustrated in the previous sections in order to deepen the investigation of the descriptional complexity of unary deterministic 1-LAs [15]. In particular, we prove that, for each integer k, there exists a unary language U_k that is accepted by a 1-LA of size polynomial in k, but for which any deterministic pushdown automaton requires a size exponential in k.

The language we consider is

$$U_k = \{a^\ell \mid \text{the } \ell\text{-th symbol of } w_k^\omega \text{ is } 1\},$$

which is the set of positions, encoded in unary, of the symbols equal to 1 in the infinite word obtained by cyclically repeating de Bruijn word w_k of order k. For example, for $k = 3$, $w_3 = 00011101$ and $U_3 = \{a^4, a^5, a^6, a^8\} \cdot \{a^8\}^*$ because the prefixes of w_k of length 4 (0001), 5 (00011), 6 (000111), and 8 (00011101) end with the symbol 1, and then they repeat after the period 00011101 having length 8: 000111010001, 0001110100011, 00011101000111, 0001110100011101, and so on.

Let us briefly show how to recognize the language U_k with a deterministic 1-LA (see Procedure 4). We can use the same approach used to recognize dB_k, with the only difference that our deterministic 1-LA *writes* the de Bruijn word on the tape (Lines 32 and 41) and, when the end of the input is reached, it accepts if the last symbol written is 1 (Lines 42 to 44). Also in this case, the obtained device has size polynomial in k.

We stress that this procedure can be implemented by a deterministic 1-LA because each cell is written during the first visit only (Lines 32 and 41).

We point out that the deterministic 1-LA described above "generates" the de Bruijn word on its tape. More precisely, receiving in input the string a^ℓ, it ends the computation with the prefix of length ℓ of w_k^ω on its tape.

Theorem 6. *For every fixed integer k, the language U_k can be recognized by a deterministic 1-LA with $O\left(k^3\right)$ states and $O\left(1\right)$ work symbols.*

As proved in [12], even each deterministic pushdown automaton accepting U_k would require an exponential size in k. Therefore we obtain the following gap in size:

Theorem 7. *Converting a unary deterministic 1-LA into an equivalent deterministic pushdown automaton can cost exponential in size.*

A natural question is whether Theorem 7 holds if we consider the conversion into *nondeterministic* pushdown automata or, equivalently, CFGs. Concerning the converse, we already proved that each unary CFG can be converted into a 1-LA of polynomial size [15].

In [15], we considered the languages $L_k = \{a^{2^k}\}$ and $J_k = \{a^{2^k}\}^*$, proving that they can be accepted by deterministic 1-LAs with $O\left(k\right)$ states and $O\left(k\right)$ work symbols. The technique used to recognize U_k developed here can be easily adapted to recognize L_k and J_k. In particular, it is sufficient to notice that, given an integer k, the word w_k computed using Procedure 4 is fixed, and so the suffix γ_k of length k of w_k is known. Moreover, we remind that w_k has length 2^k. For example, for $k = 3$, $w_3 = 00011101$ and $\gamma_3 = 101$. Hence, a deterministic 1-LA can accept J_k by writing w_k on the tape using the same approach as deBruijnUnary(k) (Lines 32 to 41) but, when it reaches the right end-marker, checks whether the suffix of the word written on the tape is γ_k. This last step can be implemented with $O\left(k\right)$ states. To accept L_k the automaton has to verify, in addition, that γ_k appears only once on the tape.

Theorem 8. *For every fixed integer k, the languages L_k and J_k can be recognized by deterministic 1-LAs with $O\left(k^3\right)$ states and $O\left(1\right)$ work symbols.*

We can observe that the size of the above-mentioned machines presented in [15] are smaller than those of the equivalent machines in Theorem 8. Indeed, the number of entries in the transition tables is quadratic in k in the former case, while it is cubic in the latter. However the work alphabet of machines in Theorem 8 is fixed, i.e., it does not depend on the parameter k.

References

1. Althöfer, I.: Tight lower bounds on the length of word chains. Inf. Process. Lett. **34**(5), 275–276 (1990). https://doi.org/10.1016/0020-0190(90)90136-L
2. Birget, J.: Intersection and union of regular languages and state complexity. Inf. Process. Lett. **43**(4), 185–190 (1992). https://doi.org/10.1016/0020-0190(92)90198-5
3. Booth, K.S.: Lexicographically least circular substrings. Inf. Process. Lett. **10**(4/5), 240–242 (1980). https://doi.org/10.1016/0020-0190(80)90149-0
4. de Bruijn, N.G.: A combinatorial problem. Nederl. Akad. Wetensch., Proc. **49**, 758–764 (1946)

5. Domaratzki, M., Pighizzini, G., Shallit, J.O.: Simulating finite automata with context-free grammars. Inf. Process. Lett. **84**(6), 339–344 (2002). https://doi.org/10.1016/S0020-0190(02)00316-2

6. Gabric, D., Sawada, J., Williams, A., Wong, D.: A framework for constructing de Bruijn sequences via simple successor rules. Discret. Math. **341**(11), 2977–2987 (2018). https://doi.org/10.1016/j.disc.2018.07.010

7. Hibbard, T.N.: A generalization of context-free determinism. Inf. Control **11**(1/2), 196–238 (1967). https://doi.org/10.1016/S0019-9958(67)90513-X

8. Hopcroft, J.E., Ullman, J.D.: Introduction to Automata Theory, Languages and Computation. Addison-Wesley (1979)

9. Knuth, D.E., Morris, J.H., Pratt, V.R.: Fast pattern matching in strings. SIAM J. Comput. **6**(2), 323–350 (1977). https://doi.org/10.1137/0206024

10. Mereghetti, C., Pighizzini, G.: Two-way automata simulations and unary languages. J. Autom. Lang. Comb. **5**(3), 287–300 (2000). https://doi.org/10.25596/jalc-2000-287

11. Meyer, A.R., Fischer, M.J.: Economy of description by automata, grammars, and formal systems. In: SWAT 1971, pp. 188–191. IEEE Computer Society (1971). https://doi.org/10.1109/SWAT.1971.11

12. Pighizzini, G.: Deterministic pushdown automata and unary languages. Int. J. Found. Comput. Sci. **20**(4), 629–645 (2009). https://doi.org/10.1142/S0129054109006784

13. Pighizzini, G.: Limited automata: properties, complexity and variants. In: Hospodár, M., Jirásková, G., Konstantinidis, S. (eds.) DCFS 2019. LNCS, vol. 11612, pp. 57–73. Springer, Cham (2019). https://doi.org/10.1007/978-3-030-23247-4_4

14. Pighizzini, G., Pisoni, A.: Limited automata and regular languages. Int. J. Found. Comput. Sci. **25**(7), 897–916 (2014). https://doi.org/10.1142/S0129054114400140

15. Pighizzini, G., Prigioniero, L.: Limited automata and unary languages. Inf. Comput. **266**, 60–74 (2019). https://doi.org/10.1016/j.ic.2019.01.002

16. Pighizzini, G., Shallit, J.O., Wang, M.: Unary context-free grammars and pushdown automata, descriptional complexity and auxiliary space lower bounds. J. Comput. Syst. Sci. **65**(2), 393–414 (2002). https://doi.org/10.1006/jcss.2002.1855

17. Rabin, M.O., Scott, D.S.: Finite automata and their decision problems. IBM J. Res. Dev. **3**(2), 114–125 (1959). https://doi.org/10.1147/rd.32.0114

18. Sakoda, W.J., Sipser, M.: Nondeterminism and the size of two way finite automata. In: STOC 1978, pp. 275–286. ACM (1978). https://doi.org/10.1145/800133.804357

19. Sawada, J., Williams, A., Wong, D.: A surprisingly simple de Bruijn sequence construction. Discret. Math. **339**(1), 127–131 (2016). https://doi.org/10.1016/j.disc.2015.08.002

20. Wagner, K.W., Wechsung, G.: Computational Complexity. D. Reidel Publishing Company, Dordrecht (1986)

Transduction of Automatic Sequences and Applications

Jeffrey Shallit[1]([✉]) [ID] and Anatoly Zavyalov[2]

[1] School of Computer Science, University of Waterloo, 200 University Ave. W.,
Waterloo, ON N2L 3G1, Canada
shallit@uwaterloo.ca
[2] University of Toronto, Toronto, ON M5S 2E4, Canada
anatoly.zavyalov@mail.utoronto.ca

Abstract. We consider the implementation of the transduction of automatic sequences, and their generalizations, in the `Walnut` software for solving decision problems in combinatorics on words. We provide a number of applications, including (a) representations of $n!$ as a sum of three squares (b) overlap-free Dyck words and (c) sums of Fibonacci representations.

1 Introduction

The k-automatic sequences form an interesting class that has been studied for more than fifty years now [1,6]. This class is defined as follows: a sequence $(a(n))_{n \geq 0}$ over a finite alphabet is k-automatic if there exists a deterministic finite automaton with output (DFAO) that, on completely processing an input of n expressed in base k, reaches a state s with output $a(n)$.

One nice property is that the first-order theory of these sequences, with addition, is algorithmically decidable [3]. `Walnut`, a free software system created and implemented by Hamoon Mousavi [14], makes it possible to state and evaluate the truth of nontrivial first-order statements about automatic sequences, often in a matter of seconds [16]. With it, we can easily reprove old theorems in a simple and uniform way, explore unresolved conjectures, and prove new theorems.

According to a famous theorem of Cobham [6], the class of k-automatic sequences is closed under deterministic t-uniform transductions. The particular model of transducer here is the following: outputs are associated with transitions, so that every input letter results in the output of exactly t letters. Cobham's theorem was later extended to the more general class of morphic sequences by Dekking [8].

Transducers make it possible to manipulate automatic sequences in useful and interesting ways. For example, it follows that the running sum and running product (taken modulo a natural number $M \geq 2$) of a k-automatic sequence taking values in $\Sigma_t = \{0, 1, \ldots, t-1\}$ is k-automatic.

Research of Jeffrey Shallit is supported by NSERC Grant number 2018-04118.

B. Nagy (Ed.): CIAA 2023, LNCS 14151, pp. 266–277, 2023.
https://doi.org/10.1007/978-3-031-40247-0_20

In this paper we report on a recent implementation of Dekking's result in Walnut, carried out by the second author. This new capability of transducers has been implemented in the latest version of Walnut, which is available for free download at

https://cs.uwaterloo.ca/~shallit/walnut.html .

With it we are able to obtain a number of results, including reproving the results of Burns [5] on the representation of $n!$ as a sum of three squares, in a simpler and more general way. We also prove a new result on overlap-free Dyck words. Lastly, we briefly discuss the transduction of Fibonacci automata and automata in other numeration systems.

Because of length restrictions, many of our results and proofs had to be omitted. The interested reader can see the full paper at [17].

2   Transducers

We assume the reader has a basic knowledge of automata theory and formal languages as given in, for example, [11]. For reasons of space we omit the usual definitions.

In this paper we deal only with 1-uniform transducers. A *1-uniform deterministic finite-state transducer* (or *transducer* for short) is a 6-tuple

$$T = \langle V, \Delta, \varphi, v_0, \Gamma, \sigma \rangle,$$

where V is a finite set of *states*, Δ is a finite set representing the *input alphabet*, $\varphi \colon V \times \Delta \to V$ is the *transition function*, $v_0 \in V$ is the *initial state*, Γ is a finite set representing the *output alphabet*, and $\sigma \colon V \times \Delta \to \Gamma$ is the *output function*. A transducer can be viewed as a deterministic finite automaton (DFA), together with an output function σ, where a single element of Γ is output at each transition.

Transducing a sequence $\mathbf{x} = (x_n)_{n \geq 0}$ means passing the sequence through the transducer symbol by symbol, obtaining an infinite sequence resulting from concatenating the single symbol output at each transition. Formally, the transduction of a sequence $\mathbf{x} = (x_n)_{n \geq 0}$ over the alphabet Δ by a transducer T is defined to be the infinite sequence

$$T(\mathbf{x}) = \sigma(v_0, x_0)\, \sigma(\varphi(v_0, x_0), x_1) \, \cdots \, \sigma(\varphi(v_0, x_0 \cdots x_{n-1}), x_n) \, \cdots .$$

Example 1. Consider the famous Thue-Morse sequence $\mathbf{t} = \mu^\omega(0)$ defined by the 2-uniform morphism $\mu \colon \Sigma^* \to \Sigma^*$ given by $\mu(0) = 01$, $\mu(1) = 10$, with $\Sigma = \{0, 1\}$. The first few terms of the sequence are:

$$\mathbf{t} = 0110100110010110 \cdots .$$

Suppose we wish to compute the running sum (mod 2) of the Thue-Morse sequence. We can do this with the transducer RUNSUM2, which transduces a sequence $a_0 a_1 a_2 \cdots$ over $\{0, 1\}$ into $b_0 b_1 b_2 \cdots$ where $b_k = (\sum_{i=0}^{k} a_i) \bmod 2$. The

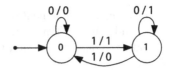

Fig. 1. Transducer RUNSUM2 for running sum mod 2.

transducer is illustrated in Fig. 1. The notation a/b means the transducer outputs b on input a. The resulting transduced sequence $0100111011100100\cdots$ is sequence A255817 in the *On-Line Encyclopedia of Integer Sequences* (OEIS) [18].

Example 2. Let $\mathbf{p} = 10111010\cdots$ be the *period-doubling sequence* [7], the fixed point of the morphism $1 \mapsto 10$, $0 \mapsto 11$, and sequence A035263 in the OEIS. Applying the transducer of Example 1 to $\mathbf{p}$ gives $\hat{\mathbf{t}} = 11010011\cdots$, the Thue-Morse sequence with its first symbol removed.

As mentioned above, the transduction of an k-automatic sequence always produces another k-automatic sequence.

Theorem 1 (Cobham). *Let $\mathbf{x} = (x_n)_{n \geq 0}$ be a k-automatic sequence over Δ, and let $T = \langle V, \Delta, \varphi, v_0, \Gamma, \sigma \rangle$ be a transducer. Then the sequence*

$$T(\mathbf{x}) = \sigma(v_0, x_0)\, \sigma(\varphi(v_0, x_0), x_1) \,\cdots\, \sigma(\varphi(v_0, x_0 \cdots x_{n-1}), x_n)\,\cdots$$

is k-automatic.

Dekking [8] provides a constructive proof of this theorem, and it is the basis for our implementation of transducers in Walnut.

3   Implementation

We give a brief overview of the construction, implementation, and complexity of the resulting automaton. See Sects. 3 and 4 of [17] for a more in-depth discussion.

3.1   Construction

We give a brief overview of the construction of the resulting DFAO that computes $T(\mathbf{x})$ as in Theorem 1, where $\mathbf{x} = (x_n)_{n \geq 0}$ is a k-automatic sequence, so $\mathbf{x} = \lambda(\mathbf{q})$ where $h\colon Q^* \to Q^*$ is a prolongable k-uniform morphism, the coding λ maps $Q^* \to \Delta^*$, $\mathbf{q} = h^\omega(q_0) = q_0 q_1 q_2 \cdots$. For all $y \in \Delta^*$, define the function $f_y\colon V \to V$ by $f_y(v) = \varphi^*(v, y)$. Note that $f_{y_1 y_2} = f_{y_2} \circ f_{y_1}$ for all $y_1, y_2 \in \Delta^*$. We need the following lemma.

Lemma 1 ([17]). *For all $w \in Q^*$, the sequences $f_{\lambda(w)}, f_{\lambda(h(w))}, \ldots, f_{\lambda(h^n(w))}, \cdots$ are ultimately periodic with the same period and preperiod, i.e., there exist integers $p \geq 1, r \geq 0$ such that $f_{\lambda(h^i(w))} = f_{\lambda(h^{p+i}(w))}$ for all $w \in Q^*$ and $i \geq r$.*

After finding p and r as in Lemma 1, define the following:

- the states $\widetilde{Q} = \{(a, \mathcal{I}(w)) : a \in Q, w \in Q^*\}$, where

$$\mathcal{I}(w) = (f_{\lambda(w)}, f_{\lambda(h(w))}, \ldots, f_{\lambda(h^{p+r-1}(w))});$$

- the k-uniform[1] morphism $\widetilde{h} \colon \widetilde{Q}^* \to \widetilde{Q}^*$ by

$$\widetilde{h}(a, \mathcal{I}(w)) = (h(a)_1, \mathcal{I}(h(w))) \cdots (h(a)_k, \mathcal{I}(h(w)h(a)_1 \cdots h(a)_{k-1})),$$

  where $h(a)_i$ denotes the i'th symbol of $h(a)$; and
- the coding $\widetilde{\lambda} \colon \widetilde{Q}^* \to \Gamma^*$ by

$$\widetilde{\lambda}(a, \mathcal{I}(w)) = \sigma(f_{\lambda(w)}(v_0), \lambda(a)) = \sigma(\varphi^*(v_0, \lambda(w)), \lambda(a)).$$

The resulting DFAO that computes $T(\mathbf{x})$ is then $M' = \langle \Omega, \Sigma, \widetilde{\delta}, \widetilde{q}_0, \Gamma, \widetilde{\lambda} \rangle$ where $\Omega \subset \widetilde{Q}$ is the set of reachable states in $\widetilde{Q}$ from $\widetilde{Q}_0 = (q_0, \mathcal{I}(\epsilon))$, the input alphabet of the DFAO is $\Sigma = \{0, \ldots, k-1\}$, and the transition function $\widetilde{\delta} \colon \widetilde{Q} \times \Sigma \to \widetilde{Q}$ is defined by

$$\widetilde{\delta}((a, \mathcal{I}(w)), d) = (h(a)_d, \mathcal{I}(h(w)h(a)_1 \cdots h(a)_{d-1})).$$

3.2 Complexity Analysis

The proof of Lemma 1 in [17] finds p and r by iterating through the finite set $(V^V)^{|Q|}$ until a repetition is found, which requires to iterate through at most $|V|^{|Q| \cdot |V|}$ maps. Consequently we get that $p, r \leq |V|^{|Q| \cdot |V|}$. The total number of states $|\Omega|$ is, in the worst case, at most $|\widetilde{Q}| \leq |Q| \cdot |V|^{(p+r)|V|} \leq |Q| \cdot |V|^{2 \cdot |V|^{|Q| \cdot |V|+1}}$. To find the reachable states Ω, we run a breadth-first search starting from the initial state, which, in the worst-case, takes time $\mathcal{O}(|Q| \cdot |V|^{(p+r)|V|} + k \cdot |Q| \cdot |V|^{(p+r)|V|}) = \mathcal{O}((k+1) \cdot |Q| \cdot |V|^{2 \cdot |V|^{|Q| \cdot |V|+1}})$.

In practice, however, we do not usually see this kind of astonishingly large worst-case complexity.[2]

4   Applications

In this section, we illustrate the usefulness of our implementation of transducers in Walnut with two applications. In the first, we find a simple and direct proof of a result of Burns [5]. In the second, we prove a new result about overlap-free Dyck words.

---

[1] Recall that the morphism $\widetilde{h} \colon \widetilde{Q}^* \to \widetilde{Q}^*$ is k-*uniform* if $|\widetilde{h}(q)| = k$ for all $q \in \widetilde{Q}$.

[2] All of the computations in this paper took at most a few seconds to run.

4.1   Factorial as a Sum of Three Squares

Let $S_3 = \{0, 1, 2, 3, 4, 5, 6, 8, \ldots\}$ be the set of natural numbers that can be expressed as the sum of three squares of natural numbers. A well-known theorem of Legendre (see, for example, [15, §1.5]) states that $N \in S_3$ if and only if it cannot be expressed in the form $4^i(8j + 7)$, where $i, j \geq 0$ are natural numbers. It follows that the characteristic sequence of the set S_3 is 2-automatic, a fact first observed by Cobham [6, Example 8, p. 172]. Indeed, $N \in S_3$ if and only if $(N)_2$ is accepted by the automaton depicted in Fig. 2. Recently a number of authors [4,5,9,10] have been interested in studying the properties of the set $S = \{n : n! \in S_3\}$. In particular, the set S is 2-automatic. In this section, we show how to determine a DFAO for (the characteristic sequence of) S by using transducers. Let $\nu_2(n)$ denote the exponent of the highest power of 2 dividing n. The basic idea is to use the fact that $n! = \prod_{1 \leq i \leq n} i$, and keep track of both the parity of the exponent of the highest power of 2 dividing i and the last 3 bits of $i/2^{\nu_2(i)}$.

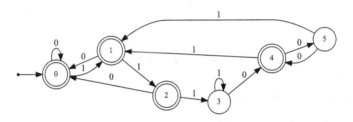

Fig. 2. Automaton accepting $(S_3)_2$.

Clearly $\nu_2(mn) = \nu_2(m) + \nu_2(n)$ for $m, n \geq 1$. It now follows that $\nu_2(n!) \bmod 2$ is the running sum (mod 2) of the sequence $\nu_2(1) \cdots \nu_2(n)$.

Let $g(n) = (n/2^{\nu_2(n)}) \bmod 8$ for $n \geq 1$ and set $g(0) = 1$. Then it is easy to see that $g(n) \in \{1, 3, 5, 7\}$ for all $n \geq 1$ and $g(mn) = (g(m)g(n)) \bmod 8$ for $m, n \geq 1$. Thus $g(n!)$ is the running product (mod 8) of the sequence $g(1)g(2) \cdots g(n)$.

Hence $\nu_2(n!) \bmod 2$ can be computed by a running-sum transducer, and $g(n!)$ can be computed by a running-product transducer. Now $n \notin S_3$ if and only if $\nu_2(n) \equiv 0 \pmod{2}$ and $g(n) = 7$, and therefore $n \notin S$ if and only if $\nu_2(n!) = \sum_{i=1}^{n} \nu_2(i) \equiv 0 \pmod{2}$ and $g(n!) \equiv \prod_{i=1}^{n} g(i) \equiv 7 \pmod{8}$.

We can now implement these ideas in `Walnut`. We first define the DFAO `NU_MOD2`, which generates the sequence $(\nu_2(n) \bmod 2)_{n \geq 1}$, illustrated in Fig. 3. This can be done with the following `Walnut` command, which defines it via a regular expression:

```
reg nu2odd msd_2 "(0|1)*10(00)*":
combine NU_MOD2 nu2odd:
```

It is easy to see that g satisfies the following identities: $g(2n) = g(n)$, and $g(8n + i) = i$ for $i \in \{1, 3, 5, 7\}$. From these identities, a least-significant-digit first automaton G8 for g is trivial to derive, as illustrated in Fig. 4. We can then reverse this automaton, using the following `Walnut` command:

`reverse G_MOD8 G8:`

and get a 12-state DFAO `G_MOD8` that computes $g(n)$ in msd-first format.

Lastly, we define the transducer `RUNPROD1357`, which transduces the sequence $g(1)g(2)\cdots g(n)$ into the sequence $h_1 h_2 \cdots h_n$ where the n'th term is the running product (mod 8) of the sequence $g(1)g(2)\cdots g(n)$, i.e., $h_k = \prod_{i=1}^{k} g(i)$ mod 8. The transducer is illustrated in Fig. 5.

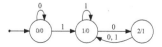

Fig. 3. DFAO `NU_MOD2` computing $\nu_2(n)$ mod 2.

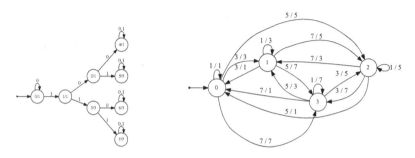

Fig. 4. DFAO computing $g(n)$ in lsd-first format.

Fig. 5. Transducer `RUNPROD1357` for running product mod 8.

We can then transduce `NU_MOD2` with `RUNSUM2` (defined in Example 1) to get a resulting DFAO `NU_RUNSUM`, depicted in Fig. 6, using the `Walnut` command

`transduce NU_RUNSUM RUNSUM2 NU_MOD2:`

We similarly transduce `G_MOD8` with `RUNPROD1357` to get a resulting DFAO `G_RUNPROD` (depicted in Fig. 7) with the following `Walnut` command:

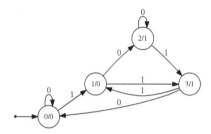

Fig. 6. DFAO `NU_RUNSUM` computing the running sum (mod 2) of `NU_MOD2`.

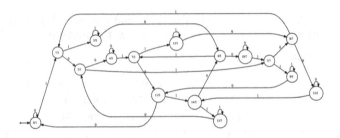

Fig. 7. DFAO G_RUNPROD computing the running product (mod 8) of G_MOD8.

```
transduce G_RUNPROD RUNPROD1357 G_MOD8:
```

Lastly, we generate the final automaton that accepts S, using the characterization above that $n \in S$ if and only if $\nu_2(n!) = \sum_{i=1}^{n} \nu_2(i) \equiv 1 \pmod 2$ or $g(n!) \equiv \prod_{i=1}^{n} g(i) \not\equiv 7 \pmod 8$, which can be directly translated into the following Walnut command:

```
def nfac_in_s "(NU_RUNSUM[i]=@1) | ~(G_RUNPROD[i]=@7)":
```

Figure 4 of Burns [5] depicts an automaton accepting n, represented in base-2 with *least-significant digit* first, if $n!$ *cannot* be written as a sum of 3 squares represented in the `lsd_k`. We can now obtain their automaton by reversing and negating the automaton `nfac_in_s` using the following Walnut command:

```
def nfac_in_s_rev_neg "~`$nfac_in_s(i)":
```

This gives us a 35-state automaton, depicted in Fig. 8, which is identical to Fig. 4 of Burns [5]. We have therefore rederived Burns' result in a simpler and more natural way.

4.2   Overlap-Free Dyck Words

In this section we cover another interesting application of transducers, to overlap-free Dyck words. We need a few definitions.

A word of the form $axaxa$, where a is a single letter, and x is a (possibly empty) word, is called an *overlap*. An example is the French word **entente**. A word is said to be *overlap-free* if it contains no block that is an overlap. A binary word is a *Dyck word* if it represents a word of balanced parentheses, if 0 is treated as a left paren and 1 as a right paren. The *nesting level* $N(x)$ of a finite Dyck word x is defined as follows: $N(\epsilon) = 0$, $N(1x0) = N(x) + 1$ if x is balanced, and $N(xy) = \max(N(x), N(y))$ if x, y are both balanced.

In the paper [13], the following result is proved: there are arbitrarily long overlap-free binary words of nesting level 3. Here we can prove the same result using a different construction involving the Thue-Morse morphism μ, where $\mu(0) = 01$ and $\mu(1) = 10$:

Theorem 2. *Define* $x_0 = 10$ *and* $x_{n+1} = \mu(101\mu(x_n)101)$ *for* $n \geq 0$. *Define* $y_n = 00x_n11$ *for* $n \geq 0$. *Then* y_n *is overlap-free for* $n \geq 0$ *and* y_n *has nesting level* 3 *for* $n \geq 1$.

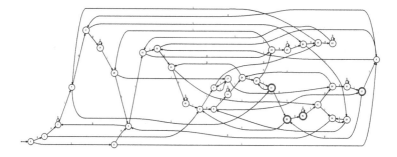

Fig. 8. LSD-first automaton accepting the complement of $S = \{n : n! \in S_3\}$.

Proof. We start by defining the infinite binary word $\mathbf{d} = 01\,y_0 y_1 y_2 \cdots$. We will prove the results about the y_n, indirectly, by proving results about $\mathbf{d}$ instead. We claim that

(a) $|y_n| = 6 \cdot 4^n$ for $n \geq 0$.
(b) $y_n = \mathbf{d}[2 \cdot 4^n .. 2 \cdot 4^{n+1} - 1]$ for $n \geq 0$.
(c) Define $I = i \in \{2\} \cup \{2 \cdot 4^n - 2 : n \geq 1\} \cup \{2 \cdot 4^n + 1 : n \geq 1\}$. Then $\mathbf{d}$ differs from $\mu^2(\mathbf{d})$ precisely at the indices in I, and furthermore $\mathbf{d}$ is the unique infinite binary word starting with 0 with this property.
(d) $\mathbf{d}$ is generated by the 10-state DFAO depicted in Fig. 9.
(e) Each y_n is overlap-free, for $n \geq 0$.
(f) Each y_n has nesting level 3, for $n \geq 1$.

We now prove each of the claims.

(a) Clearly $|x_{n+1}| = 4|x_n| + 12$ for $n \geq 0$. That, together with $|x_0| = 2$, gives $|x_n| = 6 \cdot 4^n - 4$ by an easy induction. Hence $|y_n| = 6 \cdot 4^n$.
(b) Again, an easy induction shows $\sum_{0 \leq i \leq n} |y_i| = \sum_{0 \leq i \leq n} 6 \cdot 4^i = 2 \cdot 4^{n+1} - 2$, which proves the claim.
(c) It suffices to show that $\mathbf{d}[0..2 \cdot 4^{i+1} - 1]$ differs from $\mu^2(\mathbf{d}[0..2 \cdot 4^i - 1])$ precisely at the indices $I \cap \{0, \ldots, 2 \cdot 4^{i+1} - 1\}$. To see this, note that

$$\mathbf{d}[0..2 \cdot 4^{i+1} - 1] = 01\, y_0 \prod_{1 \leq j \leq i} y_j = (01)(001011) \prod_{1 \leq j \leq i} 00\, x_j\, 11$$

$$= (01)(001011) \prod_{1 \leq j \leq i} 00100110\, \mu^2(x_{j-1})\, 10011011$$

$$= (01)(001011) \prod_{0 \leq j < i} 00100110\, \mu^2(x_j)\, 10011011.$$

On the other hand,

$$\mu^2(\mathbf{d}[0..2 \cdot 4^i - 1]) = \mu^2(01 \prod_{0 \leq j < i} y_j)$$

$$= 01101001 \prod_{0 \leq j < i} \mu^2(y_j)$$

$$= (01)(101001) \prod_{0 \leq j < i} \mu^2(00\, x_j\, 11)$$

$$= (01)(101001) \prod_{0 \leq j < i} 01100110\, \mu^2(x_j)\, 10011001.$$

Inspection now shows that the differences occur precisely at the indices mentioned. For uniqueness, assume there is some other binary word **e** starting with 01 and having the stated property. An easy induction on i now shows that $e[0..2 \cdot 4^i - 1] = d[0..2 \cdot 4^i - 1]$ for all i. Hence $e = d$.

(d) We save the automaton in `Walnut` and then verify the word generated satisfies the criterion of part (c).

```
morphism dd "0->01 1->23 2->45 3->67 4->28 5->65 6->56 7->93
    8->89 9->98":
morphism c "0->0 1->1 2->0 3->0 4->1 5->0 6->1 7->1 8->0 9->1":
promote D1 dd:
image D c D1:
# the automaton in the figure for the word d
morphism mu2 "0->0110 1->1001":
image DP mu2 D:
# the automaton for mu^2(d)
reg power4 msd_2 "0*1(00)*":
def differ "D[n]!=DP[n]":
# indices where they differ
eval test "An $differ(n) <=> (Ex $power4(x) & x>=4 &
    (n=2|n=2*x-2|n=2*x+1))":
# check if criterion satisfied
```

And `Walnut` returns TRUE.

(e) We use the following `Walnut` code. It asserts that there is a segment of **d** inside a y_n that is an overlap. When we run it, `Walnut` returns FALSE.

```
eval has_overlap "Ex,i,n $power4(x) & i>=2*x & n>=1 & i+2*n<8*x &
    At,u (t>=i & t<=i+n & u=t+n) => D[t]=D[u]":
```

(f) To check the nesting level, we first make a finite-state transducer that, on input a word x, computes the nesting level if it is ≤ 3 (and outputs 4 if it is ≥ 4 or < 0). It is depicted in Fig. 10. We now simply run our sequence **d** through this transducer and examine the output. It has no outputs of 4, so the nesting level of **d** is ≤ 3.

We can determine the positions of the 0's in the transduced sequence as follows:

```
transduce DN nest D:
eval zeros "An DN[n]=@0 <=> (Ex $power4(x) & n=2*x-1)":
eval threes "Am,n (m<n & DN[m]=@0 & DN[n]=@0 & m>=7) =>
    Et m<t & t<n & DN[t]=@3":
```

The first assertion is that 0's occur in the transduced sequence at exactly the positions corresponding to $2 \cdot 4^i - 1$. The second is that between two occurrences of 0 in the transduced sequence **dn** there is always an occurrence of 3, provided the first occurrence is at a position ≥ 7. `Walnut` returns TRUE for both. Thus each y_n, $n \geq 1$, has nesting level exactly 3.

This completes the proof.   □

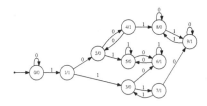

Fig. 9. DFAO for the sequence **d**.   **Fig. 10.** Nesting level transducer.

5 Going Beyond *k*-automatic sequences

So far, we have only worked with transducers for *k*-automatic sequences and automata that take input over base-*k* representations of numbers. However, some famous sequences, such as the Fibonacci word [2], are not *k*-automatic, but rather automatic with respect to a different numeration system $\mathcal{N}$, such as Fibonacci representation [12,19]. We call such sequences $\mathcal{N}$-automatic.

Dekking's construction, suitably modified, can be used to prove the following theorem.

Theorem 3. *Let* $\mathbf{a} = (a_n)_{n \geq 0}$ *be an* $\mathcal{N}$-*automatic sequence, and let* T *be a* 1-*uniform deterministic finite-state transducer. Then* $T(\mathbf{a})$ *is also* $\mathcal{N}$-*automatic.*

Proof. We provide a sketch of the construction in [17]. The $\mathcal{N}$-automatic sequence $\mathbf{a} = (a_n)_{n \geq 0}$ is computed by a finite automaton with output $M = \langle Q, \Sigma, \delta, q_0, \Delta, \lambda \rangle$, where for each $q \in Q$ and $a \in \Sigma$, we have $|\delta(q, a)| \in \{0, 1\}$ (i.e., there is *at most* one edge going out of q on input a). When M receives the $\mathcal{N}$-representation of n as input over some digit set $\Sigma = \{0, 1, \ldots, k-1\}$, the automaton computes a_n. Note that M is only defined on valid $\mathcal{N}$-representations.

The idea is to extend M to a new DFAO M' that is defined on all base-*k* representations and behaves identically to M on $\mathcal{N}$-representations. This is accomplished by defining a dead state $q_\#$ with an output of $\#$ that all inputs that are not valid $\mathcal{N}$-representations will end at. Then M' is a DFAO that accepts all inputs in Σ^*, and computes a *k*-automatic sequence $\mathbf{b} = (b_n)_{n \geq 0}$, satisfying $b_n = a_m$ where $m \in \mathbb{N}$ is such that $[m]_\mathcal{N} = [n]_k$.

We similarly define an extension T' of T where the input and output alphabets are extended to include $\#$. Upon reading $\#$ on state q, the extended transducer T' outputs $\#$ and transitions back to q. Then a new DFAO N computes $T'(\mathbf{b})$, such that for all $n, m \in \mathbb{N}$ satisfying $[m]_\mathcal{N} = [n]_k$, we have $T'(\mathbf{b}) = T(\mathbf{a})$. We may then remove all states of N that have an output of $\#$ to get a finite automaton with output N' that computes $T(\mathbf{a})$ and is only defined on valid $\mathcal{N}$-representations, completing the proof.   □

We now turn to an example. Define $s(n)$ to be the second-to-last bit of the Fibonacci representation of n (most-significant-digit first), and let $s'(n) = 1 - s(n)$. Table 1 gives the first few terms of these sequences.

Table 1. Second-to-last bit of Fibonacci representation.

n	0	1	2	3	4	5	6	7	8	9	10	11	12	13	14	15	16
$s(n)$	0	0	1	0	0	0	0	1	0	0	1	0	0	0	0	1	0
$s'(n)$	1	1	0	1	1	1	1	0	1	1	0	1	1	1	1	0	1

Then $s(n)$ is sequence <u>A123740</u> in the OEIS and $s'(n)$ is sequence <u>A005713</u>. Another characterization of $s(n)$, given in OEIS sequence <u>A123740</u>, is that $s(n) = \lfloor (n+2)\tau \rfloor - \lfloor n\tau \rfloor - 3$, where $\tau = (1 + \sqrt{5})/2$. We can verify this claim using a synchronized automaton for $\lfloor n\tau \rfloor$, given in [16, p. 278], as follows:

```
reg sldf msd_fib "0*(10|0)*10":
# second lowest bit of Fibonacci expansion is 1
combine SLDF sldf:
reg shift {0,1} {0,1} "([0,0]|[0,1][1,1]*[1,0])*":
def phin "?msd_fib (s=0 & n=0) | Ex $shift(n-1,x) & s=x+1":
def aseq "?msd_fib Ex,y x=y+3 & $phin(n+3,x) & $phin(n+1,y)":
def test1 "?msd_fib An $aseq(n) <=> SLDF[n]=@0":
```

and `Walnut` returns TRUE. We now prove the following new result:

Theorem 4. *Let* $\mathbf{f} = 01001010\cdots$ *be the Fibonacci word, fixed point of the morphism* $0 \mapsto 01$, $1 \mapsto 0$. *Then* $\mathbf{f}[n+1] = (\sum_{0 \le i \le n} s'(n)) \bmod 2$ *for all* $n \ge 0$.

Proof. We use the new capabilities of `Walnut`.

```
def nsldf "?msd_fib ~$sldf(n)":
combine NSLDF nsldf:
# sequence A005713
transduce TS RUNSUM2 NSLDF:
# A005614; this is infinite Fibonacci word shifted by 1
eval test2 "?msd_fib An TS[n]=F[n+1]":
# Walnut returns TRUE
```

□

6 Conclusion

Implementation of transducers in `Walnut` allow us to reprove existing theorems in more straightforward ways, and also prove new results. We hope readers will experiment with the new capabilities and find yet more applications.

AZ thanks Andrey Boris Khesin for useful discussions. Both authors thank the CIAA referees and Robert Burns for their helpful comments.

References

1. Allouche, J.P., Shallit, J.: Automatic Sequences: Theory, Applications, Generalizations. Cambridge University Press (2003)
2. Berstel, J.: Fibonacci words–a survey. In: Rozenberg, G., Salomaa, A. (eds.) The Book of L, pp. 13–27. Springer, Heidelberg (1986). https://doi.org/10.1007/978-3-642-95486-3_2
3. Bruyère, V., Hansel, G., Michaux, C., Villemaire, R.: Logic and p-recognizable sets of integers. Bull. Belgian Math. Soc. **1**, 191–238 (1994). Corrigendum, *Bull. Belg. Math. Soc.***1** (1994), 577
4. Burns, R.: Factorials and Legendre's three-square theorem (2021). arXiv preprint arXiv:2101.01567 [math.NT]
5. Burns, R.: Factorials and Legendre's three-square theorem: II (2022). arXiv preprint arXiv:2203.16469 [math.NT]
6. Cobham, A.: Uniform tag sequences. Math. Syst. Theor. **6**, 164–192 (1972)
7. Damanik, D.: Local symmetries in the period-doubling sequence. Disc. Appl. Math. **100**, 115–121 (2000)
8. Dekking, F.M.: Iteration of maps by an automaton. Discrete Math. **126**, 81–86 (1994)
9. Deshouillers, J.M., Luca, F.: How often is $n!$ a sum of three squares? In: Alladi, K., Klauder, J., Rao, C. (eds.) The Legacy of Alladi Ramakrishnan in the Mathematical Sciences. Springer, New York (2010). https://doi.org/10.1007/978-1-4419-6263-8_14
10. Hajdu, L., Papp, Á.: On asymptotic density properties of the sequence $(n!)_{n=0}^{\infty}$. Acta Arith. **184**, 317–340 (2018)
11. Hopcroft, J.E., Ullman, J.D.: Introduction to Automata Theory, Languages, and Computation. Addison-Wesley (1979)
12. Lekkerkerker, C.G.: Voorstelling van natuurlijke getallen door een som van getallen van Fibonacci. Simon Stevin **29**, 190–195 (1952)
13. Mol, L., Rampersad, N., Shallit, J.: Dyck words, pattern avoidance, and automatic sequences. arXiv preprint arXiv:2301.06145 [cs.DM], 15 January 2023
14. Mousavi, H.: Automatic theorem proving in Walnut (2016). arxiv preprint arXiv:1603.06017 [cs.FL]
15. Nathanson, M.B.: Additive Number Theory - The Classical Bases. Springer, New York (1996). https://doi.org/10.1007/978-1-4757-3845-2
16. Shallit, J.: The Logical Approach To Automatic Sequences: Exploring Combinatorics on Words with Walnut. London Mathematical Society Lecture Note Series, vol. 482. Cambridge University Press (2022)
17. Shallit, J., Zavyalov, A.: Transduction of automatic sequences and applications (2023). arXiv preprint arXiv:2303.15203 [cs.FL]
18. Sloane, N.J.A.: The On-line Encyclopedia of Integer Sequences. Electronic Resource, https://oeis.org
19. Zeckendorf, E.: Représentation des nombres naturels par une somme de nombres de Fibonacci ou de nombres de Lucas. Bull. Soc. Roy. Liège **41**, 179–182 (1972)

Measuring Power of Generalised Definite Languages

Ryoma Sin'ya[✉]

Akita University, Akita, Japan
ryoma@math.akita-u.ac.jp

Abstract. A language L is said to be $\mathcal{C}$-measurable, where $\mathcal{C}$ is a class of languages, if there is an infinite sequence of languages in $\mathcal{C}$ that "converges" to L. In this paper, we investigate the measuring power of GD of the class of all generalised definite languages. Although each generalised definite language only can check some local property (prefix and suffix of some bounded length), it is shown that many non-generalised-definite languages are GD-measurable. Further, we show that it is decidable whether a given regular language is GD-measurable or not.

1 Introduction

$\mathcal{C}$-measurability for a class $\mathcal{C}$ of languages is introduced by [14] and it was used for classifying non-regular languages by using regular languages. A language L is said to be $\mathcal{C}$-measurable if there is an infinite sequence of languages in $\mathcal{C}$ that converges to L. Roughly speaking, L is $\mathcal{C}$-measurable means that it can be approximated by a language in $\mathcal{C}$ with *arbitrary high precision*: the notion of "precision" is formally defined by the density of formal languages. Hence that a language L is not $\mathcal{C}$-measurable ($\mathcal{C}$-*im*measurable) means that L has a complex shape so that it can not be approximated by languages in $\mathcal{C}$. While the membership problem for a given language L and a class $\mathcal{C}$ just asks whether $L \in \mathcal{C}$, the $\mathcal{C}$-measurability asks the existence of an infinite sequence of languages in $\mathcal{C}$ that converges to L. In this sense, measurability is much more difficult than the membership problem and its analysis is a challenging task. For example, the author [15] showed that, for the class SF of all star-free languages, the class of all SF-measurable regular languages strictly contains SF but does not contain some regular languages. However, the decidability of SF-measurability for regular languages is still unknown. Only for some very restricted subclasses $\mathcal{C}$ of star-free languages, the decidability of $\mathcal{C}$-measurability is known [16]. A language L is called *locally testable* [5,9,18] if it is a finite Boolean combination of languages of the form uA^*, A^*v and A^*wA^*. Although the definition of local testability is very simple, it was shown in [16] that many non-locally-testable languages are LT-measurable, where LT is the class of all locally testable languages, and any *unambiguous polynomial* (language definable by the first-order logic with two variables) is LT-measurable. However, the decidability of LT-measurability for regular languages was left open in [16].

B. Nagy (Ed.): CIAA 2023, LNCS 14151, pp. 278–289, 2023.
https://doi.org/10.1007/978-3-031-40247-0_21

In this paper, as a continuation research of [16], we examine the measuring power of languages defined by *definiteness*, which is a natural restriction of the notion of local testability. A language L is called *definite* (*reverse definite*, respectively) [3] if it is a finite Boolean combination of languages of the form A^*u (uA^*, respectively). Also, L is called *generalised definite* [7] if it is a finite Boolean combination of languages of the form uA^* and A^*v. We consider GD-measurability and also consider D-measurability and RD-measurability where D, RD and GD is the class of all definite, reverse definite and generalised definite languages. The main results of this paper are two folds. We show:

(1) A simple automata theoretic and algebraic characterisation of RD-measurability (Theorem 1 and Theorem 3).
(2) The equivalence of the GD-measurability and the LT-measurability (Proposition 1) and a decidable characterisation of GD-measurability (Theorem 4). This decidability result answers a question posed in [16].

The structure of this paper is as follows. Section 2 provides preliminaries including density, measurability and definitions of fragments of locally testable languages. An automata theoretic characterisation of RD-measurability is given in Sect. 3, and a decidable characterisation of GD-measurability is given in Sect. 4, respectively. Related and future work are described in Sect. 5.

2 Preliminaries

This section provides the precise definitions of density, measurability and local varieties of regular languages. REG_A denotes the family of all regular languages over an alphabet A. We assume that the reader has a standard knowledge of automata theory including the concept of syntactic monoids (*cf.* [8]).

2.1 Languages and Automata

For an alphabet A, we denote the set of all words (all non-empty words, respectively) over A by A^* (A^+, respectively). We write $|w|$ for the length of w and A^n for the set of all words of length n. For a word $w \in A^*$ and a letter $a \in A$, $|w|_a$ denotes the number of occurrences of a in w. We denote by $w^r = a_k \cdots a_1$ the reverse of $w = a_1 \cdots a_k$, and denote by $L^r = \{w^r \mid w \in L\}$ the reverse of the language L. A word v is said to be a factor of a word w if $w = xvy$ for some $x, y \in A^*$. For a language $L \subseteq A^*$, we denote by $\overline{L} = A^* \setminus L$ the complement of L. A language L is said to be *dense* if $L \cap A^*wA^* \neq \emptyset$ holds for any $w \in A^*$. L is not dense means $L \cap A^*wA^* = \emptyset$ for some word w by definition, and such word w is called a *forbidden word* of L.

A deterministic automaton $\mathcal{A}$ over A is a quadruple $\mathcal{A} = (Q, \cdot, q_0, F)$ where Q is a finite set of states, $\cdot : Q \times A \to Q$ is a transition function, $q_0 \in Q$ is an initial and $F \subseteq Q$ is a set of final states. The language recognised by $\mathcal{A}$ is denoted by $L(\mathcal{A}) = \{w \in A^* \mid q_0 \cdot w \in F\}$. For a set of states $Q' \subseteq Q$ and a word w, we write $Q' \cdot w$ for the set of transition states from Q' by w: $Q' \cdot w = \{q \cdot w \mid q \in Q'\}$.

The automaton $\mathcal{A}$ is called accessible if for every state $p \in Q$ there is a word w such that $q_0 \cdot w = p$. In this paper, we only consider accessible deterministic automata. Q' is called strongly connected if for every $p, q \in Q'$, there is some word w such that $p \cdot w = q$. We say that Q' is a *sink* if it is strongly connected and there is no outgoing transition from Q', i.e., $Q' \cdot w \subseteq Q'$ for any w.

2.2 Locally Testable and Definite Languages

For a family $\mathcal{C}_A$ of languages over A, we denote by $\mathscr{B}\mathcal{C}_A$ the finite Boolean closure of $\mathcal{C}_A$. The class LT_A of all locally testable languages over A can be defined as

$$\mathrm{LT}_A = \mathscr{B}\{wA^*, A^*w, A^*wA^* \mid w \in A^*\}.$$

The class $\mathrm{D}_A, \mathrm{RD}_A$ and GD_A of all *definite, reverse definite* [3] and *generalised definite* [7] languages over A are defined as follows:

$$\mathrm{D}_A = \mathscr{B}\{A^*w \mid w \in A^*\}, \qquad \mathrm{RD}_A = \mathscr{B}\{wA^* \mid w \in A^*\},$$
$$\mathrm{GD}_A = \mathscr{B}\{A^*w, wA^* \mid w \in A^*\}.$$

Hence these classes are proper subclasses of locally testable languages.

Remark 1 (cf. [5]). In [3,7] definite languages are originally defined as follows. A language L is called:

- definite if and only if $L = E \cup A^*F$ for some finite sets $E, F \subseteq A^*$.
- reverse definite if and only if $L = E \cup FA^*$ for some finite sets $E, F \subseteq A^*$.
- generalised definite if and only if $L = E \cup \bigcup_{i \in I} F_i A^* G_i$ for some finite sets E and $F_i, G_i \subseteq A^*$ for all $i \in I$, where I is a finite index set.

For any word $w \in A^*$, the singleton $\{w\}$ can be written as the Boolean combination $wA^* \cap \overline{\bigcup_{a \in A} waA^*}$, hence any finite subset $F \subseteq A^*$ is in $\mathscr{B}\{wA^* \mid w \in A^*\}$. Conversely, for any w, the complement $\overline{wA^*}$ can be written in the form of a reverse definite language: $\{u \in A^* \mid |u| < |w|\} \cup (A^{|w|} \setminus \{w\})A^*$. Hence, these original definitions can be modified by using the finite Boolean closure as above.

2.3 Density and Measurability of Formal Languages

For a set X, we denote by $\#(X)$ the cardinality of X. We denote by $\mathbb{N}$ the set of natural numbers including 0.

Definition 1 (cf. [2]). The *density* $\delta_A(L)$ of $L \subseteq A^*$ is defined as

$$\delta_A(L) = \lim_{n \to \infty} \frac{1}{n} \sum_{k=0}^{n-1} \frac{\#(L \cap A^k)}{\#(A^k)}$$

if it exists, otherwise we write $\delta_A(L) = \perp$. The language L is called *null* if $\delta_A(L) = 0$, and dually, L is called *co-null* if $\delta_A(L) = 1$.

Example 1. It is known that every regular language has a rational density (*cf.* [11]) and it is computable. Here we explain two examples of (co-)null languages.

(1) For each word w, the language A^*wA^*, the set of all words that contain w as a factor, is of density one (co-null). This fact follows from the so-called the *infinite monkey theorem* (this is also called as "Borges's theorem", *cf.* [6, p.61, Note I.35]): take any word w. A random word of length n contains w as a factor with probability tending to 1 as $n \to \infty$.
A language L having a forbidden word w is always null: having a forbidden word w means $A^*wA^* \subseteq \overline{L}$ hence we have $\delta_A(A^*wA^*) \le \delta_A(\overline{L})$, which implies $\delta_A(\overline{L}) = 1$ by the infinite monkey theorem.
(2) The set of all palindromes $L_{\mathrm{pal}} = \{w \in A^* \mid w = w^r\}$ over $A = \{a, b\}$ is dense but null. This follows from the fact that $\#(L_{\mathrm{pal}} \cap A^n)$ equals to $2^{\lceil n/2 \rceil}$ and $2^{\lceil n/2 \rceil}/2^n < 2^{(1-n/2)}$ tends to zero if n tends to infinity.

We list some basic properties of the density as follows.

Lemma 1. *Let $K, L \subseteq A^*$ with $\delta_A(K) = \alpha, \delta_A(L) = \beta$. Then we have:*
(1) $\alpha \le \beta$ if $K \subseteq L$. (2) $\delta_A(L \setminus K) = \beta - \alpha$ if $K \subseteq L$. (3) $\delta_A(\overline{K}) = 1 - \alpha$.
(4) $\delta_A(K \cup L) \le \alpha + \beta$ if $\delta_A(K \cup L) \ne \bot$. (5) $\delta_A(K \cup L) = \alpha + \beta$ if $K \cap L = \emptyset$.
(6) $\delta_A(uL) = \delta_A(Lu) = \delta_A(L) \cdot \#(A)^{-|u|}$ for each $u \in A^*$.

For more properties of δ_A, see Chap. 13 of [2].

The notion of "measurability" on formal languages is defined by a standard measure theoretic approach as follows.

Definition 2 ([14]). Let $\mathcal{C}_A$ be a family of languages over A. For a language $L \subseteq A^*$, we define its $\mathcal{C}_A$-*inner-density* $\underline{\mu}_{\mathcal{C}_A}(L)$ and $\mathcal{C}_A$-*outer-density* $\overline{\mu}_{\mathcal{C}_A}(L)$ over A as

$$\underline{\mu}_{\mathcal{C}_A}(L) = \sup\{\delta_A(K) \mid K \subseteq L, K \in \mathcal{C}_A, \delta_A(K) \ne \bot\} \text{ and}$$
$$\overline{\mu}_{\mathcal{C}_A}(L) = \inf\{\delta_A(K) \mid L \subseteq K, K \in \mathcal{C}_A, \delta_A(K) \ne \bot\}, \text{ respectively.}$$

A language L is said to be $\mathcal{C}_A$-*measurable* if $\underline{\mu}_{\mathcal{C}_A}(L) = \overline{\mu}_{\mathcal{C}_A}(L)$ holds. We say that an infinite sequence $(L_n)_n$ of languages over A *converges to L from inner (from outer, respectively)* if $L_n \subseteq L$ ($L_n \supseteq L$, respectively) for each n and $\lim_{n \to \infty} \delta_A(L_n) = \delta_A(L)$.

We give some examples of LT_A-(im)measurable languages from [14, 16].

Example 2.

(1) The set of all palindromes $L_{\mathrm{pal}} = \{w \in A^* \mid w = w^r\}$ is LT_A-measurable. The sequence of locally testable languages $L_k = \{wA^*w^r \mid |w| = k\}$ converges to L_{pal} from outer if k tends to infinity (see [14] for the detail). The density of L_{pal} is zero as stated in Example 1, hence the constant sequence of the empty language trivially converges to L_{pal} from inner.

(2) For any real number $\alpha \in [0,1]$, there is an LT_A-measurable language L whose density is α. See [15] for the detailed construction.

(3) The language $M_k = \{w \in \{a,b\}^* \mid |w|_a = |w|_b \mod k\}$ is LT-*im*measurable for any $k \geq 2$. See [16] or Sect. 4.1 for the proof.

For a family $\mathcal{C}_A$ of languages over A, we denote by $\mathrm{Ext}_A(\mathcal{C}_A)$ ($\mathrm{RExt}_A(\mathcal{C}_A)$, respectively) the class of all $\mathcal{C}_A$-measurable languages ($\mathcal{C}_A$-measurable *regular* languages, respectively) over A. A family of regular languages over A is called a *local variety* [1] over A if it is closed under Boolean operations and left-and-right quotients.

Lemma 2 ([15]). Ext_A *is a closure operator, i.e., it satisfies the following three properties for each* $\mathcal{C} \subseteq \mathcal{D} \subseteq 2^{A^*}$: *(extensive)* $\mathcal{C} \subseteq \mathrm{Ext}_A(\mathcal{C})$, *(monotone)* $\mathrm{Ext}_A(\mathcal{C}) \subseteq \mathrm{Ext}_A(\mathcal{D})$, *and (idempotent)* $\mathrm{Ext}_A(\mathrm{Ext}_A(\mathcal{C})) = \mathrm{Ext}_A(\mathcal{C})$. *Moreover,* RExt_A *is a closure operator over the class of all local varieties of regular languages over* A, *i.e.,* $\mathcal{C}_A$-*measurability is preserved under Boolean operations and quotients for any local variety* $\mathcal{C}_A$.

The following lemma is useful and will be used in Sect. 3 and Sect. 4.

Lemma 3. *Let* $\mathcal{A} = (Q, \cdot, q_0, F)$ *be a deterministic automaton,* $Q_1, \cdots, Q_k$ *be its all sink components and let* $Q' = Q \setminus \bigcup_{i=1}^{k} Q_i$. *Then the language* $P' = \{w \in A^* \mid q_0 \cdot w \in Q'\}$ *is of density zero,* $P_i = \{w \in A^* \mid q_0 \cdot w \in Q_i\}$ *satisfies* $P_i = P_i A^*$ *and has a non-zero density for each* i.

Proof. The condition $P_i = P_i A^*$ is clear because Q_i is a sink for each i: $Q_i \cdot w \subseteq Q_i$ holds for every w. For each i, P_i is non-empty because $\mathcal{A}$ is accessible (all automata in this paper are accessible as stated in Sect. 2.1). Let w be a word in P_i. By Lemma 1, we have $\delta_A(P_i) \geq \delta_A(wA^*) = \#(A)^{-|w|} > 0$, i.e., P_i has a non-zero density. Now we show that the density of P' is zero. Let $Q' = \{q_0, q_1, \cdots, q_n\}$. For every state q_i in Q', there exists some word w_{q_i} such that $q_i \cdot w_{q_i}$ is in some sink component. Because every q_i in Q' is not in any sink component, q_i is not reachable from the state $q_i \cdot w_{q_i}$, i.e. $(q_i \cdot w_{q_i}) \cdot w \notin Q'$ for every w. Define $u_0 = w_{q_0}$ and $u_i = w_{q_i \cdot v_{i-1}}$ if $q_i \cdot v_{i-1} \in Q'$ and $u_i = \varepsilon$ otherwise for each $i \in \{1, \cdots, n\}$ where v_{i-1} is the word of the form $u_0 \cdots u_{i-1}$. By the construction, for every q_i in Q', we have $q_i \cdot u_0 \cdots u_n \notin Q'$. This means that $u_0 \cdots u_n$ is a forbidden word of P' and hence P' is of density zero by the infinite monkey theorem. $\square$

For simplicity, here after we fix an alphabet A and omit the subscript A for denoting local varieties.

3   Simple Characterisation of RD-Measurability

The next theorem gives a simple automata theoretic characterisation of RD-measurability.

Theorem 1. *For a minimal deterministic automaton $\mathcal{A}$, the followings are equivalent:*

(1) *Every sink component of $\mathcal{A}$ is a singleton.*
(2) *$L(\mathcal{A})$ is RD-measurable.*

Proof. Let $L = L(\mathcal{A}), Q_1, \cdots, Q_k$ be all sink components of $\mathcal{A} = (Q, \cdot, q_0, F)$ and let $Q' = Q \setminus \bigcup_{i=1}^{k} Q_i$. For each $i \in \{1, \cdots, k\}$, define $P_i = \{w \in A^* \mid q_0 \cdot w \in Q_i\}$ and define $P' = \{w \in A^* \mid q_0 \cdot w \in Q'\}$. Clearly, $P_1, \cdots, P_k$ and P' form the partition of A^*, and we have $\delta_A(P') = 0$ by Lemma 3.

Proof of (1) $\Rightarrow$ (2): Because each Q_i is a singleton, P_i is contained in L if the state in Q_i belongs to F and P_i is contained in $\overline{L}$ otherwise. Define

$$M = \bigcup \{P_i \mid Q_i \subseteq F\} \quad \text{and} \quad M' = \bigcup \{P_i \mid Q_i \subseteq Q \setminus F\}.$$

By the definition and Lemma 3, we have $M = MA^* \subseteq L$ and $M' = M'A^* \subseteq \overline{L}$. Because $P_1, \cdots, P_k, P'$ form the partition of A^* and the density of P' is zero, we can deduce that $\delta_A(M) + \delta_A(M') = 1$, which implies $\delta_A(M) = \delta_A(L)$ and $\delta_A(M') = \delta_A(\overline{L})$. For each $n \in \mathbb{N}$ and i, the set $M_n = \{w \in M \mid |w| \leq n\}$ and $M'_n = \{w \in M' \mid |w| \leq n\}$ are finite and hence the sequence of reverse definite languages $M_n A^*$ and $\overline{M'_n A^*}$ converges to L from inner and outer, respectively, *i.e.*(i) $M_n A^* \subseteq L$ and $\overline{M'_n A^*} \supseteq L$ holds for each n, and (ii) $\lim_{n \to \infty} \delta_A(M_n A^*) = \delta_A(L)$ and $\lim_{n \to \infty} \delta_A(\overline{M'_n A^*}) = \delta_A(L)$.

Proof of (2) $\Rightarrow$ (1): This direction is shown by contraposition. We assume that (1) is not true, *i.e.*, some sink component, say Q_j, is not a singleton. By the minimality of $\mathcal{A}$, Q_j contains at least one final state, say p, and at least one non-final state, say p' (if not, all states in Q_j are right equivalent). For each $q \in Q_j$, we write L_q for the language $L_q = \{w \in A^* \mid q_0 \cdot w = q\}$.

Because P_j is non-empty and $P_j = P_j A^*$ holds, the density of P_j is not zero. P_j has non-zero density implies that there exists at least one state q in P_j such that L_q has non-zero density. Since Q_j is a sink (strongly connected, especially), there exist some words $w_{q,p}$ and $w_{p,p'}$ such that $q \cdot w_{q,p} = p$ and $p \cdot w_{p,p'} = p'$. Thus $L_q w_{q,p} \subseteq L_p$ holds, from which we can deduce that $\delta_A(L_p) \geq \delta_A(L_q w_{q,p}) = \delta_A(L_q) \cdot \#(A)^{-|w_{q,p}|} > 0$, *i.e.*, L_p has non-zero density, say $\alpha > 0$.

We can show that, for every reverse definite language $R = E \cup FA^*$ (where E, F are finite sets) such that $R \subseteq L$, $FA^* \cap L_p = \emptyset$ holds as follows. If there is some word $w \in FA^* \cap L_p$, then $ww_{p,p'}$ is in $FA^* \cap \overline{L}$ since $ww_{p,p'} \in L_{p'}$ and p' is non-final. This violates the assumption $R \subseteq L$. This means that every reverse definite subset $R = E \cup FA^*$ of L should have density less than or equal to $\delta_A(L \setminus L_p) = \delta_A(L) - \alpha < \delta_A(L)$. Hence, no sequence of reverse definite languages converges to L from inner. $\qquad \square$

For a given automaton $\mathcal{A}$, we can construct its reverse automaton $\mathcal{A}^r$ recognising $L(\mathcal{A})^r$ by flipping final and non-final states and reversing transition relations. By the definition of definite and reverse definite languages, L is D-measurable if and only if L^r is RD-measurable. Hence, we can use Theorem 1 to deduce the decidability of D-measurability.

Corollary 1. *For a given regular language L it is decidable whether L is RD-measurable (D-measurable, respectively).*

3.1   Algebraic Characterisation

In this subsection we give an algebraic characterisation of RD-measurability, which is a natural analogy of the algebraic characterisation of RD stated as follows. Let S be a semigroup. An element $x \in S$ is called a left zero if $xS = \{x\}$ holds. An element $x \in S$ is called an idempotent if $x^2 = x$ holds.

Theorem 2 *(cf. [4]). For a regular language L and its syntactic semigroup S_L, the followings are equivalent:*

(1) *L is in RD.*
(2) *Every idempotent of S_L is a left zero.*

Let M be a monoid. For elements x and y in M, we write $x \leq_{\mathcal{R}} y$ if $xM \subseteq yM$ holds. Notice that $x \leq_{\mathcal{R}} y$ if and only if $yz = x$ for some $z \in M$. An element x is called $\mathcal{R}$-*minimal* if $y \leq_{\mathcal{R}} x$ implies $x \leq_{\mathcal{R}} y$ for every y in M.

Theorem 3. *For a regular language L and its syntactic monoid M_L, the followings are equivalent:*

(1) *L is RD-measurable.*
(2) *Every $\mathcal{R}$-minimal element of M_L is a left zero.*
(3) *Every $\mathcal{R}$-minimal idempotent of M_L is a left zero.*

Proof. Let $\mathcal{A} = (Q, \cdot, q_0, F)$ be the minimal automaton of L. Notice that M_L is isomorphic to the transition monoid $T = (\{f_w : Q \to Q \mid w \in A^*\}, \circ, f_\varepsilon)$ of $\mathcal{A}$ where f_w is the map defined by $f_w(q) = q \cdot w$, the multiplication operation $\circ$ is the composition $f_u \circ f_v = f_{uv}$ and the identity element f_ε is the identity mapping on Q. Hence, we identify M_L with T.

Proof of (1) $\Rightarrow$ (2): Let f be an $\mathcal{R}$-minimal element of T. If $f(q)$ is not in any sink component of $\mathcal{A}$ for some q, there is a some word w such that $(f \circ f_w)(q) = f(q) \cdot w$ is in some sink component. But this means that f is not $\mathcal{R}$-minimal because q is not reachable by $f(q) \cdot w$, which implies $(f \circ f_w) \circ g \neq f$ for any $g \in T$. Hence, $f(q)$ is in some sink component. By the assumption and Theorem 1, every sink component of $\mathcal{A}$ is a singleton. This means that $f(q) \cdot w = q$ holds for every w, *i.e.*, f is a left zero.

Proof of (2) $\Rightarrow$ (1): This direction is shown by contraposition. Assume (1) is not true. That is, there is a sink component $Q' \subseteq Q$ which is not a singleton by Theorem 1. Let p and q in Q' be two different states and f be an $\mathcal{R}$-minimal element in T such that $f(q_0) = p$ (such f always exists since $\mathcal{A}$ is accessible and Q' is sink). Because Q' is strongly connected, there is some word w such that $p \cdot w = q$. This means that $f \neq f \circ f_w$ (because $f(q_0) = p \neq q = (f \circ f_w)(q_0)$), *i.e.*, f is not a left zero.

Proof of (2) $\Leftrightarrow$ (3): (2) implies (3) is trivial. Assume (3). Let x be an $\mathcal{R}$-minimal element of M_L. Because M_L is finite, there is some index $i \geq 1$ such that x^i is an idempotent. By the $\mathcal{R}$-minimality of x and $x^i = x \cdot x^{i-1} \leq_{\mathcal{R}} x$, $x^i \cdot y = x$ holds for some y. But x^i is a left zero by the assumption, this means that $x = x^i$. □

4   Decidable Characterisation of GD-Measurability

In this section we consider the GD-measurability. First we show that the GD-measurability is equivalent to the LT-measurability.

Proposition 1. *A language L is* LT-*measurable if and only if L is* GD-*measurable.*

Proof. For proving the equivalence $\mathrm{Ext}_A(\mathrm{LT}) = \mathrm{Ext}_A(\mathrm{GD})$, it is enough to show that every locally testable language is GD-measurable by the monotonicity and idempotency of Ext_A (Lemma 2): $\mathrm{Ext}_A(\mathrm{GD}) \supseteq \mathrm{LT}$ implies $\mathrm{Ext}_A(\mathrm{GD}) = \mathrm{Ext}_A(\mathrm{Ext}_A(\mathrm{GD})) \supseteq \mathrm{Ext}_A(\mathrm{LT}) \supseteq \mathrm{Ext}_A(\mathrm{GD})$. Further, since GD is closed under Boolean operations, GD-measurability is closed under Boolean operations by Lemma 2 and hence we only have to show that wA^*, A^*w and A^*wA^* are all GD-measurable for every w. The languages of the form wA^* and A^*w are already in GD, thus it is enough to show that A^*wA^* is GD-measurable. This was essentially shown in [16] as follows. Since the case $w = \varepsilon$ is trivial, we assume $w = a_1 \cdots a_n$ where $a_i \in A$ and $n \geq 1$. Define $W_k = (A^k \setminus K_k)wA^*$ where $K_k = \{u \in A^k \mid ua_1 \cdots a_{n-1} \in A^*wA^*\}$ for each $k \geq 0$. Intuitively, W_k is the set of all words in which w *firstly* appears at the position $k+1$ as a factor. By definition, W_k is generalised definite (reverse definite, in particular). Clearly, $W_i \cap W_j = \emptyset$ and $\delta_A(W_i) > 0$ for each $i \neq j$, thus we have $\bigcup_{k \geq 0} W_k = A^*wA^*$ and hence $\lim_{n \to \infty} \delta_A\left(\bigcup_{k \geq 0} W_k\right) = 1$, *i.e.*, $\underline{\mu}_{\mathrm{GD}}(A^*wA^*) = 1$. Thus $A^*wA^* \in \mathrm{Ext}_A(\mathrm{GD})$. □

Next we give a decidable characterisation of GD-measurability for regular languages. The characterisation is not so much simple as the one of RD-measurability stated in Theorem 1, but the proof is a natural generalisation of the proof of Theorem 1.

Theorem 4. *Let $\mathcal{A} = (Q, \cdot, q_0, F)$ be a deterministic automaton and let $Q_1, \cdots, Q_k$ be its all sink components and let $Q' = Q \setminus \bigcup_{i=1}^k Q_i$. Define*

$$P_i = \{w \in A^* \mid q_0 \cdot w \in Q_i\} \qquad P' = \{w \in A^* \mid q_0 \cdot w \in Q'\}$$
$$S_i = \{w \in A^* \mid Q_i \cdot w \subseteq F\} \qquad S_i' = \{w \in A^* \mid Q_i \cdot w \subseteq Q \setminus F\}$$

for each $i \in \{1, \cdots, k\}$, and define

$$M = \bigcup_{i=1}^k P_i S_i \qquad \text{and} \qquad M' = \bigcup_{i=1}^k P_i S_i'.$$

Then $L = L(\mathcal{A})$ is GD-*measurable if and only if $\delta_A(L) = \delta_A(M)$ and $\delta_A(\overline{L}) = \delta_A(M')$ holds.*

Proof. By the construction, clearly $M \subseteq L$ and $M' \subseteq \overline{L}$ holds. Also, by Lemma 3, we have $M = \bigcup_{i=1}^{k} P_i A^* S_i$ and $M' = \bigcup_{i=1}^{k} P_i A^* S_i'$. Intuitively, M and M' are "largest" (with respect to the density) languages of the form PA^*S included in L and $\overline{L}$, respectively. "if" part is easy. $\delta_A(L) = \delta_A(M)$ and $\delta_A(\overline{L}) = \delta_A(M')$ implies that the two sequences of generalised definite languages $M_n = \bigcup_{i=1}^{k} \{u A^* v \mid u \in P_i, v \in S_i, |u| + |v| \leq n\}$ and the complements of $M_n' = \bigcup_{i=1}^{k} \{u A^* v \mid u \in P_i, v \in S_i', |u| + |v| \leq n\}$ converges to L if n tends to infinity from inner and outer, respectively.

Next we show "only if" part by contraposition. With out loss of generality, we can assume that $\delta_A(L) > \delta_A(M)$. For every $u, v \in A^*$, we show that

$$uA^*v \subseteq L \Rightarrow (uA^* \setminus P')v \subseteq M. \qquad (\diamondsuit)$$

This implies $\delta_A(uA^*v) = \delta_A((uA^* \setminus P')v) \leq \delta_A(M)$ (because P' has density zero by Lemma 3), from this we can conclude that every generalised definite language should have density less than or equal to the density of M. Hence, no sequence of generalised definite languages converges to L from inner by the assumption $\delta_A(L) > \delta_A(M)$. Let $u, v \in A^*$ be words satisfying $uA^*v \subseteq L$, and let uw be a word in $uA^* \setminus P'$. Because uw is not in P', uw is in P_j for some $j \in \{1, \cdots, k\}$. The condition $uA^*v \subseteq L$ implies $uwA^*v \subseteq L$ and hence we have $uww'v \in L$ for any word $w' \in A^*$. For every $q \in Q_j$, there is some word w' such that $q_0 \cdot uww' = q$ because Q_j is strongly connected. Thus we can conclude that $q \cdot v \in F$ for each $q \in Q_j$, which means that v is in S_j and hence uwv is in M (by $uw \in P_j$ and $v \in S_j$), *i.e.*, the condition $(\diamondsuit)$ is true. Let $R = E \cup \bigcup_{i \in I} F_i A^* G_i$ be a generalised definite language included in L, where E and $F_i, G_i \subseteq A^*$ are finite for all $i \in I$ and I is a finite index set. The condition $(\diamondsuit)$ and $R \subseteq L$ implies that $\bigcup_{i \in I} (F_i A^* \setminus P') G_i \subseteq M$ (note that E is density zero because it is finite). This means that any generalised definite subset of L should have a density smaller or equal to $\delta_A(M)$ which is strictly smaller than $\delta_A(L)$ by the assumption. Thus there is no convergent sequence of generalised definite languages to L from inner. $\square$

By the construction, clearly, all languages P_i, S_i, S_i' are regular and automata recognising these languages can be constructed from $\mathcal{A}$. Hence, we can effectively construct two automata recognising M and M' from $\mathcal{A}$. Also, checking the condition $\delta_A(L) = \delta_A(M)$ and $\delta_A(\overline{L}) = \delta_A(M')$ is decidable: this condition is equivalent to $\delta_A(M \cup M') = 1$, and it is decidable in linear time whether a given deterministic automaton recognises a co-null regular language (*cf.* [13]).

Corollary 2. *For a given regular language L it is decidable whether L is GD-measurable (equivalently, LT-measurable by Proposition 1).*

4.1   Remark on the Measuring Power of GD

As we stated in Example 2, the language $M_k = \{w \in \{a, b\}^* \mid |w|_a = |w|_b \bmod k\}$ is LT-*im*measurable for any $k \geq 2$. The proof of the above fact given in [16] uses an algebraic characterisation of locally testable languages. However, through Proposition 1, we can more easily prove this fact by showing that M_k is LT-immeasurable as follows.

Proposition 2. $M_k = \{w \in \{a, b\}^* \mid |w|_a = |w|_b \mod k\}$ *is GD-immeasurable for any $k \geq 2$.*

Proof. By simple calculation, we have $\delta_A(M_k) = 1/k$. By definition, every infinite generalised definite language must contain a language of the form uA^*v for some $u, v \in A^*$. Let $n = |uv|_a - |uv|_b \mod k$, define $w = b$ if $n = 0$ and $w = \varepsilon$ otherwise. Then we have $uwv \in L$ but $uwv \notin M_k$. This means that $\underline{\mu}_{\mathrm{GD}}(M_k) = 0 < \delta_A(M_k)$, *i.e.*, M_k is GD-immeasurable. $\square$

A non-empty word w is said to be *primitive* if there is no shorter word v such that $w = v^k$ for some $k \geq 2$. In [14], it is shown that the set Q of all primitive words over $A = \{a, b\}$ is REG-immeasurable where REG is the class of all regular languages. The proof given in [14] involves some non-trivial analysis of the syntactic monoid of a regular language. If we consider the more weaker notion, GD-measurability, the proof of the GD-immeasurability is almost trivial: by definition, every infinite generalised definite language must contain a language of the form uA^*v. But uA^*v contains the non-primitive word $uvuv$, hence there is no infinite generalised definite subset of Q.

From the last example, one can naturally consider that the GD-measurability is a very weaker notion than the REG-measurability. We are interested in how far the GD-measurability is from the REG-measurability: is there any natural subclass GD $\subsetneq \mathcal{C} \subsetneq$ REG of regular languages such that the $\mathcal{C}$-measurability differs from these two measurability? A possible candidate is SF the class of all star-free languages as we discussed in the next section.

5   Related and Future Work

As we stated in Sect. 1, the decidability of SF-measurability [15] for regular languages is still unknown. The decidability of LT-measurability was left open in [16], but thanks to Proposition 1 and Theorem 4, it was shown that LT-measurability (= GD-measurability) is decidable.

For some weaker fragments of star-free languages, the decidability of measurability for regular languages are known: a language L is called *piecewise testable* [12] if it can be represented as a finite Boolean combination of languages of the form $A^*a_1A^* \cdots A^*a_kA^*$ (where $a_i \in A$ for each i), and L is called *alphabet testable* if it can be represented as a finite Boolean combination of languages of the form A^*aA^* (where $a \in A$). We denote by PT and AT the class of all piecewise testable and alphabet testable languages, respectively. It was shown in [16] that AT-measurability and PT-measurability are both decidable. Moreover, AT-measurability and PT-measurability do not rely on the existence of an infinite convergent sequence, but rely on the existence of a certain *single* language [16]:

– L is AT-measurable if and only if L or its complement contains $\bigcap_{a \in A} A^*aA^*$.

- L is PT-measurable if and only if L or its complement contains a language of the form $A^*a_1A^* \cdots A^*a_kA^*$

In [17] the tight complexity bounds of AT-measurability and PT-measurability for regular languages was given: AT-measurability is co-**NP**-*complete* and PT-measurability is *decidable in linear time*, if an input regular language is given by a deterministic automaton. Even though AT is a very restricted subclass of PT, the complexity of AT-measurability is much higher than PT-measurability. This contrast is interesting. Thanks to Theorem 1, RD-measurability is decidable in linear time, if an input regular language is given by a minimal automaton.

Our future work are three kinds.

(1) Give the tight complexity bound of D- and GD-measurability.
(2) Prove or disprove $\mathrm{Ext}_A(\mathrm{GD}) \subsetneq \mathrm{Ext}_A(\mathrm{SF})$.
(3) If $\mathrm{Ext}_A(\mathrm{GD}) \subsetneq \mathrm{Ext}_A(\mathrm{SF})$, prove or disprove the decidability of SF-measurability.

As demonstrated in the proof of Theorem 4, GD-measurability heavily relies on the existence of an *infinite sequence* of different generalised definite languages. Hence the situation is essentially different with AT-measurability and PT-measurability. One might naturally consider that GD-measurability has a more higher complexity than AT-measurability.

To tackle the problem (2) and (3), perhaps we can use some known techniques related to star-free languages, for example, the so-called *separation problem* for a language class $\mathcal{C}$: for a given pair of regular languages (L_1, L_2), is there a language L in $\mathcal{C}$ such that $L_1 \subseteq L$ and $L \cap L_2 = \emptyset$ (L "separates" L_1 and L_2)? It is known that the separation problem for SF is decidable [10].

Acknowledgements. I am grateful to Mark V. Lawson whose helpful discussions were extremely valuable. The author also thank to anonymous reviewers for many valuable comments. This work was supported by JST ACT-X Grant Number JPM-JAX210B, Japan.

References

1. Adámek, J., Milius, S., Myers, R.S.R., Urbat, H.: Generalized Eilenberg theorem I: local varieties of languages. In: Muscholl, A. (ed.) FoSSaCS 2014. LNCS, vol. 8412, pp. 366–380. Springer, Heidelberg (2014). https://doi.org/10.1007/978-3-642-54830-7_24
2. Berstel, J., Perrin, D., Reutenauer, C.: Codes and Automata, Encyclopedia of mathematics and its applications, vol. 129. Cambridge University Press (2010)
3. Brzozowski, J.: Canonical regular expressions and minimal state graphs for definite events. Math. Theor. Automata **12**, 529–561 (1962)
4. Brzozowski, J.: On Aperiodic I-monoids. Department of Computer Science. University of Waterloo. Research Report CS-75-28 (1975)
5. Brzozowski, J., Simon, I.: Characterizations of locally testable events. Discrete Math. **4**, 243–271 (1973)

6. Flajolet, P., Sedgewick, R.: Analytic Combinatorics, 1st edn. Cambridge University Press, New York (2009)

7. Ginzburg, A.: About some properties of definite, reverse-definite and related automata. IEEE Trans. Electron. Comput. **EC–15**(5), 806–810 (1966)

8. Lawson, M.V.: Finite Automata. Chapman and Hall/CRC (2004)

9. McNaughton, R.: Algebraic decision procedures for local testability. Math. Syst. Theor. **8**, 60–76 (1974)

10. Place, T., Zeitoun, M.: Separating regular languages with first-order logic. Log. Meth. Comput. Sci. **12**(1) (2016)

11. Salomaa, A., Soittola, M.: Automata Theoretic Aspects of Formal Power Series. Springer, New York (1978). https://doi.org/10.1007/978-1-4612-6264-0

12. Simon, I.: Piecewise testable events. In: Automata Theory and Formal Languages, pp. 214–222 (1975)

13. Sin'ya, R.: An automata theoretic approach to the zero-one law for regular languages. In: Games, Automata, Logics and Formal Verification, pp. 172–185 (2015)

14. Sin'ya, R.: Asymptotic approximation by regular languages. In: Current Trends in Theory and Practice of Computer Science, pp. 74–88 (2021)

15. Sin'ya, R.: Carathéodory extensions of subclasses of regular languages. In: Developments in Language Theory, pp. 355–367 (2021)

16. Sin'ya, R.: Measuring power of locally testable languages. In: Diekert, V., Volkov, M. (eds.) Developments in Language Theory, DLT 2022. LNCS, vol. 13257, pp. 274–285. Springer, Cham (2022). https://doi.org/10.1007/978-3-031-05578-2_22

17. Sin'ya, R., Yamaguchi, Y., Nakamura, Y.: Regular languages that can be approximated by testing subword occurrences. Comput. Softw. **40**(2), 49–60 (2023). (written in Japanese)

18. Zalcstein, Y.: Locally testable languages. J. Comput. Syst. Sci. **6**(2), 151–167 (1972)

Smaller Representation of Compiled Regular Expressions

Sicheol Sung[1], Sang-Ki Ko[2], and Yo-Sub Han[1(✉)]

[1] Department of Computer Science, Yonsei University, Seoul, Republic of Korea
{sicheol.sung,emmous}@yonsei.ac.kr
[2] Department of Computer Science and Engineering, Kangwon National University,
Chuncheon, Gangwon-do, Republic of Korea
sangkiko@kangwon.ac.kr

Abstract. We consider the problem of running the regex pattern matching in a space-efficient manner. Given a regex, we suggest a bit-packing scheme for representing a compiled regex in a compressed way, which is its position automaton. Our scheme reduces its representation size further by relying on the homogeneous property of position automata and practical features of regexes. We implement the proposed scheme and evaluate the memory consumption using a practical regex benchmark dataset. Our approach produces a much smaller representation compared to two common FA representations. In addition, experimental results show that our bit-packing regex engine is effective for matching regexes that have large compiled forms, by showing less memory consumption compared to the current state-of-the-art regex engine (RE2).

Keywords: Regex engine · Position automata · Bit-packing scheme · Regular expressions

1 Introduction

Regular expressions (regexes in short) are useful to represent patterns in various applications such as natural language processing, intrusion detection systems, and bioinformatics [9,17,21]. Once we write a desirable pattern P as a regex, the regex pattern matching engine (regex engine in short) finds corresponding pattern occurrences in a text string T with respect to the pattern regex. A typical regex engine runs in two steps: compilation and matching. In the compilation step, the regex engine converts P into an equivalent finite-state automaton called a *compiled regex*. Then, in the matching step, the engine determines whether or not the compiled regex accepts T (membership test), or reports where P occurs as a substring of T (searching test). [10] Most programming languages allow regexes and speed up the matching process by storing compiled regexes in memory. For example, the regex engine of C# keeps a compiled regex in main memory and avoids the re-compilation of the same regex repeatedly [8]. A C++ library named ctre eliminates the regex compilation overhead in runtime,

B. Nagy (Ed.): CIAA 2023, LNCS 14151, pp. 290–301, 2023.
https://doi.org/10.1007/978-3-031-40247-0_22

by compiling a program with regexes [10]. As the volume of data increases, the regex becomes larger and more complicated. These larger regexes may cause a speed issue when the amount of memory is limited; for example, IoT or sensor devices. This motivates us to study an effective way of reducing the memory consumption for storing compiled regexes.

We present a bit-packing scheme for space-efficient representation of position automata [16,20], which are one of the most popular compiled regex representations. Our method facilitates the compression methods for word dictionaries by Daciuk and Weiss [11] with additional features to support practical regexes.

Our bit-packing regex engine (BP-engine) compiles a regex to its position automaton, and converts the automaton to a bit-packed regex (BP-regex) according to our bit-packing scheme. Our engine matches the BP-regex directly for memory efficiency, bypassing the need for the unpacking process that converts a BP-regex back into the corresponding position automaton. This gives rise to memory-efficient processing. For instance, an intrusion detection system embedded in an IoT device has limited computing resources. The system often processes large regexes of malicious patterns, which are hard to process in low-resource devices. We may convert such regexes to BP-regexes, then our BP-engine directly processes BP-regexes without redundant unpacking steps. Figure 1 shows an overview of our matching process.

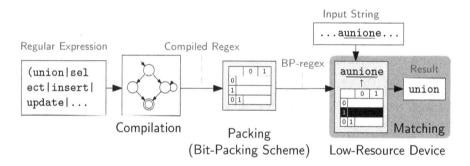

Fig. 1. An overview of our BP-regex matching process with a BP-regex.

We demonstrate the effectiveness of our memory-saving approach by answering the following questions.

RQ1: How much does the proposed scheme reduce the size of compiled regexes compared to existing formats? (Sect. 5.2)

RQ2: How much does the precompiled BP-regex reduce the memory usage during pattern matching compared to the engines with compilation? (Sect. 5.3)

RQ3: How long does it take for BP-regex to pack a compiled regex according to our scheme? (Sect. 5.4)

We present a bit-packing scheme for converting practical regexes to smaller representations. Our scheme relies on the homogeneous property [5] of the position automata (Sect. 3.1) and some practical features of regexes (Sect. 3.2) for reducing the representation size. We implement our BP-engine based on the proposed scheme, and demonstrate its effectiveness for addressing the memory shortage problem through experiments.

The rest of the paper is organized as follows. In Sect. 2, we define the terms including special labels in practical regexes. We propose a bit-packing scheme and present major features of the scheme in Sect. 3. Then, in Sect. 4, we recall the Grail format and dumped regexes for comparison. We evaluate our scheme and compare its performance with existing solutions in Sect. 5 and conclude the paper in Sect. 6.

2 Background

An alphabet Σ is a set of symbols and a string w is a finite sequence of symbols. A regex E over Σ is defined as follows. The empty set $\emptyset$, the empty string λ, and a single symbol $\sigma \in \Sigma$ are regexes. For regexes E_1 and E_2, the concatenation $(E_1 E_2)$, the union $(E_1 | E_2)$ and Kleene star (E_1^*) are regexes. A nondeterministic finite-state automaton (NFA) is a quintuple $A = (Q, \Sigma, \delta, q_0, F)$, where Q is a set of states, Σ is a finite alphabet, $\delta : Q \times \Sigma \to 2^Q$ is a transition function, $q_0 \in Q$ is the initial state, and $F \subseteq Q$ is a set of final states. The NFA A recognizes the language $\{w \in \Sigma^* \mid \delta^*(q_0, w) \cap F \neq \emptyset\}$, where $\delta^* : Q \times \Sigma^* \to 2^Q$ is defined by $\delta^*(q, \lambda) = \{q\}$ and $\delta^*(p, w\sigma) = \{r \in Q \mid r \in \delta(q, \sigma) \text{ for some } q \in \delta^*(p, w)\}$. An NFA is a deterministic finite-state automaton (DFA) if $|\delta(q, \sigma)| \leq 1$ for any $q \in Q$ and $\sigma \in \Sigma$.

We can think of δ as a set of labeled transitions (p, σ, q), where $q \in \delta(p, \sigma)$. For each transition $(p, \sigma, q) \in \delta$, we say states p and q as the source state and the target state of the transition, respectively. We also say that the transition is an out-transition of p and an in-transition of q. An NFA has the *homogeneous property* if the in-transitions of each state have the same label, and position automata have the homogeneous property [6]. The density of A is the ratio $\frac{|\delta|}{|Q|^2}$ of the number of transitions to the number of state pairs. Note that the density of position automata is between 0 and 1 because each state of a position automaton has at most $|Q|$ out-transitions. For any symbol σ, we say that a state $q \in \delta(p, \sigma)$ is an adjacent state of p. We do not use the term next state to avoid confusion with the state that follows in the memory.

Real-world regexes have several features for practical implementation. For instance, anchors denote empty strings at constrained positions; anchor ^ and anchor \$ denote start-of-line and end-of-line, respectively. A character class is another special feature that denotes a set of characters by enclosing the characters or ranges of characters in square brackets. One can also denote the inverted character set by the caret following the opening bracket. For example, [^aeiou0-9] is the set of characters that are neither vowels nor digits.

Literal denotes non-special symbols. In practice, most regex engines have a priority mechanism that enables an engine to choose one out-transition over another out-transitions to facilitate the matching preference [2].

3 Proposed Bit-Packing Scheme

Daciuk and Weiss [12] proposed an effective bit-packing scheme for storing a word dictionary in an acyclic DFA. They represented each state by listing its out-transitions together with their labels and the addresses of target states (target addresses). They also introduced a *variable-length coding* (v-coding) of addresses. The first bit of each byte indicates whether or not the byte is the last byte of the encoded address, and the remaining seven bits denote the value of the address. For a state q, let *in-memory next state* be the state following q in memory. The representation of a transition also includes N-, L-, and F- flags as follows:

- N: **n**o address; we do not need a target state address since the target state is the in-memory next state.
- L: this is the **l**ast transition of the source state.
- F: this is a transition to a **f**inal state.

This scheme produces the CFSA2 format, which is a dictionary representation that enables an acyclic DFA matching against a CFSA2 input. The scheme is suitable for a finite set of strings but cannot handle an infinite set of strings or character classes.

Based on their work, we propose a bit-packing scheme for compiled regexes (= NFAs) with practical regex features. Figure 2 shows the change in the transition layout; the left is in the CFSA2 format and the right one is our scheme.

Fig. 2. An illustration for (a) a transition in the CFSA2 format and (b) a state in the BP-regex format.

While v-coding in the CFSA2 format represents a single target address only, we extend v-coding to handle multiple target addresses that share the source state and the same label. In addition to the first bit indicating whether the byte is at the end of an encoded address, we use the second bit of the last byte to indicate that the address is the last address of the label. Then, we arrange the transitions in decreasing order of priority.

We also support special labels for practical regexes. The first byte denotes a label type, and the following bytes represent a multi-byte encoded label whose length and encoding method depend on its label type. The NS-value denotes <u>n</u>o addresses and a <u>s</u>elf-loop. If there is an adjacent state that is neither an in-memory next state nor the state itself, then the NS-value is 0. Otherwise, we have three possible cases for the NS-value. The value 1 indicates that the state has only one adjacent state, which is the in-memory next state. The values 2 and 3 both indicate that the state also has a self-loop; the only difference between these two values is the priority between the self-loop and the transition targeting the in-memory next state. For these three cases, we do not need the address of a target state. We use two more optimizations for saving space—the homogeneous property optimization and the consecutive literal optimization.

3.1 Homogeneous Property Optimization

Once we represent a state in the BP-regex format depicted in Fig. 2(b), each out-transition of a state consumes memory for its label. We utilize the homogeneous property of position automata—all in-transitions of a state have the same label—to reduce the memory consumption. Thus, instead of keeping all out-transition labels for each state, we only keep its unique in-transition label. We represent the initial state as the sequence of the next addresses without labels because it has no in-transitions. Figure 3 illustrates the homogeneous property optimization.

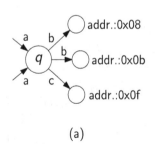

(a)

(b)

0		literal
0		ASCII code of 'b'
0	1	address 0x08
0	0	address 0x08
0		literal
0		ASCII code of 'c'
0	0	address 0x0f

(c)

0		literal
0		ASCII code of 'a'
0	1	address 0x08
0	1	address 0x0b
0	0	address 0x0f

Fig. 3. (a) A graph representation of a state, (b) storing all out-transition labels and (c) storing its unique in-transition label based on the homogeneous property.

3.2   Consecutive Literal Optimization

Literal labels often appear in sequence as strings within patterns in practical regexes. For example, a pattern ^(\w+:)?tape-record-admin$[1] has consecutive literals tape-record-admin. Sung et al. [28] demonstrated the high usage of concatenation compared to other regular operations. After the compilation, symbols in the middle of these strings become states with a single in-transition and a single out-transition—in the previous example, each symbol of ape-record-admi does. In such cases, it is efficient to have all bytes within a certain interval represent the ASCII code of the literal label rather than spending extra bytes to indicate that the state label is literal. We optimize the consecutive literal label case by utilizing the first bit of the ASCII byte. Figure 4 depicts the consecutive literal optimization.

	bit							
	7	6	5	4	3	2	1	0
0	NS-value		0		1			
1	1	ASCII code of the 1st literal						
2	1	ASCII code of the 2nd literal						
			...					
k	0	ASCII code of the last literal						
⋮	addresses of the last state							
			...					

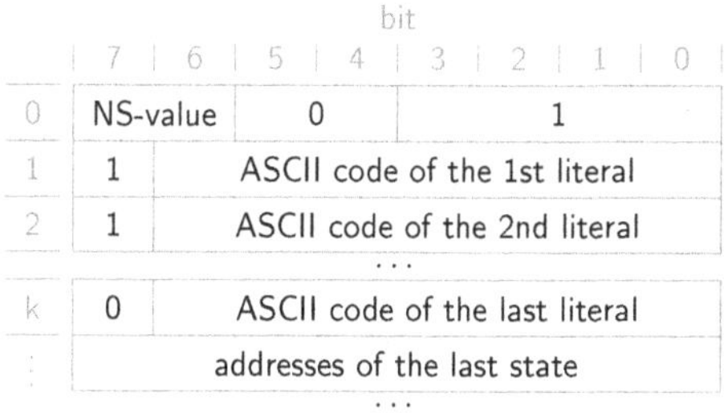

Fig. 4. The layout of a sequence of states with literal labels after the consecutive literal optimization.

4   NFA Structure Formats

There are several formats to represent NFAs including FAdo [1], Grail [26], JFLAP [27] or memory dumping [25]. We consider Grail and memory dumping for comparison.

4.1   Grail Format

Grail is a C++ library for automata programming, which is developed for supporting formal language and automata research. The library uses a newline-separated list of instructions for input automata. Each instruction is a transition triple of a source state, a label and a target state. Grail denotes the initial state 0 by the instruction '(START) |- 0'. Final states F are denoted using a label -| and a target state (FINAL); the instruction is 'F -| (FINAL)'.

---

[1] We bring this example from polyglot [13]. Here, \w is a character class for alphanumeric characters and an underscore.

4.2   NFA Structure Dumping

Structure dumping is an object serializing technique that saves an entire data instance in the main memory as a file without deforming. We call the format obtained by dumping such NFA objects *Dump*, which represents a state as a transition map where the key is an out-transition label of the state and the value is an array of pointers to target states of each transition [7,19].

Although the exact size in memory of the NFA state might vary depending on its implementations, the size of each transition map cannot be less than the sum of the sizes of its transition label σ and the pointers to the target states in $\delta(q,\sigma)$. For a fair comparison, we underestimate the size of a dumped file of an NFA by adding up the sizes of essential components: target-state pointers, out-transition labels for each state, and pointers to initial and final states. We assume 4-byte pointers, and the byte size of the character class label is

$$1 + (\text{\# of characters in the class}) + 2 \times (\text{\# of ranges in the class}),$$

where the first one-byte term is a flag value indicating the inversion of the classes. We also consider another format called *HDump*, which is similar to Dump, where each state has only an in-transition label of the state instead of out-transition labels. This is for the homogeneous property of position automata.

5   Experimental Results and Analysis

5.1   Experiment Setting

Our BP-engine is based on the Thompson matching algorithm [29] for the matching. Note that the current state-of-the-art regex engine, RE2, also uses the Thompson algorithm. We evaluate the matching time by the membership test instead of the searching test since we focus on the peak memory consumption for smaller FA representations while reading the whole input. Note that it is straightforward to extend the membership test to the search test by prepending and appending Σ^* to an input regex.

We use the polyglot corpus [13], a large-scale regex corpus for statistical analysis of real-world regexes, for our evaluation dataset. The polyglot corpus contains a vast collection of regexes from eight different programming languages: JavaScript, Java, PHP, Python, Ruby, Go, Perl, and Rust. These languages are widely used in different fields, and the corpus is a common benchmark dataset for practical regex tasks [3,14,22,28].

Since the polyglot dataset has only regexes and no matching strings, we construct input strings using Xeger [30] that generates strings by performing a random walk. From 537,806 regexes in the dataset, we remove 171,111 regexes that RE2 and our engine cannot process due to unsupported features such as lookaround or backreference. The number of the remaining regexes is 366,695. We use Massif of Valgrind [23] for tracking the memory footprint of engines and reporting the peak memory usage. For each experiment, we exclude outliers that do not fall within 1.5 times the interquartile range to prevent the statistical analysis from being biased by few extreme cases [31].

5.2   Size Reduction of Our Scheme

We compare BP-regexes to Grail and Dump formats to verify the possible advantages of using BP-regexes from a storage perspective (RQ1). We use the Dump format for the underestimation of the dumped files. Figure 5 shows the size of each NFA format obtained by compiling real-world regexes. The result shows that regexes in the Grail format are the largest, representing NFAs using 121.73 bytes on average. The Dump format uses 92.0 bytes, and its optimized version, HDump, compresses the size to 83.62 bytes. Our BP-regex achieves the smallest size 23.17 bytes.

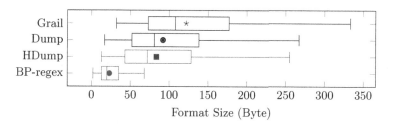

Fig. 5. The box-and-whisker plot of each NFA format size; (1) the whiskers indicate the minimum and maximum values, (2) the box indicates first second and third quartile values, and (3) the dot indicates the average value.

The comparison between Dump and HDump demonstrates the usefulness of the homogeneous optimization—reduces the size of NFA by 8.38 bytes on average. The homogeneous optimization essentially reduces the size of states with more than one out-transition label. In general, as the density of the transition function increases, the number of states with more than one out-transition label increases as well. This implies that the homogeneous optimization is beneficial for high-density NFAs. Table 1 supports our claim by showing that the optimization reduces the size more when the NFA has a higher density.

Table 1. The average size of the memory dump method with or without the homogeneous optimization for different NFA densities. We divide the density ranges non-linearly to keep the number of regexes in each row similar. The last column shows the average of the size reduction $((1 - \text{HDump}/\text{Dump}) \times 100\%)$.

Density of NFA $(\rho = \lvert \delta \rvert / \lvert Q \rvert^2)$	# of Regexes	Size (Byte)		Reduction (%)
		Dump	HDump	
$0 < \rho \leq 1/16$	135,730	255.19	224.16	4.81 ↓
$1/16 < \rho \leq 1/8$	115,082	76.84	68.67	9.73 ↓
$1/8 < \rho \leq 1$	115,883	54.80	44.32	18.19 ↓

A small NFA representation is important for fast matching as well as memory efficiency. During the matching process, we can reduce cache misses and page faults by loading a large number of FA transitions into memory. However, we notice that RE2 outperforms our BP-engine in our experiment. This is because of the several optimizations that the current BP-engine does not support.

5.3 Memory Saved by Precompiled BP-Regex

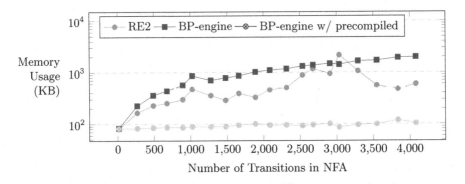

Fig. 6. The peak memory usage for regex pattern matching with respect to the number of transitions.

We examine how much memory is saved by precompiling the BP-regex, instead of compiling a regex at runtime (RQ2). We take the peak memory usage for evaluations as it is a common metric for evaluating memory-efficient algorithms [18,24]. We analyze our BP-engine matching process with the compilation and the matching process that uses precompiled BP-regexes. For comparison, we also measure the peak memory usage of RE2, which compiles regexes at runtime.

Figure 6 shows that the BP-engine with precompiled BP-regex outperforms the other engines as the NFA is larger. In other words, compiling the regex before the runtime is more efficient for long regexes that potentially generate more transitions when compiled, such as malicious pattern rules in intrusion detection systems. For example, our BP-engine is suitable for real-time virus packet monitoring on an IoT device by using BP-regexes.

Observe that the memory consumption of our approach is stable, whereas the other approaches increase the consumption substantially as the number of transitions increases. For example, the benefit of our approach becomes notable when the number of transitions is 20 or more. This is because the compilations consume more memory for constructing larger NFAs and, therefore, our bit-packing scheme has more room for improvement. Overall, based on the experimental results in Sects. 5.2 and 5.3, we claim that using precompiled regex formatted in BP-regex is the most suitable approach when the amount of memory is limited.

5.4   Packing Time of BP-Regex

RQ3 examines the impact of bit-packing on overall time consumption for pre-matching regex processing. We measure the compilation time and the bit-packing time to answer the question. We study RE2 as well as BP-regex to compare the compilation step of the real-world regex engine. Figure 7 shows the time spent on the compilation of RE2, BP-engine without packing, and BP-engine with packing.

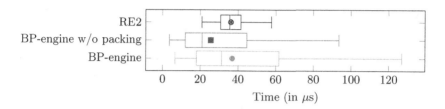

Fig. 7. The compilation time of different engines.

The compilation of BP-regex takes 25.6 ms on average, and adding the packing increases the preprocessing time by 11.3 ms. Comparing the average 25.6 and 36.9 ms of BP-regex and BP-regex with packing, the packing increases the preprocessing time by 1.35 times.

The result shows a trade-off for the compactness of BP-regex shown in Sect. 5.2. Note that, in most cases, the compilation time of BP-regex is shorter than RE2 but has a significantly higher variance. This is due to the NFA construction algorithm; BP-regex uses the position construction whose resulting automaton has $O(n^2)$ transitions in the worst-case [4], where n is the number of symbols and operations in a regex. On the other hand, RE2 uses a modified Thompson construction that produces $O(n)$ transitions [15]. This gives rise to a large variance in the compilation time of the BP-engine compared to RE2.

6   Conclusions

We have proposed a bit-packing scheme that packs compiled regexes into space-efficient BP-regexes. Our scheme compresses compiled regexes by using (1) extended v-coding that supports multiple target addresses, (2) a smaller representation of states using the homogeneous property, and (3) an optimization for literals labels that occur consecutively. Our BP-engine performs the regex matching with low memory consumption by using BP-regexes.

The experiments demonstrate the effectiveness of our scheme. We have compared the resulting BP-regex with the Grail format and memory dumping, and demonstrated that the BP-regex has a much smaller size. Then, we have presented an advantage of our approach for loading a bit-packed NFA in terms of peak memory consumption. Our approach is more beneficial when the input regex has a larger compiled form. The last experiment has shown that the additional time is relatively small compared to the reduction of memory usage.

Acknowledgments. We wish to thank the referees for the careful reading of the paper and valuable suggestions. This research was supported by the NRF grant funded by MIST (RS-2023-00208094).

References

1. Almeida, A., Almeida, M., Alves, J., Moreira, N., Reis, R.: FAdo and GUItar. In: Maneth, S. (ed.) CIAA 2009. LNCS, vol. 5642, pp. 65–74. Springer, Heidelberg (2009). https://doi.org/10.1007/978-3-642-02979-0_10
2. Berglund, M., van der Merwe, B.: On the semantics of regular expression parsing in the wild. In: Proceedings of the 20th International Conference on Implementation and Application of Automata, pp. 292–304 (2015)
3. Berglund, M., van der Merwe, B., van Litsenborgh, S.: Regular expressions with lookahead. J. Univ. Comput. Sci. **27**(4), 324–340 (2021)
4. Brüggemann-Klein, A.: Regular expressions into finite automata. Theoret. Comput. Sci. **120**(2), 197–213 (1993)
5. Caron, P., Ziadi, D.: Characterization of Glushkov automata. Theoret. Comput. Sci. **233**(1–2), 75–90 (2000)
6. Champarnaud, J., Coulon, F., Paranthoën, T.: Compact and fast algorithms for safe regular expression search. Int. J. Comput. Math. **81**(4), 383–401 (2004)
7. Chang, C., Paige, R.: From regular expressions to DFA's using compressed NFA's. Theoret. Comput. Sci. **178**(1–2), 1–36 (1997)
8. Contributors: Compilation and reuse in regular expressions, September 2021. https://learn.microsoft.com/en-us/dotnet/standard/base-types/compilation-and-reuse-in-regular-expressions. Accessed 25 Apr 2023
9. Cortes, C., Mohri, M.: Learning with weighted transducers. In: Proceedings of the 7th International Workshop on Finite-State Methods and Natural Language Processing, pp. 14–22 (2008)
10. CTRE: Compile time regular expression in C++, January 2023. https://github.com/hanickadot/compile-time-regular-expressions. Accessed 25 Apr 2023
11. Daciuk, J.: Experiments with automata compression. In: Proceedings of the 5th International Conference on Implementation and Application of Automata, pp. 105–112 (2001)
12. Daciuk, J., Weiss, D.: Smaller representation of finite state automata. In: Proceedings of the 16th International Conference on Implementation and Application of Automata, pp. 118–129 (2011)
13. Davis, J.C., Michael IV, L.G., Coghlan, C.A., Servant, F., Lee, D.: Why aren't regular expressions a lingua franca? An empirical study on the re-use and portability of regular expressions. In: Proceedings of the 27th ACM Joint Meeting on European Software Engineering Conference and Symposium on the Foundations of Software Engineering, pp. 443–454 (2019)
14. Davis, J.C., Moyer, D., Kazerouni, A.M., Lee, D.: Testing regex generalizability and its implications: a large-scale many-language measurement study. In: Proceedings of the 34th IEEE/ACM International Conference on Automated Software Engineering, pp. 427–439 (2019)

15. Giammarresi, D., Ponty, J., Wood, D., Ziadi, D.: A characterization of Thompson digraphs. Discret. Appl. Math. **134**(1–3), 317–337 (2004)
16. Glushkov, V.M.: The abstract theory of automata. Russ. Math. Surv. **16**(5), 1–53 (1961)
17. Hossain, S.: Visualization of bioinformatics data with dash bio. In: Proceedings of the 18th Python in Science Conference, pp. 126–133 (2019)
18. Lin, J., Chen, W.M., Lin, Y., Cohn, J., Gan, C., Han, S.: MCUNet: tiny deep learning on IoT devices. In: Advances in Neural Information Processing System, vol. 33, pp. 11711–11722 (2020)
19. Luo, B., Lee, D., Lee, W., Liu, P.: QFilter: fine-grained run-time XML access control via NFA-based query rewriting. In: Proceedings of the 13th ACM International Conference on Information and Knowledge Management, pp. 543–552 (2004)
20. McNaughton, R., Yamada, H.: Regular expressions and state graphs for automata. IRE Trans. Electron. Comput. **EC-9**(1), 39–47 (1960)
21. Meiners, C.R., Patel, J., Norige, E., Torng, E., Liu, A.X.: Fast regular expression matching using small TCAMs for network intrusion detection and prevention systems. In: Proceedings of 19th USENIX Security Symposium, pp. 111–126 (2010)
22. van der Merwe, B., Mouton, J., van Litsenborgh, S., Berglund, M.: Memoized regular expressions. In: Maneth, S. (ed.) CIAA 2021. LNCS, vol. 12803, pp. 39–52. Springer, Cham (2021). https://doi.org/10.1007/978-3-030-79121-6_4
23. Nethercote, N., Seward, J.: Valgrind: a framework for heavyweight dynamic binary instrumentation. In: Proceedings of the ACM SIGPLAN 2007 Conference on Programming Language Design and Implementation, pp. 89–100 (2007)
24. Nunes, D.S.N., Ayala-Rincón, M.: A compressed suffix tree based implementation with low peak memory usage. Electron. Notes Theoret. Comput. Sci. **302**, 73–94 (2014)
25. Ramey, R.: November 2004. https://www.boost.org/doc/libs/1_82_0/libs/serialization/doc/index.html. Accessed 25 Apr 2023
26. Raymond, D.R., Wood, D.: Grail: a C++ library for automata and expressions. J. Symb. Comput. **17**(4), 341–350 (1994)
27. Rodger, S.H., Finley, T.W.: JFLAP: An Interactive Formal Languages and Automata Package. Jones & Bartlett Learning (2006)
28. Sung, S., Cheon, H., Han, Y.S.: How to settle the ReDoS problem: back to the classical automata theory. In: Proceedings of the 26th International Conference on Implementation and Application of Automata, pp. 34–49 (2022)
29. Thompson, K.: Programming techniques: regular expression search algorithm. Commun. ACM **11**(6), 419–422 (1968)
30. Xeger: Xeger (2019). https://pypi.org/project/xeger/. Accessed 25 Apr 2023
31. Zani, S., Riani, M., Corbellini, A.: Robust bivariate boxplots and multiple outlier detection. Comput. Stat. Data Anal. **28**(3), 257–270 (1998)

Author Index

B. Nagy (Ed.): CIAA 2023, LNCS 14151, pp. 303–304, 2023.
https://doi.org/10.1007/978-3-031-40247-0

Printed in the United States
by Baker & Taylor Publisher Services